Polymer-Based Flexible Materials

Polymer-Based Flexible Materials

Editors

Jiangtao Xu
Sihang Zhang

Basel • Beijing • Wuhan • Barcelona • Belgrade • Novi Sad • Cluj • Manchester

Editors

Jiangtao Xu
College of Materials
and Energy
South China Agricultural
University
Guangzhou
China

Sihang Zhang
School of Food Science
and Engineering
Hainan University
Haikou
China

Editorial Office
MDPI
St. Alban-Anlage 66
4052 Basel, Switzerland

This is a reprint of articles from the Special Issue published online in the open access journal *Polymers* (ISSN 2073-4360) (available at: www.mdpi.com/journal/polymers/special_issues/577699743V).

For citation purposes, cite each article independently as indicated on the article page online and as indicated below:

Lastname, A.A.; Lastname, B.B. Article Title. *Journal Name* **Year**, *Volume Number*, Page Range.

ISBN 978-3-7258-1166-3 (Hbk)
ISBN 978-3-7258-1165-6 (PDF)
doi.org/10.3390/books978-3-7258-1165-6

Cover image courtesy of Sihang Zhang

© 2024 by the authors. Articles in this book are Open Access and distributed under the Creative Commons Attribution (CC BY) license. The book as a whole is distributed by MDPI under the terms and conditions of the Creative Commons Attribution-NonCommercial-NoDerivs (CC BY-NC-ND) license.

Contents

About the Editors vii

Preface ix

Chenkai Jiang and Bin Sheng
Linear Capacitive Pressure Sensor with Gradient Architecture through Laser Ablation on MWCNT/Ecoflex Film
Reprinted from: *Polymers* **2024**, *16*, 962, doi:10.3390/polym16070962 1

Chunai Dai, Yang Shi, Zhen Li, Tingting Hu, Xiao Wang and Yi Ding et al.
The Design, Synthesis, and Characterization of Epoxy Vitrimers with Enhanced Glass Transition Temperatures
Reprinted from: *Polymers* **2023**, *15*, 4346, doi:10.3390/polym15224346 14

Lianli Deng, Zehua Wang, Bailu Qu, Ying Liu, Wei Qiu and Shaohe Qi
A Comparative Study on the Properties of Rosin-Based Epoxy Resins with Different Flexible Chains
Reprinted from: *Polymers* **2023**, *15*, 4246, doi:10.3390/polym15214246 25

Jian Qiu, Jusha Ma, Wenjia Han, Xiao Wang, Xunchun Wang and Maliya Heini et al.
Effects of Electron Irradiation and Temperature on Mechanical Properties of Polyimide Film
Reprinted from: *Polymers* **2023**, *15*, 3805, doi:10.3390/polym15183805 40

Yuan Yang, Bo Yang, Zhengping Chang, Jihao Duan and Weihua Chen
Research Status of and Prospects for 3D Printing for Continuous Fiber-Reinforced Thermoplastic Composites
Reprinted from: *Polymers* **2023**, *15*, 3653, doi:10.3390/polym15173653 53

Luyao Gao, Fuwei Liu, Qinru Wei, Zhiwei Cai, Jiajia Duan and Fuqun Li et al.
Fabrication of Highly Conductive Porous Fe_3O_4@RGO/PEDOT:PSS Composite Films via Acid Post-Treatment and Their Applications as Electrochemical Supercapacitor and Thermoelectric Material
Reprinted from: *Polymers* **2023**, *15*, 3453, doi:10.3390/polym15163453 70

Yan Li, Hongwei Hu, Teddy Salim, Guanggui Cheng, Yeng Ming Lam and Jianning Ding
Flexible Wet-Spun PEDOT:PSS Microfibers Integrating Thermal-Sensing and Joule Heating Functions for Smart Textiles
Reprinted from: *Polymers* **2023**, *15*, 3432, doi:10.3390/polym15163432 83

Jiayi Li, Shangbi Chen, Jingyu Zhou, Lei Tang, Chenkai Jiang and Dawei Zhang et al.
Flexible $BaTiO_3$-PDMS Capacitive Pressure Sensor of High Sensitivity with Gradient Micro-Structure by Laser Engraving and Molding
Reprinted from: *Polymers* **2023**, *15*, 3292, doi:10.3390/polym15153292 93

Wei Wang, Yang Liu and Zongwu Xie
A Modified Constitutive Model for Isotropic Hyperelastic Polymeric Materials and Its Parameter Identification
Reprinted from: *Polymers* **2023**, *15*, 3172, doi:10.3390/polym15153172 109

Ying Su, Xiaoming Zhao and Yue Han
Phase Change Microcapsule Composite Material with Intelligent Thermoregulation Function for Infrared Camouflage
Reprinted from: *Polymers* **2023**, *15*, 3055, doi:10.3390/polym15143055 141

Fan Xiao, Shunyu Jin, Wan Zhang, Yingxin Zhang, Hang Zhou and Yuan Huang
Wearable Pressure Sensor Using Porous Natural Polymer Hydrogel Elastomers with High Sensitivity over a Wide Sensing Range
Reprinted from: *Polymers* **2023**, *15*, 2736, doi:10.3390/polym15122736 . . . . **158**

Chun Wei, Xiaofei Hao, Chaoying Mao, Fachun Zhong and Zhongping Liu
The Mechanical Properties of Silicone Rubber Composites with Shear Thickening Fluid Microcapsules
Reprinted from: *Polymers* **2023**, *15*, 2704, doi:10.3390/polym15122704 . . . . **169**

Jiahao Liang, Jianxin Nie, Haijun Zhang, Xueyong Guo, Shi Yan and Ming Han
Interaction Mechanism of Composite Propellant Components under Heating Conditions
Reprinted from: *Polymers* **2023**, *15*, 2485, doi:10.3390/polym15112485 . . . . **180**

About the Editors

Jiangtao Xu

Dr. Xu Jiangtao received his B.S. and M.S. degrees from Qingdao University in 2013 and 2016, respectively, and his Ph.D. degree from The Hong Kong Polytechnic University in 2021. He is currently working as an Associate Professor in the School of Materials and Energy, South China Agricultural University. He has published more than 20 refereed papers, including ACS Applied Materials and Interfaces, Nano Research, Composites Part B, etc. He has also published a number of papers in the field of materials and energy. His research interests are mainly in flexible electronics, novel nanofiber materials and surface-enhanced Raman spectroscopy.

Sihang Zhang

Dr. Sihang Zhang graduated from Sichuan University with a master's degree in 2018. He graduated from the State Key Laboratory of Polymer Materials and Engineering at Sichuan University and obtained his Doctor's degree in 2021. He was selected for the Hong Kong Innovation and Technology Commission's Research Talent Hub (RTH-ITF) Program and worked as a postdoctoral fellow in the School of Fashion and Textiles at the Hong Kong Polytechnic University during 2021.09–2023.11. He joined the Department of Food Quality and Safety, School of Food Science and Engineering, Hainan University as a Research Associate in 2023. His research interests include the design and development of rapid food safety detection technologies, the preparation of surface-enhanced Raman scattering sensors, and the utilization of natural polymers in food. He is currently a Guest Editor of *Polymers*. He has authored more than 15 published SCI papers. He has participated in the National Natural Science Foundation of China, Sichuan Provincial Natural Science Foundation and Hong Kong Innovation and Technology Commission Projects.

Preface

Polymer-based flexible materials play a significant role in modern technology and daily life, from flexible electronics, sensors, and wearable devices to biomedical applications, and their wide range of applications highlights their outstanding performance and versatility. The purpose of this reprint is to collect the recent advancements in the research and potential applications of polymer-based flexible materials in various fields. In addition, it will present some innovative application examples that not only demonstrate the practical value of polymer-based flexible materials, but also highlight their potential in advancing science and technology and improving the quality of human life. Finally, we will discuss the major challenges we are currently facing in this field and future research directions, with the aim of inspiring more researchers and engineers to pursue the study of polymer-based flexible materials, encouraging further developments in the field. Through the introduction and discussion in this reprint, readers will gain a comprehensive understanding of the importance of polymer-based flexible materials and their significant impact on society. We expect that this reprint will inform and inspire researchers, engineers, and students in related fields, and promote further research and applications in this area. This collection was assembled and organized by Associate Professor Xu Jiangtao of South China Agricultural University, and Associate Professor Zhang Sihang of Hainan University contributed as a Guest Editor. We are very grateful to all of the experts and scholars who actively contributed to this reprint.

Jiangtao Xu and Sihang Zhang
Editors

Article

Linear Capacitive Pressure Sensor with Gradient Architecture through Laser Ablation on MWCNT/Ecoflex Film

Chenkai Jiang [1,2] and Bin Sheng [1,2,*]

1. School of Optical-Electrical and Computer Engineering, University of Shanghai for Science and Technology, Shanghai 200093, China; 15003438539@163.com
2. Shanghai Key Laboratory of Modern Optical Systems, Engineering Research Center of Optical Instruments and Systems, Shanghai 200093, China
* Correspondence: bsheng@usst.edu.cn

Abstract: The practical application of flexible pressure sensors, including electronic skins, wearable devices, human–machine interaction, etc., has attracted widespread attention. However, the linear response range of pressure sensors remains an issue. Ecoflex, as a silicone rubber, is a common material for flexible pressure sensors. Herein, we have innovatively designed and fabricated a pressure sensor with a gradient micro-cone architecture generated by CO_2 laser ablation of MWCNT/Ecoflex dielectric layer film. In cooperation with the gradient micro-cone architecture and a dielectric layer of MWCNT/Ecoflex with a variable high dielectric constant under pressure, the pressure sensor exhibits linearity (R^2 = 0.990) within the pressure range of 0–60 kPa, boasting a sensitivity of 0.75 kPa^{-1}. Secondly, the sensor exhibits a rapid response time of 95 ms, a recovery time of 129 ms, hysteresis of 6.6%, and stability over 500 cycles. Moreover, the sensor effectively exhibited comprehensive detection of physiological signals, airflow detection, and Morse code communication, thereby demonstrating the potential for various applications.

Keywords: flexible capacitive pressure sensor; gradient micro-cone architecture; laser ablation; dielectric constant; linear response; silicone rubber

1. Introduction

The wide use of flexible sensors, including use in applications such as electronic skins [1–4], human–computer interaction [5,6], cardiovascular monitoring [7], body joint detection [8–10], breathing tests [11–13], and information communication [14,15], has attracted academic attention. According to the working principles, flexible pressure sensors can be classified as piezoresistive [16,17], capacitive [18], triboelectric [19], and piezoelectric [20], which can effectively convert mechanical deformation into quantifiable electrical signals. However, the main problem with existing pressure sensors is the nonlinear change in sensor sensitivity under pressure, making it imperative to devise a capacitive pressure sensor with both good sensitivity and linearity.

Solid silicone rubber (PDMS, Ecoflex) is often used in conventional capacitive sensors, but the sensitivity is limited due to the lack of a micro-structured dielectric layer and poor deformation capacity. Therefore, microstructure dielectrics with higher compressibility have been developed to raise sensitivity, including micro-domes [21], micro-cones [22], micro-pyramids [23–25], and porous structures [26–28]. Although microstructure dielectrics can be manufactured using various molds [29,30], the expense and inefficiency of the technology limits its development. To overcome these constraints, Xue et al. [31] introduced the method of obtaining microstructures in dielectric layers through laser ablation, which is of high efficiency and which is capable of producing large manufacturing areas, providing a new way to prepare microstructures.

In addition, several studies have already proved that increasing the dielectric constant can improve the sensitivity of sensors [32–34], which is usually achieved by mixing silicone rubber with conductive fillers (carbon nanotubes [33] and graphene [35]). Meanwhile, carbon nanotubes, due to their excellent conductivity and high aspect ratio, can significantly reduce dielectric permeation thresholds and have become the most commonly used conductive filler in research.

To improve the linearity of the sensor over the working range, Zhou et al. [36] successfully presented a capacitive sensor with the MWCNT/PDMS gradient architecture dielectric layer, achieving a high level of linearity (R^2 = 0.993) and sensitivity of 0.065 kPa^{-1} at a pressure of 1600 kPa. Xie et al. [37] proposed a linear capacitive pressure sensor with wrinkled PDMS/CNT dielectric layers that exhibits good linearity (R^2 = 0.975) and sensitivity of 1.448 kPa^{-1} at 20 kPa. However, the above studies failed to analyze the variations in the dielectric constant of dielectric layer composite films when film is subject to pressure. This crucial factor ensures linearity response in pressure sensors, which is the issue explored in our study.

In this work, we present an MWCNT/Ecoflex pressure sensor with a gradient micro-cone architecture achieved through laser ablation. The response in the relative dielectric constant of the MWCNT/Ecoflex film under pressure was investigated. The sensor exhibited good linearity and sensitivity by coordinating the gradient micro-cone architecture and the film, reaching sensitivity of 0.75 kPa^{-1} and significantly improving linearity (R^2 = 0.990) with the 2.5 wt% MWCNT under 0–60 kPa. Additionally, the sensor demonstrated excellent repeatability with a fast response time of 95 ms and a recovery time of 129 ms over 500 cycles. Laser ablation significantly accelerates the microstructure manufacturing process, and the gradient micro-cone-structured dielectric layer can be formed within 5 min, which supports being customized separately. Finally, we demonstrate the applications of pressure sensors in human signal detection, airflow detection, and Morse code communication.

2. Materials and Methods

2.1. Materials

MWCNTs (purity: >95%; diameter: <8 nm; length: 0.5–2 μm) were purchased from Suzhou Tanfeng Material Technology. Ecoflex 00-30 was obtained from Smooth-On. Hexane (≥97.0%) purchased from the China Pharmaceutical Group Corporation was used as the diluent.

2.2. Preparation of Ecoflex Film

The beaker, glass rod, and Petri dish were cleaned with hexane and then dried in a heated oven at 60 °C for 30 min. Subsequently, they were removed from the oven and cooled. Ecoflex parts A and B were mixed at a ratio of 1:1 and stirred for 15 min. The resulting mixture was then poured into a Petri dish and put into a vacuum oven for 15 min to eliminate air bubbles. The Ecoflex fluid was then cured at room temperature for 3 h.

2.3. MWCNT/Ecoflex Film Preparation

The MWCNT and dispersant (polyvinylpyrrolidone, PVP) were precisely quantified to provide a mass ratio of 10:1 [38]. The MWCNT and PVP were dispersed into the hexane solution by an ultrasonic cleaner (30 min) (GW0303, GW Ultrasonic Instruments Co., Ltd., Shenzhen, China). Ecoflex part A was then incorporated into the solution and agitated at 1000 rpm for 30 min at high speed. Ecoflex part B was then added and the mixture was stirred for an additional 30 min. The resulting liquid mixture was poured into a mold. After the evaporation of hexane, the uncured MWCNT and Ecoflex mixture was then vacuum-dried in an oven for 30 min and heated at 60 °C for 1 h to achieve complete curing.

2.4. Experimental Setup

The sensor was placed on a pressure test platform (ZQ-990B, ZhiTuo, Ltd., Dongguan, Guangdong, China) that is capable of applying a pressure of 0–60 kPa. The capacitance data were collected using an LCR instrument (TH2822D, Tonghui, Jiangsu, China) with a copper wire connecting the sensor and the LCR operating at 1 kHz. All data were collected at a room temperature of 25 °C. The morphologies of the samples were analyzed using a scanning electron microscope (JSM-IT500HR, Japan Electronics, Tokyo, Japan). Fourier Transform Infrared (FTIR) spectra of the samples were obtained using a Spectrum FTIR spectrometer manufactured by PerkinElmer (L1280127, PerkinElmer, Waltham, MA, USA). X-ray photoelectron spectroscopy (XPS) (AXIS Ultra DLD, Shimadzu, Kyoto, Japan) was used to analyze the elemental composition of samples. Raman spectra were acquired through use of a Raman microscope (RAMANtouch, Nanophoton, Kyoto, Japan) with 532 nm laser excitation. The Tri-Strong TIDE Industrial Camera Microscope was used to measured optical microscope images.

3. Results and Discussion

3.1. Fabrication of the Sensor

Figure 1a shows the structure of the sensor. The sensor comprises electrodes and a gradient micro-cone architecture dielectric layer. It can be attached to the skin to measure physiological parameters. Among them, the electrodes and dielectric layer were glued together with Ecoflex. Copper foil (5 μm) and PI tape (0.02 mm) were combined to form the electrodes. The fabrication schematic for the dielectric layer is shown in Figure 1b. First, the MWCNT/Ecoflex solution was poured into a mold for curing. Second, an 8 × 8 (8 mm × 8 mm) array was designed using CAD and downloaded to a CO_2 laser (K3020, Julong Laser Co, Ltd., Liaocheng, Shandong, China). Finally, we used a CO_2 laser to ablate the surface of the MWCNT/Ecoflex film to form a gradient micro-cone architecture after cleaning the carbide material produced during the ablation process. The finished sensor is shown in Figure S10.

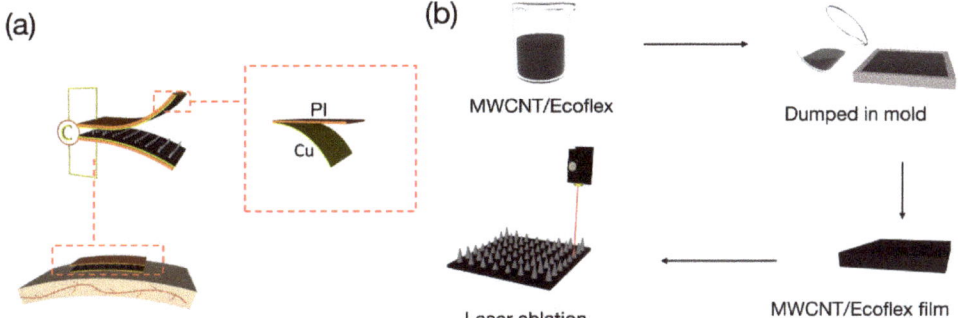

Figure 1. (**a**) Structure of a flexible pressure sensor on human skin. (**b**) Dielectric layer fabrication step for flexible pressure sensors.

To investigate the effect of microstructures on sensor performance, the Ecoflex film was ablated using different power levels and different numbers of ablations. The ablation region is shown in black in Figure S1, in which an 8 × 8 array of micro-cones was fabricated. The low heights of the micro-cones (Figure S2) can be attributed to the utilization of low power when we ablated the films using the parameters in Table S1. Conversely, when the Ecoflex film was ablated utilizing the parameters in Table S2, the high power destroyed the micro-cones (see Figure S3). Consequently, we used the appropriate power parameters in Table S3 for the following study, which allowed us to ablate high micro-cones (see Figure S4) with increasing micro-cone heights as the number of ablations increased.

Figure 2a shows the relative capacitive response of the Ecoflex dielectric sensor with different numbers of ablations at 0–60 kPa. Sensor sensitivity increases as the number of ablations increases, owing to the improved deformability of the dielectric layer caused by the increased height of the micro-cones. Sensor sensitivity was 0.041 kPa^{-1} at 0–60 kPa when the ablations were repeated four times. As described in [36], it has been shown in recent years that sensors with linear gradient architecture dielectric layers exhibit more linearity and sensitivity than those without a gradient architecture dielectric layer. Therefore, to fabricate a micro-cone dielectric layer with linear gradient micro-cones, small areas were ablated (Table S4) after the micro-cones were fabricated, as shown in Table S3. Finally, four levels of linear gradient micro-cones (see Figure S5) were fabricated in the dielectric layer. Two ablation schemes with different numbers of height distributed micro-cones are shown in Figures S6 and S7. Figure 2b shows the sensor's relative capacitive response for the three ablation schemes. The sensitivity of the ablation scheme shown in Figure S7 (0.055 kPa^{-1}) is higher than that of the scheme shown in Figure S6 (0.047 kPa^{-1}) and the scheme shown in Figure S1 without the linear gradient architecture (0.041 kPa^{-1}). The linear range of Figure S7 is also greater than that of Figure S1 without a gradient architecture. In summary, we chose Figure S7 for the ablation in the present study. However, sensor sensitivity still dropped, resulting in a nonlinear response over the pressure range (0–60 kPa). To solve these issues, Ecoflex film was mixed with MWCNTs.

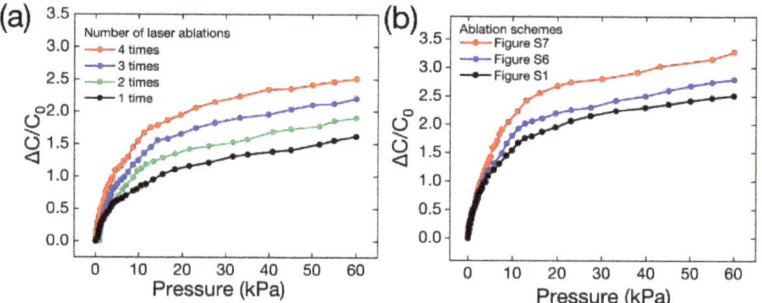

Figure 2. Relative capacitive response of the Ecoflex dielectric pressure sensor under 0–60 kPa. (**a**) Relative capacitive response of the Ecoflex pressure sensor with different numbers of ablations. (**b**) Relative capacitive response of Ecoflex pressure sensors with three different laser ablation schemes.

3.2. Characterization

The height of the micro-cone ablation for MWCNT/Ecoflex was slightly lower than that of the micro-cone ablation for Ecoflex alone. Figure 3a–d show side views of the MWCNT/Ecoflex micro-cones. The height of the micro-cone structure increases as the number of ablations increases (see Table S5). Figure 3e,f show SEM images of the MWCNT/Ecoflex dielectric layer with gradient micro-cones, enabling distinct observations of the laser ablation micro-structured at four different height levels (see Table S6). Figure 3f–i show the SEM images of the cross-section of the MWCNT/Ecoflex film. The presence of micron-scale porous cracks in the cross-section, which has a porosity of 2.7%, was attributed to the volatilization of hexane.

Figure 4 displays the ATR-FTIR and Raman spectra, and Figure 4a illustrates the ATR-FTIR spectrum acquired for the MWCNT/Ecoflex film, the MWCNT/Ecoflex film after laser ablation, and the ablation product of the MWCNT/Ecoflex film. The absorption peaks of the MWCNT/Ecoflex film are almost consistent with those of the MWCNT/Ecoflex film after laser ablation because the Si-C and SiO_2 peaks are located at a position similar to that of Si-O-Si, and the signals of its peak are hidden in the Si-O-Si peak. The spectrum exhibits an absorption peak at 2963 cm^{-1}, which can be attributed to the asymmetric vibration of CH_3. The CH_3 group exhibited an asymmetric stretching vibration, leading to a relatively weak peak at 1412 cm^{-1} and a strong absorption peak at 1258 cm^{-1}. The asymmetric stretching vibration and deformation of Si-O-Si caused significant absorption peaks at 1066 cm^{-1} and 1010 cm^{-1}. The prominent absorption peak at 788 cm^{-1} was related to Si-CH$_3$ [15]. The ATR-FTIR spectra of the ablation product of MWCNT/Ecoflex film are depicted in Figure 4a, which shows an absorption peak at 1066 cm^{-1} from the asymmetric tension vibration of Si-O-Si. The weak peak at 800 cm^{-1} was related to the asymmetric and symmetric deformation of Si-O and Si-C. Therefore, we considered that SiO_2 and SiC were produced during laser ablation. These analyses above were similar to the findings of previous studies [15,39]. In addition, Figure 4b shows the Raman spectra of the Ecoflex film, MWCNT film, and the MWCNT/Ecoflex film, as well as the MWCNT/Ecoflex film after laser ablation. Here, the MWCNT film, MWCNT/Ecoflex film, and laser-ablated MWCNT/Ecoflex film showed three peaks of carbon in the D band at ~1346 cm^{-1}, the G band at ~1582 cm^{-1}, and the 2D band at ~2703 cm^{-1}. Meanwhile, for the MWCNT/Ecoflex film, the laser-ablated MWCNT/Ecoflex film also exhibited the characteristic peaks of the Ecoflex signals, and the Si-C signals of the laser-ablated films are adjacent to the peaks of the Ecoflex signals. These peaks are consistent with the conclusions of a previous study [40].

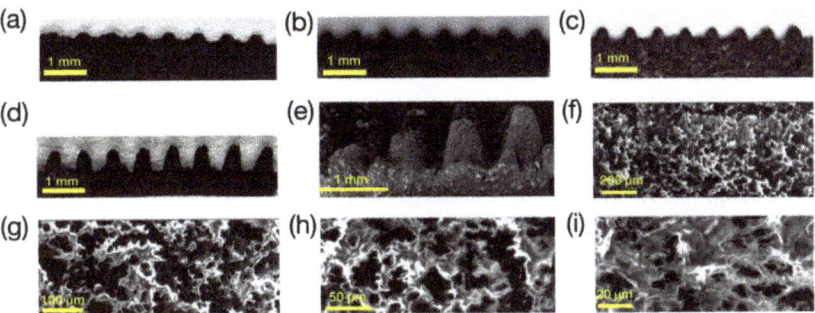

Figure 3. The surface morphology of MWCNT/Ecoflex dielectric layers. (a–d) Side view optical images of the MWCNT/Ecoflex micro-cones obtained using the parameters of Table S5 with different ablation numbers: (a) one time, (b) two times, (c) three times, (d) four times. (e,f) SEM images of the side view of the dielectric layer with gradient micro-cones. (g–i) Cross-sectional SEM images of the MWCNT/Ecoflex film. The black area is MWCNTs and the white area is Ecoflex.

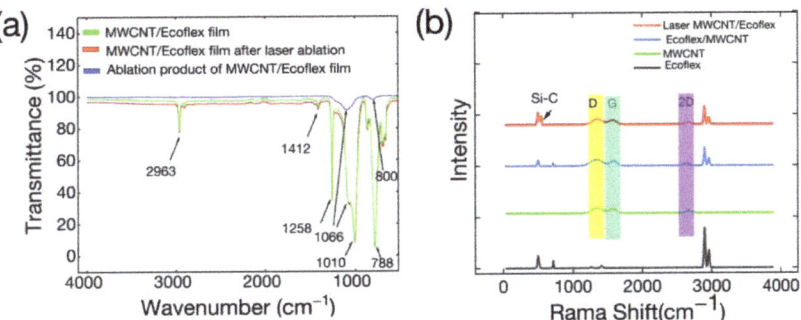

Figure 4. The ATR-FTIR spectrum and Raman spectrum. (a) The ATR-FTIR spectrum of MWCNT/Ecoflex film, the ATR-FTIR spectrum of MWCNT/Ecoflex film after laser ablation, and the ATR-FTIR spectrum of the ablation product of MWCNT/Ecoflex film. (b) The Raman spectrum of the MWCNT/Ecoflex film after laser ablation, the Raman spectrum of MWCNT/Ecoflex film, the Raman spectrum of MWCNT film, and the Raman spectrum of Ecoflex film.

To comprehensively analyze the composition and functional groups present in MWCNT/Ecoflex film, the XPS spectra of C1s, O1s, and Si 2p in these films were collected. The fitted spectra of the C1s peaks of the MWCNT/Ecoflex film and the laser-ablated MWCNT/Ecoflex film (Figure 5a,d) can be deconvolved into two distinct peaks at 284.4 and 284.8 eV, which are attributed to C-C(sp2) and C-C(sp3). The fitted spectra of the O1s peaks of the MWCNT/Ecoflex film and the laser-ablated MWCNT/Ecoflex film (Figure 5b,e) exhibit two distinct peaks at 531.9 eV and 532.7 eV, signifying C-Si-O and Si-O_x. The fitted spectra of the Si 2p peaks of the MWCNT/Ecoflex and laser-ablated MWCNT/Ecoflex films (Figure 5c,f) can be decomposed into three distinct peaks at 101.8 eV, 102.5 eV, and 100.9 eV, respectively, which are attributed to C-Si-O, Si-O$_x$, and Si-C. Notably, the x values of the Si-O$_x$ peaks in the O 1s and Si 2p spectra of the MWCNT/Ecoflex film and the MWCNT/Ecoflex film after laser ablation are estimated to be in the range of 1 to 2. These peaks are consistent with previous findings [40–42]. The XPS findings are consistent with the findings obtained from the Raman and ATR-FTIR spectroscopy analyses.

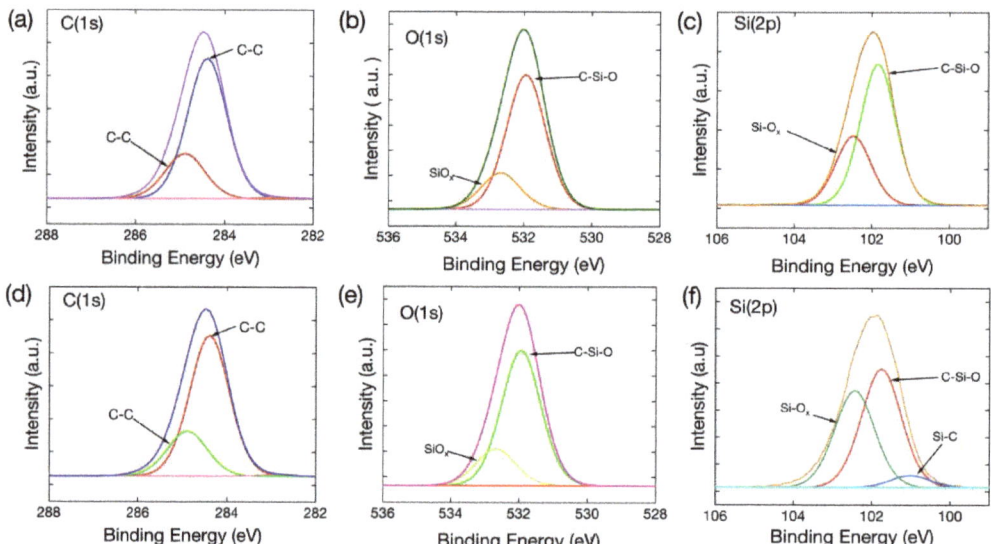

Figure 5. XPS spectra of the films. (**a**–**c**) XPS spectra of MWCNT/Ecoflex film for C 1s, O 1s, and Si 2p. (**d**–**f**) XPS spectra of MWCNT/Ecoflex film after laser ablation for C 1s, O 1s, and Si 2p.

The dielectric constant of the MWCNT/Ecoflex film was also investigated as shown in Equation (1), where ε_0, ε_r, A, and d represent the dielectric constant in vacuum, the relative dielectric constant, the area directly opposite electrodes, and the distance between the electrodes, respectively. c is the capacitance. Figure 6a shows the dielectric constant of the MWCNT/Ecoflex film at different mass fractions. The MWCNT/Ecoflex film exhibited an optimal relative dielectric constant of 650 at 2.5 wt%.

$$\varepsilon_r = \frac{cd}{\varepsilon_0 A} \qquad (1)$$

$$\sigma = \frac{l}{RA} \qquad (2)$$

The conductivity and percolation thresholds are shown in Figure 6b. The equation for electrical conductivity is outlined in Equation (2). σ represents conductivity, R denotes the resistance of the MWCNT/Ecoflex film, A represents the film's cross-sectional area, and l indicates the spacing between electrodes. Owing to the exceptional electrical conductivity and high aspect ratios exhibited by MWCNTs, the film was able to swiftly establish conductive pathways. Hence, percolation theory exhibits a significant increase in conductivity at the critical mass fraction of MWCNTs. Percolation theory establishes the relation between conductivity and the mass fraction of MWCNTs as follows:

$$\sigma_a \propto \sigma_0 (p - p_a)^t \qquad (3)$$

As shown in Equation (3), σ_a represents the film's conductivity and σ_0 is the initial conductivity, where p denotes the MWCNT mass fraction, p_a signifies the percolation threshold, and t is the threshold index. For the MWCNT/Ecoflex films, Equation (3) was used to calculate p_a and t as 0.7 wt% and 2.4, respectively. These data are similar to the findings of previous studies [43,44]. Figure 6c illustrates the Young's modulus of the MWCNT/Ecoflex films with different mass fractions of MWCNTs, ranging from 0.40 MPa to 0.54 MPa. Figure 6d depicts the response in permittivity under pressure (0–60 kPa) with different mass fractions of the MWCNT/Ecoflex films. We collected the value of the deformation of the film under pressure on the test platform, and the initial thickness value minus the deformation value is the true height of the film under pressure. Using Equation (1), we then obtained the exact dielectric constant of the film under pressure. All of the above capacitance data were measured at an LCR frequency of 1 kHz. The response of the relative dielectric constant of the films increases with increasing MWCNT mass ratio under 60 kPa and reaches a maximum 3.2 at 2.5 wt%.

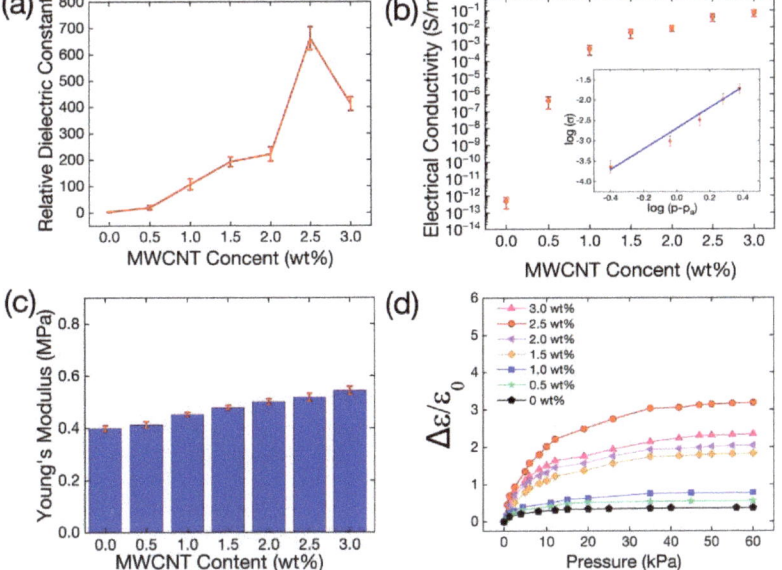

Figure 6. The characteristics of MWCNT/Ecoflex film. (**a**) Relative dielectric constant of MWCNT/Ecoflex films with different mass fractions of MWCNTs. (**b**) The relationship between the mass fraction of MWCNTs and the conductivity of MWCNT/Ecoflex films. (**c**) Young's modulus of MWCNT/Ecoflex films with different mass fractions of MWCNTs. (**d**) The response of the relative dielectric constant of the MWCNT/Ecoflex films with different mass fractions of MWCNTs under 60 kPa.

3.3. The Performance of the Flexible Pressure Sensor

A classical parallel plate capacitor was used for our capacitive pressure sensor, as shown in Equation (4). In our sensor, the dielectric constant under vacuum and the area directly opposite the area directly opposite the two electrodes was fixed. In this case, the capacitance change is caused by ε_r and d. Figure 7a shows a working flow diagram and the equivalent circuit.

$$c = \frac{\varepsilon_0 \varepsilon_r A}{d} \quad (4)$$

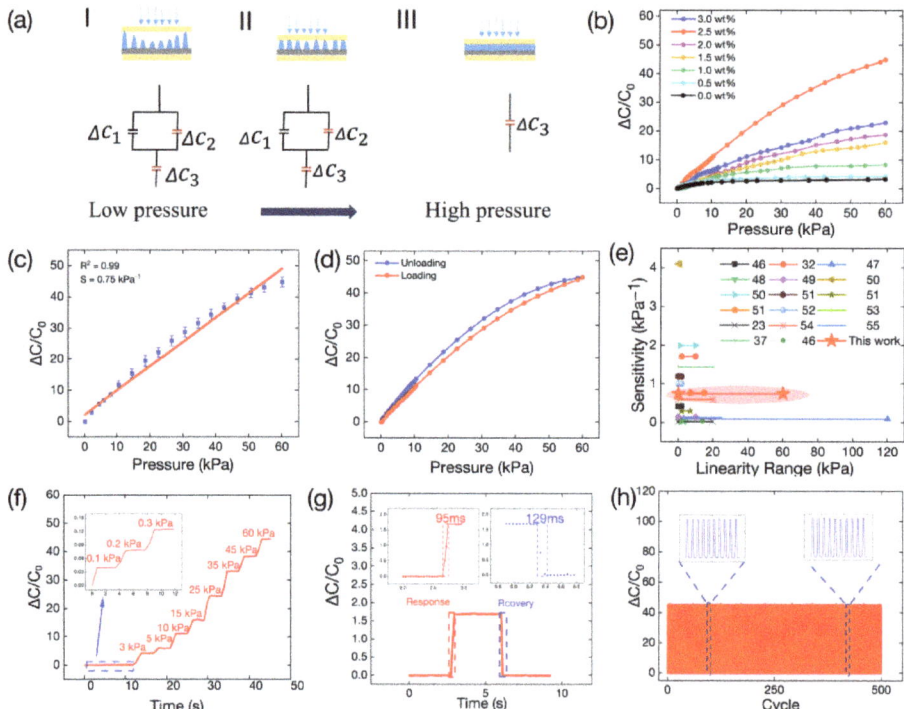

Figure 7. A working flow diagram and the performance of the sensor. (**a**) Working flow diagram of the MWCNT/Ecoflex capacitive pressure sensor with gradient architecture. (**b**) Relative capacitive response of pressure sensors with different mass fractions of MWCNTs under 60 kPa. (**c**) Linear fitting of capacitance response for the sensor under 60 kPa. (**d**) Relative capacitive response of the pressure sensor during a 60 kPa hysteresis test. (**e**) Sensor sensitivity and linearity range compared to recently published capacitive pressure sensors. (**f**) Relative capacitive response of the sensor during a step pressure test at 0–60 kPa. (**g**) Response time of the sensor at a pressure of 1.3 kPa. (**h**) The stability test of the sensor for 500 cycles at 60 kPa.

The sensor's dielectric layer consists of the micro-cone layer and the thin film layer. The micro-cone layer contains two parallel capacitors denoted by c_1 and c_2, where c_1 represents a capacitor with air in the micro-cone layer and c_2 represents the MWCNT/Ecoflex micro-cones. The MWCNT/Ecoflex dielectric film capacitor is indicated by c_3. As shown in Figure 7a-I, the three capacitors maintain their initial capacitance under unpressured conditions. In Figure 7a-II, it can be observed that the gradient architecture micro-cones in the sensor are gradually compressed when subjected to pressure. This can be attributed to the air and lower Young's modulus in the micro-cone layer compared to the film layer, while the Δd value of the micro-cone layer exhibits a rapid increase during this stage. Meanwhile, the Δd value of the film layer is almost negligible. Equation (5) shows the Lichtenecker rule, where ε_a, ε_c, v_a, and v_c represent the relative dielectric constant of air, the relative dielectric constant of the MWCNT/Ecoflex film, and the volume fractions of air and the MWCNT/Ecoflex film, respectively. When the sensor's dielectric layer is compressed, the air is replaced by the MWCNT/Ecoflex material. In this study, the dielectric constant of ε_a was equal to 1, while the dielectric constant of ε_c for the 2.5 wt% MWCNT/Ecoflex composite was measured to be 650. As the sensor is compressed, the MWCNT/Ecoflex micro-cones progressively replace the air, resulting in an increase in the relative dielectric constant, which is indicated by ε_t. In Figure 7a-II, the sensor's response to pressure is related to changes in c_1 and c_2 in the micro-cone dielectric layer, while the change in c_3 is negligible. The micro-cone layer was pressed into the film as the pressure increased, as shown in Figure 7a-III. The response for c_3 is that of sensor capacitance. As the dielectric constant of the film increases under pressure, the capacitance of the sensor continues to increase. Therefore, the micro-cone dielectric layer, in combination with a thin film, results in a sensor with high sensitivity and linearity.

$$\varepsilon_t = \varepsilon_a v_a + \varepsilon_c v_c \tag{5}$$

$$s = \frac{\partial\left(\frac{\Delta c}{c_0}\right)}{\partial p} \tag{6}$$

Optimization of dielectric layer film thickness is also considered (Figure S8). When the thickness of the film layer is 200 μm, the compression range of the dielectric layer is smaller, resulting in a working range that is narrower than that of a sensor with a thickness of 400 μm. When the thickness of the film layer reaches 600 μm, the linear range of the sensor is smaller than that of a sensor with a substrate thickness of 400 μm, which means that the microstructures become less useful in the pressure range when the thickness of film increases. This result is consistent with the findings of previous related work [23]. Figure 7b illustrates the relative capacitive response of the pressure sensor with different mass fractions of MWCNTs at 60 kPa. The relative capacitance change is expressed as $\Delta c/c_0$, in which Δc and c_0 represent the change capacitance and the starting capacitance, respectively, while ∂p is the range of pressure changes. Compared to the other MWCNT mass fractions, the relative dielectric constant reaches an optimum value of 650 at 2.5 wt% (see Figure 6a). Meanwhile, the variation in the MWCNT/Ecoflex film's relative dielectric constant under pressure also shows the highest change at 2.5 wt% (see Figure 6d). These two factors are responsible for the highest response of the sensor occurring at 2.5 wt%. The sensitivity of the sensor is defined by Equation (6). When the mass fraction of MWCNTs was 2.5 wt%, the capacitive sensor exhibited a sensitivity 0.75 kPa^{-1} at 60 kPa, which is 14 times higher than that with the Ecoflex dielectric layer. The linearity of the sensor ($R^2 = 0.990$) is shown in Figure 7c. As shown in Figure 7d, a maximum hysteresis of 6.6% is obtained by loading and unloading pressure. Figure 7e compares the performance of our sensor and other capacitive sensors in terms of sensitivity and linearity. Detailed information is provided in Table S7. Figure 7f displays the capacitance response of the sensor when a step pressure of 0–60 kPa was applied to it. The limit response of the sensor at 1 Pa was also investigated (Figure S9). Figure 7g shows a response time of 95 ms and a recovery response time of 129 ms at a pressure of 1.3 kPa. Due to its close proximity to human skin's response time, it can be applied to applications involving real-time signal processing on the surface of humans [45]. To examine the stability of the sensor under pressure, a 500-cycle test was performed at 60 kPa, as shown in Figure 7h where 10 cycles between 90 and 100 cycles and 420 and 430 cycles are shown.

3.4. Applications of the Sensor

Pulse signals indicate an individual's health because of their association with cardiac systole and diastole. Our sensor can detect small signals from carotid arteries, as shown in Figure 8a. Swallowing is a complicated muscular process in the human body, and it is noteworthy that various digestive disorders can impede swallowing. Consequently, the monitoring of swallowing signals becomes imperative in the course of treatment for such disorders. In Figure 8b, throat swallowing is monitored by attaching the sensor. Figure 8c shows that the sensor can detect a bending elbow. The capacitive signal, which increases with the degree of elbow flexion, also increases. This indicates that the sensor has the potential to be a reliable instrument for recording arm activities. Figure 8d shows the pressure sensor attached to the body's abdomen. It can be observed that the capacitance changes regularly with breathing. The sensor can thus detect muscle activity. Assessment of the condition of the calf muscles is imperative for sprinters during the development of a scientific training program. The sensor can be affixed to the calf region, where it monitors the tension and relaxation of the calf muscles, as shown in Figure 8e. Additionally, the sensor was able to detect the perceived degree of airflow (Figure 8f). Morse code is a form of communication that can be emitted not only by radio but also by other means, such as sound and gestures. This type of communication plays an irreplaceable role in emergency rescue and disaster preparedness. Figure 8g–i show an application of Morse code, where the user presses a finger on the sensor causing a change in capacitance, which outputs the Morse code for "SOS" and "USST".

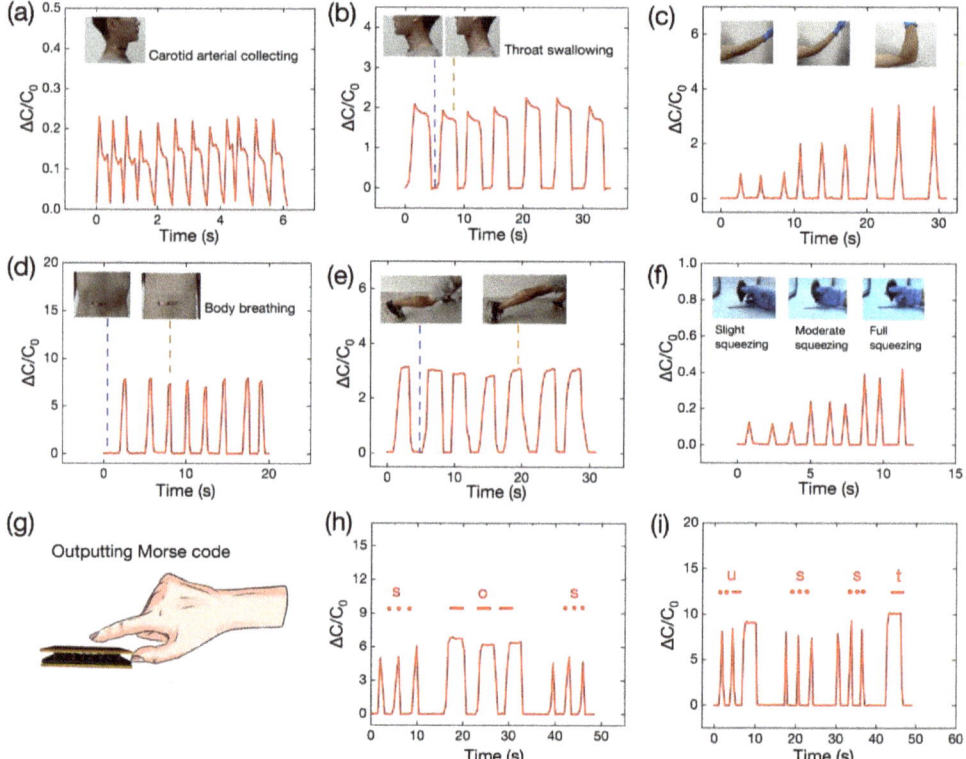

Figure 8. Relevant applications of the sensor. (**a**) The detected signal from carotid artery pulsation when the sensor is bonded to the neck. (**b**) The detection of swallowing signals when the sensor is bonded to the throat surface. (**c**) The sensor attached to arm joints to detect different degrees of flexion. (**d**) The sensor detecting breathing in the human abdomen. (**e**) The detection of muscle activity when the sensor is bonded to the calf. (**f**) Different levels of squeeze applied to an air blow ball result in different degrees of airflow, which can be detected by the sensor. (**g**–**i**) Morse code output from a finger pressing on the sensor.

4. Conclusions

In this work, we proposed the fabrication of a new manufacturing method for capacitive pressure sensors whose MWCNT/Ecoflex dielectric layer includes a thin film and a gradient micro-cone structure achieved by CO_2 laser ablation. This fabrication method offers the advantages of high manufacturing efficiency and customizability. We have optimized the power and speed of the laser ablation to determine the optimum ablation parameters, as well as the optimal height of the microstructure. Variation in the MWCNT/Ecoflex film's relative dielectric constant under pressure and with different mass fractions was also investigated. In addition, we studied the effect of doped MWCNTs on sensor performance and ultimately confirmed that the optimal configuration was MWCNTs with a mass ratio of 2.5 wt%. The sensor exhibited a sensitivity of 0.75 kPa^{-1} and demonstrated exceptional linearity (R^2 = 0.990) from 0 to 60 kPa with a response time of 95 ms, a low lag time of 6.6%, and a recovery time of 129 ms. Finally, this study elucidates the potential applications of this sensor, including the detection of body signals such as pulse signals, swallowing signals, and arm joint flexion signals, the testing of calf muscle strength, and outputting Morse code messages entered by finger presses.

Supplementary Materials: The following supporting information can be downloaded at: https://www.mdpi.com/article/10.3390/polym16070962/s1, Figure S1: Design drawing of the CO_2 laser ablation array (black areas are ablated areas); Table S1: CO_2 laser ablation parameters; Figure S2: Side view of the Ecoflex micro-cones using the parameters of Table S1; Table S2: CO_2 laser ablation parameters; Figure S3: Side view of the Ecoflex micro-cones using the parameters of Table S2; Table S3: CO_2 laser ablation parameters; Figure S4: Side view of the Ecoflex micro-cones using the parameters of Table S3; Table S4: CO_2 laser ablation parameters; Figure S5: Side view of the Ecoflex micro-cones using the parameters of Table S4; Table S5: CO_2 laser ablation parameters; Figure S6: Scheme with gradient micro-cone architecture with 20 W of ablation power in the black area, 6 W in the dark blue area, 7 W in the light blue area, and 8 W in the green area; Figure S7: Scheme with gradient micro-cone architecture with 20 W of ablation power in the black area, 6 W in the dark blue area, 7 W in the light blue area, and 8 W in the green area; Table S6: CO_2 laser ablation parameters; Figure S8: Relative capacitive response of sensors with different dielectric layer thicknesses; Table S7: Comparison of work in recent years. Figure S9: The limit of detection of sensor. Figure S10: Top and side views of the finished sensor. References [46–55] are relevant capacitive flexible pressure works and are cited in Table S7.

Author Contributions: Conceptualization, B.S.; Formal analysis, B.S.; Writing—original draft, C.J.; Writing—review & editing, B.S.; Supervision, B.S.; Funding acquisition, B.S. All authors have read and agreed to the published version of the manuscript.

Funding: We gratefully acknowledge the support of the Natural Science Foundation of Shanghai (19ZR1436100), the Shanghai Municipal Innovation and Entrepreneurship Training Program for College Students (SH2023045), and the National Natural Science Foundation of China (11105149).

Institutional Review Board Statement: Not applicable.

Data Availability Statement: Data are contained within the article.

Acknowledgments: The authors express gratitude to the editors and the reviewers for their constructive and helpful review comments.

Conflicts of Interest: The authors declare no conflicts of interest.

References

1. Jung, S.; Lee, J.; Hyeon, T.; Lee, M.; Kim, D. Fabric-Based Integrated Energy Devices for Wearable Activity Monitors. *Adv. Mater.* **2014**, *26*, 6329–6334. [CrossRef] [PubMed]
2. Bae, G.; Han, J.; Lee, G.; Lee, S.; Kim, S.; Park, S.; Kwon, J.; Jung, S.; Cho, K. Pressure/Temperature Sensing Bimodal Electronic Skin with Stimulus Discriminability and Linear Sensitivity. *Adv. Mater.* **2018**, *30*, e1803388. [CrossRef] [PubMed]
3. Wen, Z.; Zhou, J.; Zhao, S.; Chen, S.; Zhang, D.; Sheng, B. Recyclable EGaIn/TPU Sheath–Core Fibres for Superelastic Electronics and Sensing Applications. *J. Mater. Chem. C* **2023**, *11*, 12163–12173. [CrossRef]
4. Yan, Z.; Liu, Y.; Xiong, J.; Wang, B.; Dai, L.; Gao, M.; Pan, T.; Yang, W.; Lin, Y. Hierarchical Serpentine-Helix Combination for 3D Stretchable Electronics. *Adv. Mater.* **2023**, *35*, 2210238. [CrossRef] [PubMed]
5. Xiong, Y.; Shen, Y.; Tian, L.; Hu, Y.; Zhu, P.; Sun, R.; Wong, C.-P. A Flexible, Ultra-Highly Sensitive and Stable Capacitive Pressure Sensor with Convex Microarrays for Motion and Health Monitoring. *Nano Energy* **2020**, *70*, 104436. [CrossRef]
6. Lin, F.; Zhu, Y.; You, Z.; Li, W.; Chen, J.; Zheng, X.; Zheng, G.; Song, Z.; You, X.; Xu, Y. Ultrastrong and Tough Urushiol-Based Ionic Conductive Double Network Hydrogels as Flexible Strain Sensors. *Polymers* **2023**, *15*, 3219. [CrossRef] [PubMed]
7. Boutry, C.M.; Nguyen, A.; Lawal, Q.O.; Chortos, A.; Rondeau-Gagné, S.; Bao, Z. A Sensitive and Biodegradable Pressure Sensor Array for Cardiovascular Monitoring. *Adv. Mater.* **2015**, *27*, 6954–6961. [CrossRef]
8. Xiong, Y.; Xiao, J.; Chen, J.; Xu, D.; Zhao, S.; Chen, S.; Sheng, B. A Multifunctional Hollow TPU Fiber Filled with Liquid Metal Exhibiting Fast Electrothermal Deformation and Recovery. *Soft Matter* **2021**, *17*, 10016–10024. [CrossRef] [PubMed]

9. Shen, T.; Liu, S.; Yue, X.; Wang, Z.; Liu, H.; Yin, R.; Liu, C.; Shen, C. High-Performance Fibrous Strain Sensor with Synergistic Sensing Layer for Human Motion Recognition and Robot Control. *Adv. Compos. Hybrid Mater.* **2023**, *6*, 127. [CrossRef]
10. Xiao, J.; Xiong, Y.; Chen, J.; Zhao, S.; Chen, S.; Xu, B.; Sheng, B. Ultrasensitive and Highly Stretchable Fibers with Dual Conductive Microstructural Sheaths for Human Motion and Micro Vibration Sensing. *Nanoscale* **2022**, *14*, 1962–1970. [CrossRef]
11. Wang, Q.; Tong, J.; Wang, N.; Chen, S.; Sheng, B. Humidity Sensor of Tunnel-Cracked Nickel@polyurethane Sponge for Respiratory and Perspiration Sensing. *Sens. Actuators B Chem.* **2021**, *330*, 129322. [CrossRef]
12. Wang, J.; Wang, N.; Xu, D.; Tang, L.; Sheng, B. Flexible Humidity Sensors Composed with Electrodes of Laser Induced Graphene and Sputtered Sensitive Films Derived from Poly(Ether-Ether-Ketone). *Sens. Actuators B Chem.* **2023**, *375*, 132846. [CrossRef]
13. Wang, Y.; Hou, S.; Li, T.; Jin, S.; Shao, Y.; Yang, H.; Wu, D.; Dai, S.; Lu, Y.; Chen, S.; et al. Flexible Capacitive Humidity Sensors Based on Ionic Conductive Wood-Derived Cellulose Nanopapers. *ACS Appl. Mater. Interfaces* **2020**, *12*, 41896–41904. [CrossRef] [PubMed]
14. Zhou, J.; Zhao, S.; Tang, L.; Zhang, D.; Sheng, B. Programmable and Weldable Superelastic EGaIn/TPU Composite Fiber by Wet Spinning for Flexible Electronics. *ACS Appl. Mater. Interfaces* **2023**, *15*, 57533–57544. [CrossRef]
15. Tang, L.; Zhou, J.; Zhang, D.; Sheng, B. Laser-Induced Graphene Electrodes on Poly(Ether–Ether–Ketone)/PDMS Composite Films for Flexible Strain and Humidity Sensors. *ACS Appl. Nano Mater.* **2023**, *6*, 17802–17813. [CrossRef]
16. Sang, Z.; Ke, K.; Manas-Zloczower, I. Design Strategy for Porous Composites Aimed at Pressure Sensor Application. *Small* **2019**, *15*, 1903487. [CrossRef] [PubMed]
17. Arruda, L.M.; Moreira, I.P.; Sanivada, U.K.; Carvalho, H.; Fangueiro, R. Development of Piezoresistive Sensors Based on Graphene Nanoplatelets Screen-Printed on Woven and Knitted Fabrics: Optimisation of Active Layer Formulation and Transversal/Longitudinal Textile Direction. *Materials* **2022**, *15*, 5185. [CrossRef] [PubMed]
18. Niu, H.; Gao, S.; Yue, W.; Li, Y.; Zhou, W.; Liu, H. Highly Morphology-Controllable and Highly Sensitive Capacitive Tactile Sensor Based on Epidermis-Dermis-Inspired Interlocked Asymmetric-Nanocone Arrays for Detection of Tiny Pressure. *Small* **2020**, *16*, 1904774. [CrossRef] [PubMed]
19. Zhu, G.; Yang, W.Q.; Zhang, T.; Jing, Q.; Chen, J.; Zhou, Y.S.; Bai, P.; Wang, Z.L. Self-Powered, Ultrasensitive, Flexible Tactile Sensors Based on Contact Electrification. *Nano Lett.* **2014**, *14*, 3208–3213. [CrossRef]
20. Bai, N.; Wang, L.; Wang, Q.; Deng, J.; Wang, Y.; Lu, P.; Huang, J.; Li, G.; Zhang, Y.; Yang, J.; et al. Graded Intrafillable Architecture-Based Iontronic Pressure Sensor with Ultra-Broad-Range High Sensitivity. *Nat. Commun.* **2020**, *11*, 209. [CrossRef]
21. Park, J.; Lee, Y.; Hong, J.; Ha, M.; Jung, Y.-D.; Lim, H.; Kim, S.Y.; Ko, H. Giant Tunneling Piezoresistance of Composite Elastomers with Interlocked Microdome Arrays for Ultrasensitive and Multimodal Electronic Skins. *ACS Nano* **2014**, *8*, 4689–4697. [CrossRef] [PubMed]
22. Palaniappan, V.; Masihi, S.; Panahi, M.; Maddipatla, D.; Bose, A.K.; Zhang, X.; Narakathu, B.B.; Bazuin, B.J.; Atashbar, M.Z. Laser-Assisted Fabrication of a Highly Sensitive and Flexible Micro Pyramid-Structured Pressure Sensor for E-Skin Applications. *IEEE Sens. J.* **2020**, *20*, 7605–7613. [CrossRef]
23. Ruth, S.R.A.; Beker, L.; Tran, H.; Feig, V.R.; Matsuhisa, N.; Bao, Z. Rational Design of Capacitive Pressure Sensors Based on Pyramidal Microstructures for Specialized Monitoring of Biosignals. *Adv. Funct. Mater.* **2020**, *30*, 1903100. [CrossRef]
24. Choong, C.; Shim, M.; Lee, B.; Jeon, S.; Ko, D.; Kang, T.; Bae, J.; Lee, S.H.; Byun, K.; Im, J.; et al. Highly Stretchable Resistive Pressure Sensors Using a Conductive Elastomeric Composite on a Micropyramid Array. *Adv. Mater.* **2014**, *26*, 3451–3458. [CrossRef] [PubMed]
25. Deng, W.; Huang, X.; Chu, W.; Chen, Y.; Mao, L.; Tang, Q.; Yang, W. Microstructure-Based Interfacial Tuning Mechanism of Capacitive Pressure Sensors for Electronic Skin. *J. Sens.* **2016**, *2016*, 1–8. [CrossRef]
26. Kim, J.-O.; Kwon, S.Y.; Kim, Y.; Choi, H.B.; Yang, J.C.; Oh, J.; Lee, H.S.; Sim, J.Y.; Ryu, S.; Park, S. Highly Ordered 3D Microstructure-Based Electronic Skin Capable of Differentiating Pressure, Temperature, and Proximity. *ACS Appl. Mater. Interfaces* **2019**, *11*, 1503–1511. [CrossRef] [PubMed]
27. Jang, Y.; Jo, J.; Lee, S.-H.; Kim, I.; Lee, T.-M.; Woo, K.; Kwon, S.; Kim, H. Fabrication of Highly Sensitive Capacitive Pressure Sensors Using a Bubble-Popping PDMS. *Polymers* **2023**, *15*, 3301. [CrossRef] [PubMed]
28. Zhang, Q.; Jia, W.; Ji, C.; Pei, Z.; Jing, Z.; Cheng, Y.; Zhang, W.; Zhuo, K.; Ji, J.; Yuan, Z.; et al. Flexible Wide-Range Capacitive Pressure Sensor Using Micropore PE Tape as Template. *Smart Mater. Struct.* **2019**, *28*, 115040. [CrossRef]
29. Park, C.W.; Moon, Y.G.; Seong, H.; Jung, S.W.; Oh, J.-Y.; Na, B.S.; Park, N.-M.; Lee, S.S.; Im, S.G.; Koo, J.B. Photolithography-Based Patterning of Liquid Metal Interconnects for Monolithically Integrated Stretchable Circuits. *ACS Appl. Mater. Interfaces* **2016**, *8*, 15459–15465. [CrossRef]
30. Su, B.; Gong, S.; Ma, Z.; Yap, L.W.; Cheng, W. Mimosa-Inspired Design of a Flexible Pressure Sensor with Touch Sensitivity. *Small* **2015**, *11*, 1886–1891. [CrossRef]
31. Du, Q.; Liu, L.; Tang, R.; Ai, J.; Wang, Z.; Fu, Q.; Li, C.; Chen, Y.; Feng, X. High-Performance Flexible Pressure Sensor Based on Controllable Hierarchical Microstructures by Laser Scribing for Wearable Electronics. *Adv. Mater. Technol.* **2021**, *6*, 2100122. [CrossRef]
32. Choi, J.; Kwon, D.; Kim, K.; Park, J.; Orbe, D.D.; Gu, J.; Ahn, J.; Cho, I.; Jeong, Y.; Oh, Y.; et al. Synergetic Effect of Porous Elastomer and Percolation of Carbon Nanotube Filler toward High Performance Capacitive Pressure Sensors. *ACS Appl. Mater. Interfaces* **2020**, *12*, 1698–1706. [CrossRef] [PubMed]

33. Zhang, Z.; Zhang, Q.; Zhang, H.; Li, B.; Zang, J.; Zhao, X.; Zhao, X.; Xue, C. A Novel MXene-Based High-Performance Flexible Pressure Sensor for Detection of Human Motion. *Smart Mater. Struct.* **2023**, *32*, 065007. [CrossRef]
34. Sharma, S.; Chhetry, A.; Sharifuzzaman, M.; Yoon, H.; Park, J.Y. Wearable Capacitive Pressure Sensor Based on MXene Composite Nanofibrous Scaffolds for Reliable Human Physiological Signal Acquisition. *ACS Appl. Mater. Interfaces* **2020**, *12*, 22212–22224. [CrossRef] [PubMed]
35. Kou, H.; Zhang, L.; Tan, Q.; Liu, G.; Dong, H.; Zhang, W.; Xiong, J. Wireless Wide-Range Pressure Sensor Based on Graphene/PDMS Sponge for Tactile Monitoring. *Sci. Rep.* **2019**, *9*, 3916. [CrossRef] [PubMed]
36. Ji, B.; Zhou, Q.; Lei, M.; Ding, S.; Song, Q.; Gao, Y.; Li, S.; Xu, Y.; Zhou, Y.; Zhou, B. Gradient Architecture-Enabled Capacitive Tactile Sensor with High Sensitivity and Ultrabroad Linearity Range. *Small* **2021**, *17*, 2103312. [CrossRef] [PubMed]
37. Lv, C.; Tian, C.; Jiang, J.; Dang, Y.; Liu, Y.; Duan, X.; Li, Q.; Chen, X.; Xie, M. Ultrasensitive Linear Capacitive Pressure Sensor with Wrinkled Microstructures for Tactile Perception. *Adv. Sci.* **2023**, *10*, 2206807. [CrossRef] [PubMed]
38. Huang, Y.; Zheng, Y.; Song, W.; Ma, Y.; Wu, J.; Fan, L. Poly(Vinyl Pyrrolidone) Wrapped Multi-Walled Carbon Nanotube/Poly(Vinyl Alcohol) Composite Hydrogels. *Compos. Part Appl. Sci. Manuf.* **2011**, *42*, 1398–1405. [CrossRef]
39. Tehrani, F.; Goh, F.; Goh, B.T.; Rahman, M.; Abdul Rahman, S. Pressure Dependent Structural and Optical Properties of Silicon Carbide Thin Films Deposited by Hot Wire Chemical Vapor Deposition from Pure Silane and Methane Gases. *J. Mater. Sci. Mater. Electron.* **2012**, *24*, 1361–1368. [CrossRef]
40. Armyanov, S.; Stankova, N.E.; Atanasov, P.A.; Valova, E.; Kolev, K.; Georgieva, J.; Steenhaut, O.; Baert, K.; Hubin, A. XPS and μ-Raman Study of Nanosecond-Laser Processing of Poly(Dimethylsiloxane) (PDMS). *Nucl. Instrum. Methods Phys. Res. Sect. B Beam Interact. Mater. Atoms* **2015**, *360*, 30–35. [CrossRef]
41. Hantsche, H. High Resolution XPS of Organic Polymers, the Scienta ESCA300 Database. By G. Beamson and D. Briggs, Wiley, Chichester 1992, 295 pp., Hardcover, £ 65.00, ISBN 0-471-93592-1. *Adv. Mater.* **1993**, *5*, 778. [CrossRef]
42. Xu, D.; Liu, B.; Wang, N.; Zhou, J.; Tang, L.; Zhang, D.; Sheng, B. Ultrasensitive and Flexible Humidity Sensors Fabricated by Ion Beam Sputtering and Deposition from Polydimethylsiloxane. *Vacuum* **2023**, *213*, 112125. [CrossRef]
43. Oh, J.; Kim, D.-Y.; Kim, H.; Hur, O.-N.; Park, S.-H. Comparative Study of Carbon Nanotube Composites as Capacitive and Piezoresistive Pressure Sensors under Varying Conditions. *Materials* **2022**, *15*, 7637. [CrossRef] [PubMed]
44. Hur, O.-N.; Ha, J.-H.; Park, S.-H. Strain-Sensing Properties of Multi-Walled Carbon Nanotube/Polydimethylsiloxane Composites with Different Aspect Ratio and Filler Contents. *Materials* **2020**, *13*, 2431. [CrossRef]
45. Yang, J.; Luo, S.; Zhou, X.; Li, J.; Fu, J.; Yang, W.; Wei, D. Flexible, Tunable, and Ultrasensitive Capacitive Pressure Sensor with Microconformal Graphene Electrodes. *ACS Appl. Mater. Interfaces* **2019**, *11*, 14997–15006. [CrossRef] [PubMed]
46. Luo, Y.; Shao, J.; Chen, S.; Chen, X.; Tian, H.; Li, X.; Wang, L.; Wang, D.; Lu, B. Flexible Capacitive Pressure Sensor Enhanced by Tilted Micropillar Arrays. *ACS Appl. Mater. Interfaces* **2019**, *11*, 17796–17803. [CrossRef] [PubMed]
47. Liu, Q.; Liu, Z.; Li, C.; Xie, K.; Zhu, P.; Shao, B.; Zhang, J.; Yang, J.; Zhang, J.; Wang, Q.; et al. Highly Transparent and Flexible Iontronic Pressure Sensors Based on an Opaque to Transparent Transition. *Adv. Sci.* **2020**, *7*, 2000348. [CrossRef] [PubMed]
48. Qiu, J.; Guo, X.; Chu, R.; Wang, S.; Zeng, W.; Qu, L.; Zhao, Y.; Yan, F.; Xing, G. Rapid-Response, Low Detection Limit, and High-Sensitivity Capacitive Flexible Tactile Sensor Based on Three-Dimensional Porous Dielectric Layer for Wearable Electronic Skin. *ACS Appl. Mater. Interfaces* **2019**, *11*, 40716–40725. [CrossRef]
49. Kim, J.; Chou, E.-F.; Le, J.; Wong, S.; Chu, M.; Khine, M. Soft Wearable Pressure Sensors for Beat-to-Beat Blood Pressure Monitoring. *Adv. Healthc. Mater.* **2019**, *8*, 1900109. [CrossRef]
50. Su, Q.; Zou, Q.; Li, Y.; Chen, Y.; Teng, S.Y.; Kelleher, J.T.; Nith, R.; Cheng, P.; Li, N.; Liu, W.; et al. A Stretchable and Strain-Unperturbed Pressure Sensor for Motion Interference–Free Tactile Monitoring on Skins. *Sci. Adv.* **2021**, *7*, eabi4563. [CrossRef]
51. Wan, Y.; Qiu, Z.; Hong, Y.; Wang, Y.; Zhang, J.; Liu, Q.; Wu, Z.; Guo, C.F. A Highly Sensitive Flexible Capacitive Tactile Sensor with Sparse and High-Aspect-Ratio Microstructures. *Adv. Electron. Mater.* **2018**, *4*, 1700586. [CrossRef]
52. Jin, T.; Pan, Y.; Jeon, G.-J.; Yeom, H.-I.; Zhang, S.; Paik, K.-W.; Park, S.-H.K. Ultrathin Nanofibrous Membranes Containing Insulating Microbeads for Highly Sensitive Flexible Pressure Sensors. *ACS Appl. Mater. Interfaces* **2020**, *12*, 13348–13359. [CrossRef] [PubMed]
53. Lee, K.; Lee, J.; Kim, G.; Kim, Y.; Kang, S.; Cho, S.; Kim, S.; Kim, J.-K.; Lee, W.; Kim, D.-E.; et al. Rough-Surface-Enabled Capacitive Pressure Sensors with 3D Touch Capability. *Small* **2017**, *13*, 1700368. [CrossRef] [PubMed]
54. Kwon, D.; Lee, T.-I.; Shim, J.; Ryu, S.; Kim, M.S.; Kim, S.; Kim, T.-S.; Park, I. Highly Sensitive, Flexible, and Wearable Pressure Sensor Based on a Giant Piezocapacitive Effect of Three-Dimensional Microporous Elastomeric Dielectric Layer. *ACS Appl. Mater. Interfaces* **2016**, *8*, 16922–16931. [CrossRef]
55. Chhetry, A.; Sharma, S.; Yoon, H.; Ko, S.; Park, J.Y. Enhanced Sensitivity of Capacitive Pressure and Strain Sensor Based on $CaCu_3Ti_4O_{12}$ Wrapped Hybrid Sponge for Wearable Applications. *Adv. Funct. Mater.* **2020**, *30*, 1910020. [CrossRef]

Disclaimer/Publisher's Note: The statements, opinions and data contained in all publications are solely those of the individual author(s) and contributor(s) and not of MDPI and/or the editor(s). MDPI and/or the editor(s) disclaim responsibility for any injury to people or property resulting from any ideas, methods, instructions or products referred to in the content.

Article

The Design, Synthesis, and Characterization of Epoxy Vitrimers with Enhanced Glass Transition Temperatures

Chunai Dai [1,†], Yang Shi [1,†], Zhen Li [2,*], Tingting Hu [1], Xiao Wang [1], Yi Ding [1], Luting Yan [1], Yaohua Liang [3], Yingze Cao [4,*] and Pengfei Wang [2,*]

1. School of Physical Science and Engineering, Beijing Jiaotong University, Beijing 100044, China; chadai@bjtu.edu.cn (C.D.); 19126255@bjtu.edu.cn (Y.S.); 23126690@bjtu.edu.cn (X.W.); 23121824@bjtu.edu.cn (Y.D.); ltyan@bjtu.edu.cn (L.Y.)
2. China Academy of Aerospace Science and Innovation, Beijing 100088, China
3. Department of Agricultural and Biosystems Engineering, South Dakota State University, Brookings, SD 57007, USA
4. China Academy of Space Technology, Beijing 100094, China
* Correspondence: chemlizhen@gmail.com (Z.L.); caoyingze@163.com (Y.C.); hvhe@163.com (P.W.)
† These authors contributed equally to this work.

Abstract: A series of epoxy vitrimers (EVs) with enhanced glass transition temperatures (T_gs) were synthesized by curing epoxy resin E51 with different ratios of phthalic anhydride and sebacic acid as curing agents, and 1,5,7-triazabicyclic [4.4.0] dece-5-ene as a transesterification catalyst, and their curing dynamics, rheological properties, mechanical properties, and thermal stability were comprehensively investigated. By adjusting the molar ratio of the anhydride to the carboxylic acid in the curing agent, the T_gs of the EVs increased from 79 to 143 °C with the increase in the anhydride content. In particular, the material EV-5.5 with a high usable T_g of 98 °C could undergo stress relaxation through the transesterification reaction when exposed to high temperatures (160 to 200 °C), and the correlation between the relaxation time and temperature follows the Arrhenius equation. Moreover, EV-5.5 exhibited elastomeric behavior, where brittle fractures occurred before yielding, which demonstrated a tensile strength of 52 MPa. EV-5.5 also exhibited good thermal stability with a decomposition temperature (T_{d5}) of 322 °C. This study introduces new possibilities for practical applications of thermoset epoxy resins under special environmental conditions.

Keywords: epoxy vitrimer; glass transition temperature; epoxy resin E51; curing agent; phthalic anhydride; sebacic acid

Citation: Dai, C.; Shi, Y.; Li, Z.; Hu, T.; Wang, X.; Ding, Y.; Yan, L.; Liang, Y.; Cao, Y.; Wang, P. The Design, Synthesis, and Characterization of Epoxy Vitrimers with Enhanced Glass Transition Temperatures. Polymers 2023, 15, 4346. https://doi.org/10.3390/polym15224346

Academic Editors: Jiangtao Xu and Sihang Zhang

Received: 14 September 2023
Revised: 17 October 2023
Accepted: 27 October 2023
Published: 7 November 2023

Copyright: © 2023 by the authors. Licensee MDPI, Basel, Switzerland. This article is an open access article distributed under the terms and conditions of the Creative Commons Attribution (CC BY) license (https://creativecommons.org/licenses/by/4.0/).

1. Introduction

Epoxy resins are frequently used as coatings, flame retardant materials, structural materials, and semiconductor component packaging materials in the application fields of construction, electronics, automotives, aerospace, and so on, due to their excellent mechanical properties, good heat resistance and chemical resistance, and good adhesion to a variety of substrates [1,2]. However, since epoxy resins are thermosets with internal polymer chain cross-linked structures that are three-dimensional and permanent, making them insoluble and infusible, it is difficult for them to be recycled and reused once they have been shaped, which limits their applications to a large extent and results in a waste of resources [3].

In 2011, Leibler and coworkers [4] used zinc acetate and zinc acetyl acetonate as catalysts to catalyze the classic epoxy chemistry reaction of the diglycidyl ether of bisphenol A and fatty dicarboxylic and tricarboxylic acids, and obtained a cure containing a large number of dynamic exchange ester bonds. Thanks to the transesterification reaction (TER), the classic thermoset epoxy resin was fluid while maintaining the integrity and insolubility of its cross-linked network at high temperatures and could be reshaped and reprocessed.

Such polymers with covalently cross-linked networks and glass-like fluid properties are therefore called vitrimers [5]. After Leibler's group proposed the TER-based epoxy vitrimer (EV), numerous researchers developed vitrimers based on disulfide bonds [6–10], amino exchange [11,12], olefin metathesis [13,14] and dynamic imine bonds [15–17], which greatly enriched the vitrimer system. However, due to the fact that the introduction of dynamic bonds mostly requires special functional groups, and the preparation processes are complex or costly, the research on EVs continues to focus on the creation of dynamic covalent bonds via TER [18–20].

Most TER-based EV materials are prepared on the basis of acid anhydride- or carboxylic acid-cured epoxy resins. The raw materials for epoxy resins are diverse and so are the curing agents, and the key to the preparation of EVs is the addition of the appropriate catalysts to initiate the TER and catalyze curing reactions. The present catalysts mainly include organic salts (zinc acetate, zinc acetyl acetonate, dibutyltin diacetate, etc.), strong bases (1,5,7-triazide bicyclic (4.4.0) dece-5-ene (TBD), etc.) [20,21] and so on. Currently, most researchers in this field are focusing on the effects of bio-based resin feedstocks [22,23], curing agent types, catalyst types or catalyst-free conditions, epoxy group functionality, and material grain size on the cross-linking process and the mechanical properties of EV materials [21], which have facilitated the development of diverse applications, such as chemical recycling and self-healing, energy storage, electronic devices, shape-shifting materials and devices, artificial muscles, and microfabrication [24–29].

The ultimate potential of vitrimers, however, is limited by the ability to tune the glass transition temperature (T_g) for the target application. At elevated temperatures, vitrimers exhibit the characteristics of viscoelastic fluids, displaying flow behavior. Conversely, at lower temperatures, the exchange reactions within vitrimers occur at a significantly slower rate, giving them characteristics resembling those of conventional thermoset materials. The transition from a liquid-like state to a solid state is reversible and corresponds to a glass transition phenomenon [30]. The T_g is the critical parameter that demarcates the transition between a glassy state and a rubbery state, which is of significant relevance in determining the temperature at which vitrimers become activated [31]. Advanced and smart EV materials have sparked our interest in their application to aerospace vehicles. Aerospace vehicles are a special target application case. Because the exposed components of aerospace vehicles frequently reach temperatures beyond 90 °C when operating continuously under the special environment conditions of air resistance and solar radiation [32], the suitable EV materials are suggested to have an appropriate T_g of above 90 °C while maintaining certain mechanical properties. For this purpose, we concentrate on the design, preparation, and characterization of EVs with enhanced T_gs exceeding 90 °C.

T_g is the temperature at which the chain segments in the polymer change from the frozen state to the moving state, and the movement of the chain segments is realized through the internal rotation of single bonds on the main chain. Hence, the flexibility of the polymer chain has an influence on the T_g of the polymer. Binary and ternary fatty acids tend to have long flexible chains connected via multiple single bonds, so the T_gs of epoxy resins or vitrimers cured with them are typically low, mostly less than 50 °C [30]. Currently, researchers primarily employ two methods to produce epoxy resins or vitrimers with elevated T_gs. The first approach involves the use of epoxy monomers and curing agents that incorporate rigid groups, such as phenyl and biphenyl, or multifunctional groups. This strategy aims to enhance the rigidity of the molecular chains and the cross-linking density of the cured products, thereby augmenting the resistance to molecular chain segment movement [33–36]. For example, Wu and coworkers [37] used such an effective strategy, where natural glycyrrhizic acid (GL) with sebacic acid (SA) were used as curing agents to prepare TER-based EVs. The prepared V4 exhibited a fast stress relaxation (a relaxation time of 130 s at 180 °C) and a usable T_g of 61 °C. An alternative approach is to combine epoxy with other resins with high T_gs, such as benzocaine and cyanate. Through copolymerization, the resulting product could exhibit a T_g exceeding 150 °C [38,39]. However, improving the T_g, network rearrangement rate, and mechanical properties simultaneously to meet material requirements in the special environment of aerospace vehicles remains a big challenge for TER-based EVs [40].

Based on previous reports, our group investigated the use of some curing agents containing rigid groups in the original epoxy–carboxylic acid vitrimer system. The preliminary experiments revealed that aromatic carboxylic acids such as isophthalic acid and terephthalic acid demonstrated higher melting points (greater than 200 °C) and lower reaction temperatures with the epoxy group (the reaction could occur at 130 °C under the condition of TBD catalyst), which resulted in poor process performance; 4,4′-diaminodiphenylmethane with a melting point of 90 °C could be used to cure the diglycidyl ether of bisphenol A or E51 to enhance the T_g of the products, but high-T_g systems suffered from insufficient TERs due to the lack of ester bonds and the low network mobility.

Finally, in this study, a determined curing agent containing rigid groups employed to cure epoxy resin oligomers is reported. Specifically, epoxy resin E51 was utilized as the monomer, while a mixture of phthalic anhydride (PA) and sebacic acid (SA) in varying proportions served as the curing agent. The curing reaction took place in the presence of the transesterification catalyst of TBD. As a result, EVs with elevated T_gs were successfully synthesized (Figure 1), which introduce new possibilities for practical applications of thermosets such as healing or convenient processability in a wider temperature range. Additionally, the created EV-5.5 with a T_g of 98 °C had a suitable stress relaxation rate as well as strong mechanical and thermal stability, making it an intriguing candidate material for use in aerospace applications under unique environmental circumstances.

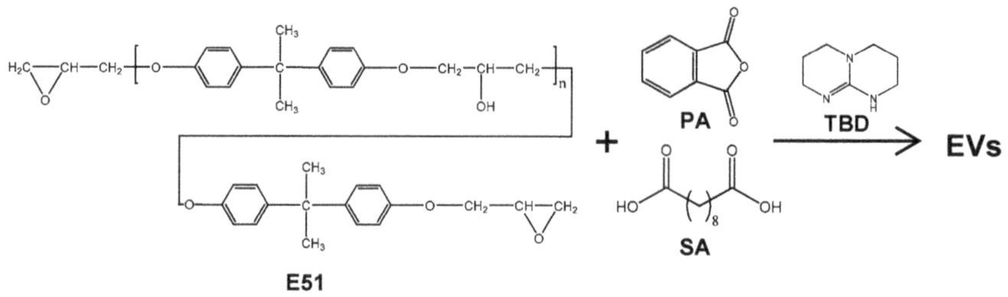

Figure 1. Curing reaction of EVs synthesized from E51, PA, and SA using TBD as the catalyst.

2. Materials and Methods

2.1. Chemicals

The petroleum-based epoxy resin E51 was from Nantong Xingchen Synthetic Material Co., Ltd. (Jiangsu, China); the TBD, and SA came from Tianjin HEOWNS Biochemical Technology Co., Ltd. (Tianjin, China); and the PA came from Shanghai Aladdin Biochemical Technology Co., Ltd. (Shanghai, China). All chemicals used were of analytical grade.

2.2. Synthesis

At ambient temperature, a total of 4 g of epoxy resin E51 and a varying amount of SA (ranging from 0.21 to 1.23 g) were carefully measured and placed into a container. The container, which had a metal bowl coated with polytetrafluoroethylene on the inner wall, was then positioned in a magnetic stirring heating sleeve and heated to a constant temperature of 135 °C for a duration of 5 min. This heating process facilitated the melting of the epoxy resin E51 and the SA. Subsequently, the heating was discontinued, and a quantity of PA (ranging from 1.21 to 2.72 g) was added to the mixture. Continuous agitation was maintained throughout the cooling process to ensure that the PA melted and formed a eutectic with the E51 and SA. Once the eutectic temperature reached 125 °C, 0.28 g of TBD was introduced into the container. After 30 s of vigorous stirring, the container was promptly removed to prevent the excessive polymerization of the prepolymer. The resulting prepolymer mass of 2 to 3 g was then weighed and placed on a die specifically designed for plate pressing. The EV material was obtained by subjecting the prepolymer to hot pressing at a temperature of 180 °C and a pressure of 4 MPa for a duration of 4 h using a hot press [3]. The quantities of the raw materials used for each EV are given in Table 1.

Table 1. Feed compositions of each EV (the relative relationship between the moles of groups or molecules and the mole number of the epoxy group is set to 10).

Label	Epoxy Group (E51)	Anhydride Group (PA)	Carboxyl Group (SA)	TBD
EV-4	10	4	6	1
EV-5	10	5	5	1
EV-5.5	10	5.5	4.5	1
EV-6	10	6	4	1
EV-7	10	7	3	1
EV-8	10	8	2	1
EV-9	10	9	1	1

2.3. Characterization

Differential scanning calorimetry (DSC) was performed on a DSC instrument (Q2000, TA Instruments, Newark, DE, USA), utilizing a standard aluminum crucible. The sample weight was carefully controlled within the 6–10 mg range. The temperature range for the DSC test was set from 0 to 200 °C, with a heating rate of 10 °C/min. The entire test was carried out under a nitrogen atmosphere.

Stress–relaxation experiments were conducted on a USA TA Instruments rheometer (AR-G2, TA Instruments, Newark, DE, USA) in torsion geometry with 8 mm diameter samples. An axial force of -0.01 N and a deformation of 1% were applied. The relaxation times were measured for 63% relaxation.

The tensile properties were evaluated using an electronic universal material testing machine (5943, Instron, Boston, MA, USA). Rectangular samples measuring $40 \times 10 \times 0.5$ mm^3 were cut from the EV material and placed on the machine. The test procedure involved gradually increasing the axial tension from 0 N at a rate of 30 N/min until the sample fractured.

The surface morphology was examined using a scanning electron microscope (SEM, Quattro C, Thermo Fisher Scientific, Waltham, MA, USA).

Thermogravimetric (TG) and differential thermal gravimetric (DTG) analyses were performed using a thermogravimeter (TGA-Q50, TA Instruments, Newark, Delaware, USA). The EV material was subjected to a gradual heating process in a nitrogen atmosphere, reaching a maximum temperature of 740 °C at a heating rate of 10 °C/min.

3. Results and Discussion

3.1. Proposed Mechanisms of the Curing Reactions

The potential mechanisms underlying the reactions between epoxy groups and anhydrides or carboxylic acids, catalyzed by bases, are depicted in Figure 2. Figure 2a illustrates various reaction steps involving the epoxy group and the anhydride, with a base (take R_3N as the example) as the catalyst. Firstly, the anhydride reacts with R_3N, resulting in the formation of a carboxylate anion. Subsequently, the carboxylate anion reacts with the epoxy group, causing the opening of the epoxy group and the formation of an alkoxide anion. Finally, the alkoxide anion reacts with another anhydride, leading to the formation of a new carboxylate anion [41]. Similarly, Figure 2b demonstrates the reaction steps between the epoxy group and the carboxylic acid, catalyzed by a base (abbreviated as B). Initially, the carboxylic acid reacts with the B, resulting in the formation of a carboxylate anion. Subsequently, the carboxylate anion reacts with the epoxy group, leading to the opening of the epoxy group and the formation of an alkoxide anion. Finally, a new reaction occurs between the alkoxide anion and another carboxylic acid.

However, upon comparing step 3 in Figure 2a with step 3 in Figure 2b, it becomes evident that the use of anhydrides as a curing agent leads to the formation of carboxylate anions (marked with the red box), which will subsequently react with epoxy groups. This results in the repetition of steps 2 and 3, leading to the creation of new branch structures on the polymer chain. Consequently, the cross-link density of the cured product is increased. Conversely, when carboxylic acids are employed as a curing agent, the reaction with epoxy groups does not yield such branch structures. Therefore, by adjusting the ratio of anhydrides to carboxylic acids in the curing agent, it is possible to modify the cross-link density and, consequently, the T_g of the resulting EV material.

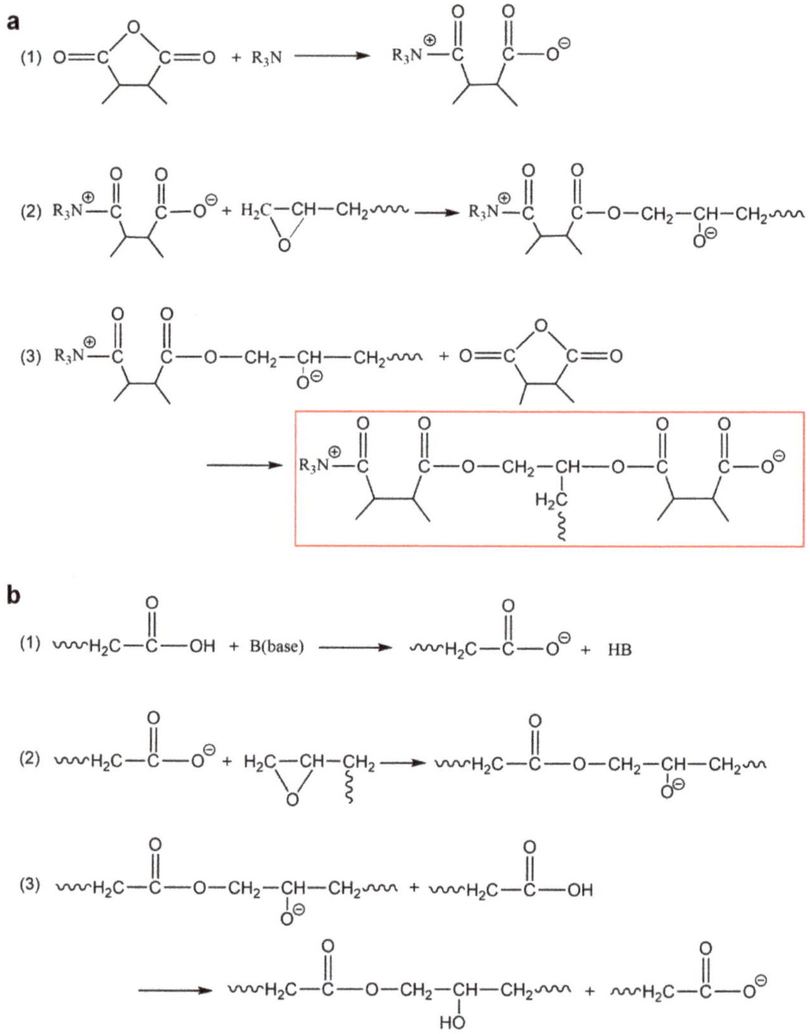

Figure 2. Proposed reaction steps between epoxy and (**a**) anhydrides catalyzed by bases (take R$_3$N as the example); (**b**) carboxylic acids catalyzed by bases (abbreviated as B).

3.2. T_gs of EV Materials

To investigate the effect of the curing agent on the T_g of the vitrimer, DSC was performed on the EV materials cured with varying levels of anhydride content in the curing agent. Figure 3a displays the DSC curves for the different EV materials, and Figure 3b illustrates their detailed T_gs. As depicted in Figure 3, the T_gs of the EVs gradually increased from 79 to 143 °C as the proportion of anhydrides in the curing agents increased. This can be attributed to two primary factors. First, the presence of rigid benzene rings in

PA leads to significant steric hindrance, thereby raising the energy barrier for molecular chain rotation. Second, in contrast to carboxylic acids, the reaction of anhydride with epoxy groups, facilitated by a base catalyst, enhances the cross-linkage density of cured EV materials. Consequently, when the total molar amount of anhydride and carboxyl groups in the curing agent and the molar amount of epoxy groups in the epoxy oligomer were maintained at a 1:1 ratio, the higher the proportion of PA in the curing agent, the higher the T_g of the cured EV. The T_g of the cured EV increased beyond 90 °C when the molar ratio of anhydride to carboxyl groups in the curing agent was 5:5, which met the basic material requirements for low-orbit spaceflight [32].

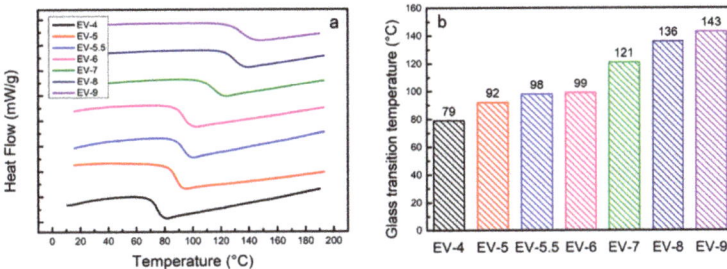

Figure 3. (**a**) DSC curves and (**b**) T_gs of EVs prepared with different ratios of anhydride to carboxylic acid as curing agents.

3.3. Rheological Properties of EV Materials

In order to investigate the rheological properties of EV materials at elevated temperatures, shear stress relaxation experiments were conducted on EV materials with T_gs exceeding 90 °C at 150 and 180 °C. The results presented in Figure 4 and Table 2 demonstrate that the majority of EV materials exhibited stress relaxation within the duration of the test (40,000 s) at both temperatures. This suggests that the EV materials were capable of undergoing a TER at higher temperatures, leading to changes in the cross-linked network's topology. Notably, the relaxation times of EV-5, EV-6, and EV-7 at 180 °C were significantly shorter than those at 150 °C. This observation can be attributed to the accelerated TER rate at elevated temperatures, which expedites the rearrangement process of the cross-linked network.

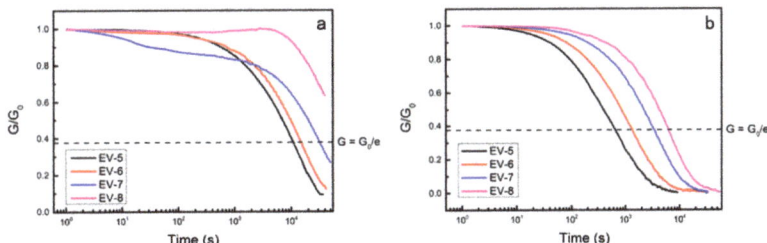

Figure 4. Stress–relaxation curves of several EV materials with T_g > 90 °C at different temperatures: (**a**) 150 °C, (**b**) 180 °C.

Table 2. Relaxation time of each EV (T_g > 90 °C) in the shear rheological test at different temperatures.

Test Temperature (°C)	Relaxation Time (s)			
	EV-5	EV-6	EV-7	EV-8
150	10,872	15,170	32,800	>40,000
180	660	1334	3460	6170

Furthermore, the relaxation curves at the same test temperature reveal that the higher the content of PA in the curing agent, the longer the relaxation time of the cured EV material. This phenomenon can be attributed to the molecular structure of PA and its reaction characteristics with the epoxy group. In the presence of the base catalyst, the reaction between carboxylic acid and epoxy oligomers efficiently generates two reactants, namely the ester group and the hydroxyl group, facilitating transesterification. Conversely, the reaction between anhydride and epoxy oligomers, facilitated by the base catalyst, only produces ester groups. Consequently, a higher content of anhydride in the curing agent leads to an unfavorable group composition in the curing product for the TER. Additionally, PA possesses a rigid benzene ring structure with significant steric hindrance, which hinders contact between the ester groups and hydroxyl groups in EV materials. Based on the aforementioned findings, it can be concluded that a higher amount of PA in the curing agent hampered the TER of the prepared EV material and slowed down the stress relaxation process. One thing to note here is that the principle of shear rheology testing is based on some basic assumptions; i.e., only when the input or output strain or stress is applied to the sample, and the flow field is a pure shear flow field, the test results are reliable, so different samples often need to adopt their own specific test conditions and techniques to ensure that these basic assumptions are valid [42]. The set of samples (EV-5, EV-6, EV-7, and EV-8) used here to determine the relaxation times were different, but in order to ensure the comparability of the test results, the same test conditions were used, which may affect the accuracy of the absolute values of the relaxation times for the samples. Therefore, the absolute values of the relaxation times here are suggested to be neglected and their relative values are considered. Comprehensively considering the performance requirements of a T_g above 90 °C and a shorter stress relaxation time, the material EV-5.5 should be able to meet the criteria. Thus, EV-5.5 was primarily evaluated in the follow-up work.

Shear stress relaxation experiments were conducted on EV-5.5 at various temperatures, ranging from 160 to 200 °C. The stress relaxation curves of the EV-5.5 material at different temperatures are depicted in Figure 5a, while the corresponding relaxation times are presented in Table 3. The experimental findings indicate that, as the shear stress relaxation test temperature increased from 160 to 200 °C, the relaxation time of the EV-5.5 material gradually decreased from 23,354 s to 1350 s. This observation suggests that the rate of the topological transition of the cross-linked network within the material accelerates with increasing temperature. The test results further indicate that the correlation between the relaxation time (τ) and the shear rheological test temperature (T) for the EV-5.5 material adheres to the Arrhenius equation, $\ln\tau = \ln\tau_0 + E_a/RT$. Specifically, the $\ln\tau$ at each temperature exhibits a linear relationship with $1000/T$, as depicted in Figure 5b. This finding provides further evidence that the EV materials possessed the rheological characteristics that are commonly observed in vitrimer materials at elevated temperatures [30]. By analyzing the slope of the fitted curve and employing the Arrhenius equation, the activation energy (E_a) associated with the relaxation process of the EV-5.5 material was determined to be 123.8 kJ/mol, which was in good agreement with those reported by Leibler and other researchers (69–150 kJ mol^{-1}) [35]. Compared with V4 with glycyrrhizic acid as the component curing agent, as reported by Wu et al. [37], EV-5.5 exhibited a higher T_g and a slower relaxation rate, which was capable of meeting the requirements for our target applications in aerospace.

Table 3. Relaxation times for the EV-5.5 material at various temperatures.

Test temperature (°C)	160	170	180	190	200
Relaxation time, τ (s)	23,354	10,536	4664	2231	1350

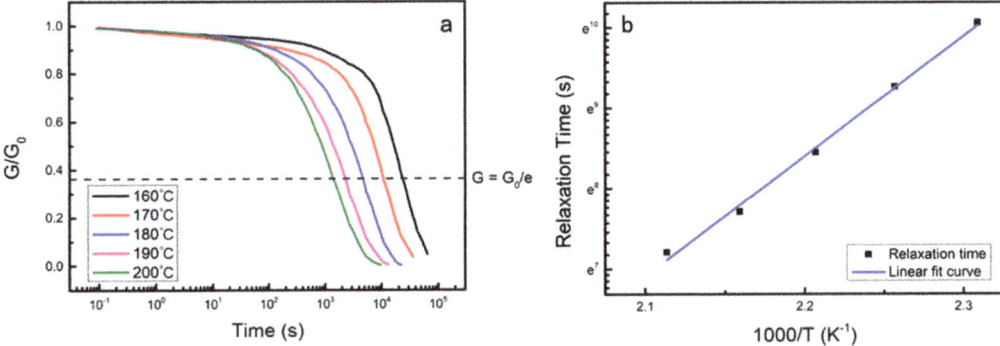

Figure 5. (a) Normalized stress–relaxation curves of the EV-5.5 material at different temperatures; (b) fitting of the relaxation times to the Arrhenius equation.

3.4. Mechanical Properties of EV-5.5

To assess the mechanical properties of the EV material, EV-5.5 was subjected to a tensile test at room temperature. The resulting tensile curve is depicted in Figure 6a. The graph illustrates the linear relationship between stress and strain with increasing applied load. Upon reaching an applied load of 52 ± 6 MPa, the material experienced a rupture, exhibiting a 12.5 ± 1.9% elongation at the rupture and a rupture energy of 3.83 ± 1.02 MJ/m^3.

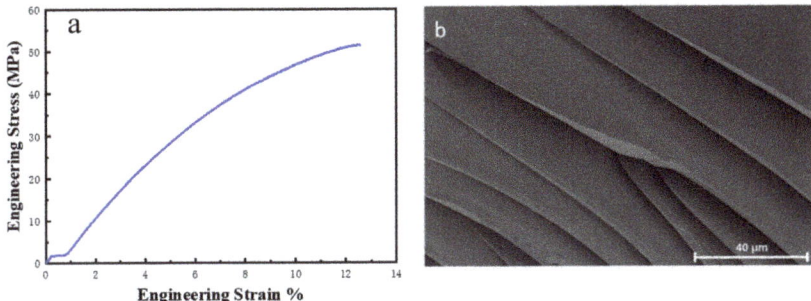

Figure 6. (a) Tensile curve of EV-5.5; (b) SEM image of the rupture cross-section.

The tensile curve does not display any stress peaks or plateaus prior to the material's rupture, implying the absence of significant yield phenomena; namely, the EV-5.5 material exhibited typical elastomeric behavior throughout the stretching process, with brittle breaking occurring before yielding. This observation is further supported in the SEM image of the rupture's cross-section, as displayed in Figure 6b. The fracture surface of EV-5.5 exhibited smooth and well-ordered streamer-like lines, indicating the occurrence of stress concentrations during tensile processes and confirming the brittle fracture behavior. Compared with the reported V4 and a few bio-based vitrimers, EV-5.5 showed a significantly better tensile strength [37].

3.5. Thermal Stabilities of EV-5.5

The thermal stability of EV-5.5 was characterized using TG analysis and the results are presented in Figure 7. The red TG curve shows how the weight of the EV-5.5 sample decreased when it was heated. The initial decomposition temperature of EV-5.5 was determined to be 260 °C. Within the temperature range of 0–260 °C, the sample experienced a weight loss of 0.5%, which is attributed to the volatilization of the TBD catalyst, the unreacted curing agent, and other substances. As the temperature increased, the decomposition rate of EV-5.5 accelerated, with decomposition temperatures of 322, 341, and 437 °C corresponding to weight losses of 5%, 10%, and 50%, respectively. The EV-5.5 material demonstrated good thermal stability, and the temperature at 5% weight loss (T_{d5}) was substantially higher than 263 °C [37]. By performing differential calculations on the TG curve, the DTG curve (blue line) was obtained, which provides insight into the decomposition rate of the materials at different temperatures. The DTG curve reveals that EV-5.5 decomposed rapidly within the temperature range of 300–500 °C, with the decomposition rates peaking at 354 and 415 °C.

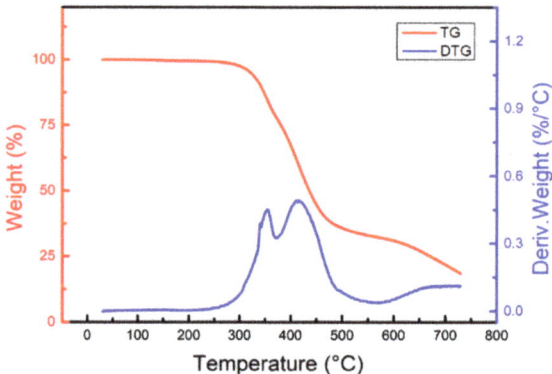

Figure 7. TG curve (red line) and DTG curve (blue line) of the EV-5.5 material.

4. Conclusions

The EV materials were synthesized using epoxy E51 as the base resin, with PA and SA as the curing agents and TBD as the catalyst. Shear rheology tests revealed that the EV materials exhibited stress relaxation behavior at high temperatures, and the relaxation time was found to be consistent with that of vitrimer materials. DSC and shear rheology tests demonstrated that the T_g of the cured EV material increased, and the fluidity decreased with higher anhydride contents in the curing agent. This can be attributed to the larger steric hindrance of the molecular structure of PA, which leads to an increased cross-linking density in the cured product when reacted with epoxy under TBD catalysis. Tensile tests revealed that the EV-5.5 sample displayed elastomeric properties, with brittle breaking occurring before yielding, which exhibited a tensile strength of 52 MPa, an elongation at fracture of 12.5%, and a fracture energy of 3.83 MJ/m^3. The initial decomposition temperature of EV-5.5 was 260 °C, according to the TG analysis data. Two decomposition peaks appeared as the temperature rose, at 354 and 415 °C, respectively, and the decomposition temperatures of 5%, 10%, and 50% were 322 °C, 341 °C, and 437 °C, respectively. This study, aiming to prepare EV materials with enhanced and tunable T_gs, opens up new opportunities for thermosets' practical applications.

Author Contributions: Conceptualization, Z.L., Y.C. and P.W.; methodology, C.D.; validation, T.H., X.W. and Y.D.; formal analysis, Y.S.; investigation, Y.S.; resources, Z.L.; data curation, C.D.; writing—original draft preparation, Y.S.; writing—review and editing, C.D. and Y.L.; visualization, Y.L.; supervision, L.Y.; project administration, Z.L. and P.W.; funding acquisition, Z.L. and P.W. All authors have read and agreed to the published version of the manuscript.

Funding: This research was funded by the National Natural Science Foundation of China (Grant No. 52003291).

Institutional Review Board Statement: Not applicable.

Data Availability Statement: The authors confirm that the data supporting the findings of this study are available within the article.

Conflicts of Interest: The authors declare no conflict of interest.

References

1. Chen, P.; Liu, S.; Wang, D. *Epoxy Resin and Its Application*; Chemical Industry Press: Beijing, China, 2011.
2. Yang, S.; Huo, S.; Wang, J.; Zhang, B.; Wang, H. A highly fire-safe and smoke-suppressive single-component epoxy resin with switchable curing temperature and rapid curing rate. *Compos. Part B-Eng.* 2021, 207, 108601. [CrossRef]
3. Yang, Y. Epoxy Based Vitrimer Composites. Ph.D. Dissertation, Tsinghua University, Beijing, China, 2017.
4. Montarnal, D.; Capelot, M.; Tournilhac, F.; Leibler, L. Silica-like malleable materials from permanent organic networks. *Science* 2011, 334, 965–968. [CrossRef] [PubMed]
5. Denissen, W.; Winne, J.M.; Du Prez, F.E. Vitrimers: Permanent organic networks with glass-like fluidity. *Chem. Sci.* 2016, 7, 30–38. [CrossRef] [PubMed]
6. Martin, R.; Rekondo, A.; de Luzuriaga, A.R.; Cabañero, G.; Grande, H.J.; Odriozola, I. The processability of a poly(urea-urethane) elastomer reversibly crosslinked with aromatic disulfide bridges. *J. Mater. Chem. A* 2014, 2, 5710–5715. [CrossRef]
7. Tsarevsky, N.V.; Matyjaszewski, K. Reversible redox cleavage/coupling of polystyrene with disulfide or thiol groups prepared by atom transfer radical polymerization. *Macromolecules* 2002, 35, 9009–9014. [CrossRef]
8. Michal, B.T.; Jaye, C.A.; Spencer, E.J.; Rowan, S.J. Inherently photohealable and thermal shape-memory polydisulfide networks. *ACS Macro Lett.* 2013, 2, 694–699. [CrossRef]
9. Pepels, M.; Filot, I.; Klumperman, B.; Goossens, H. Self-healing systems based on disulfide-thiol exchange reactions. *Polym. Chem.* 2013, 4, 4955–4965. [CrossRef]
10. Canadell, J.; Goossens, H.; Klumperman, B. Self-healing materials based on disulfide links. *Macromolecules* 2011, 44, 2536–2541. [CrossRef]
11. Stefani, H.A.; Costa, I.M.; Silva, D.D.O. ChemInform abstract: An easy synthesis of enaminones in water as solvent. *ChemInform* 2000, 32, 1526–1528. [CrossRef]
12. Denissen, W.; Rivero, G.; Nicolaÿ, R.; Leibler, L.; Winne, J.M.; Du Prez, F.E. Vinylogous urethane vitrimers. *Adv. Funct. Mater.* 2015, 25, 2451–2457. [CrossRef]
13. Lu, Y.; Guan, Z. Olefin metathesis for effective polymer healing via dynamic exchange of strong carbon-carbon double bonds. *J. Am. Chem. Soc.* 2012, 134, 14226–14231. [CrossRef] [PubMed]
14. Vougioukalakis, G.C.; Grubbs, R.H. Ruthenium-based heterocyclic carbene-coordinated olefin metathesis catalysts. *Chem. Rev.* 2010, 110, 1746–1787. [CrossRef] [PubMed]
15. Taynton, P.; Yu, K.; Shoemaker, R.K.; Jin, Y.; Qi, H.J.; Zhang, W. Heat- or water-driven malleability in a highly recyclable covalent network polymer. *Adv. Mater.* 2014, 26, 3938–3942. [CrossRef] [PubMed]
16. Yang, B.; Zhang, Y.; Zhang, X.; Tao, L.; Li, S.; Wei, Y. Facilely prepared inexpensive and biocompatible self-healing hydrogel: A new injectable cell therapy carrier. *Polym. Chem.* 2012, 3, 3235–3238. [CrossRef]
17. Deng, G.; Li, F.; Yu, H.; Liu, F.; Liu, C.; Sun, W.; Jiang, H.; Chen, Y. Dynamic hydrogels with an environmental adaptive self-healing ability and dual responsive sol-gel transitions. *ACS Macro Lett.* 2012, 1, 275–279. [CrossRef]
18. Zhou, L.; Liu, J.; Wu, S.; Chen, G.; Yang, S.; Yang, L. Research Progress of Vitrimer Materials. *Mater. Rep.* 2020, 34, 585–591.
19. Liu, H.; Wei, L.; Sun, Z.; Liu, C.; Liu, Y. Effect of Catalysts on the Performance of Vitrimers Based on Dynamic Ester Exchange. *Trans. China Electrotech. Soc.* 2023, in press.
20. Liu, H.; Wei, L.; Sun, Z.; Liu, C.; Liu, Y. Preparation and Properties of Recyclable Vitrified Epoxy Resin Based on Transesterification. *Trans. China Electrotech. Soc.* 2023, 38, 4019–4029.
21. Wang, Y.; Liu, X.; Jing, X.; Li, Y. Research Progress in Epoxy Vitrimer. *Chem. Bull.* 2021, 84, 313–321.
22. Liu, L.; Ju, B. Research progress of bio-based vitrimer materials. *Eng. Plast. Appl.* 2021, 49, 135–140.
23. Yang, X.; Guo, L.; Xu, X.; Shang, S.; Liu, H. A fully bio-based epoxy vitrimer: Self-healing, triple-shape memory and reprocessing triggered by dynamic covalent bond exchange. *Mater. Des.* 2019, 186, 108248. [CrossRef]
24. Zheng, N.; Xu, Y.; Zhao, Q.; Xie, T. Dynamic covalent polymer networks: A molecular platform for designing functions beyond chemical recycling and self-healing. *Chem. Rev.* 2021, 121, 1716–1745. [CrossRef] [PubMed]
25. Obadia, M.M.; Mudraboyina, B.P.; Serghei, A.; Montarnal, D.; Drockenmuller, E. Reprocessing and recycling of highly cross-linked ion-conducting networks through transalkylation exchanges of C–N bonds. *J. Am. Chem. Soc.* 2015, 137, 6078–6083. [CrossRef] [PubMed]
26. Zheng, N.; Fang, Z.; Zou, W.; Zhao, Q.; Xie, T. Thermoset shape-memory polyurethane with intrinsic plasticity enabled by transcarbamoylation. *Angew. Chem. Int. Ed.* 2016, 128, 11593–11597. [CrossRef]
27. Yan, P.; Zhao, W.; Jiang, L.; Wu, B.; Hu, K.; Yuan, Y.; Lei, J. Reconfiguration and shape memory triggered by heat and light of carbon nanotube–polyurethane vitrimer composites. *J. Appl. Polym. Sci.* 2018, 135, 45784. [CrossRef]
28. Huang, X.; Liu, H.; Fan, Z.; Wang, H.; Huang, G.; Wu, J. Hyperbranched polymer toughened and reinforced self-healing epoxy vitrimer. *Polym. J.* 2019, 50, 535–542.
29. Huang, L.; Yang, Y.; Niu, Z.; Wu, R.; Fan, W.; Dai, Q.; He, J.; Bai, C. Catalyst-free vitrimer cross-linked by biomass-derived compounds with mechanical robustness, reprocessability and multi-shape memory effects. *Macromol. Rapid Commun.* 2021, 42, 2100432. [CrossRef] [PubMed]

30. Capelot, M.; Unterlass, M.M.; Tournilhac, F.; Leibler, L. Catalytic control of the vitrimer glass transition. *ACS Macro Lett.* **2012**, *1*, 789–792. [CrossRef]
31. Dao, T.D.; Ha, N.S.; Goo, N.S.; Yu, W.R. Design, fabrication, and bending test of shape memory polymer composite hinges for space deployable structures. *J. Intell. Mater. Syst. Struct.* **2018**, *29*, 1560–1574. [CrossRef]
32. Cao, X.; Jiang, C.; Feng, H.; Kong, J.; Xu, Y.; Wang, L.; Wu, S. Design and Implementation of the Deployable Sunshield on GF-7 Satellite Remote Sensing Camera. *Spacecr. Recovery Remote Sens.* **2020**, *41*, 67–77.
33. Ren, H.; Sun, J.; Wu, B.; Zhou, Q. Synthesis and characterization of a novel epoxy resin containing naphthyl/dicyclopentadiene moieties and its cured polymer. *Polymer* **2006**, *47*, 8309–8316. [CrossRef]
34. Ren, H.; Sun, J.; Zhao, Q.; Zhou, Q.; Ling, Q. Synthesis and characterization of a novel heat resistant epoxy resin based on N, N′-bis (5-hydroxy-1-naphthyl) pyromellitic diimide. *Polymer* **2008**, *49*, 5249–5253. [CrossRef]
35. Park, S.; Jin, F.; Lee, J. Thermal and mechanical properties of tetrafunctional epoxy resin toughened with epoxidized soybean oil. *Mater. Sci. Eng. A Struct.* **2004**, *374*, 109–114. [CrossRef]
36. Lin, C.H.; Feng, Y.R.; Dai, K.H.; Chang, H.C.; Juang, T.Y. Synthesis of a benzoxazine with precisely two phenolic OH linkages and the properties of its high-performance copolymers. *J. Polym. Sci. Polym. Chem.* **2013**, *51*, 2686–2694.
37. Wu, J.; Gao, L.; Guo, Z.; Zhang, H.; Zhang, B.; Hu, J.; Li, M.H. Natural glycyrrhizic acid: Improving stress relaxation rate and glass transition temperature simultaneously in epoxy vitrimers. *Green Chem.* **2021**, *23*, 5647–5655. [CrossRef]
38. Tanpitaksit, T.; Jubsilp, C.; Rimdusit, S. Effects of benzoxazine resin on property enhancement of shape memory epoxy: A dual function of benzoxazine resin as a curing agent and a stable network segment. *Express Polym. Lett.* **2015**, *9*, 824–837. [CrossRef]
39. Ariraman, M.; Sasikumar, R.; Alagar, M. Shape memory effect on the formation of oxazoline and triazine rings of BCC/DGEBA copolymer. *RSC Adv.* **2015**, *5*, 69720–69727. [CrossRef]
40. Giebler, M.; Sperling, C.; Kaiser, S.; Duretek, I.; Schlögl, S. Epoxy-Anhydride Vitrimers from Aminoglycidyl Resins with High Glass Transition Temperature and Efficient Stress Relaxation. *Polymers* **2020**, *12*, 1148. [CrossRef]
41. Liu, Y.; Zhao, J.; Zhao, L.; Li, W.; Zhang, H.; Yu, X.; Zhang, Z. High performance shape memory epoxy/carbon nanotube nanocomposites. *ACS Appl. Mater. Interfaces* **2016**, *8*, 311–320. [CrossRef]
42. Liu, S.; Cao, X.; Zhang, J.; Han, Y.; Zhao, X.; Chen, Q. Toward Correct Measurements of Shear Rheometry. *Acta Polym. Sin.* **2021**, *52*, 406–422.

Disclaimer/Publisher's Note: The statements, opinions and data contained in all publications are solely those of the individual author(s) and contributor(s) and not of MDPI and/or the editor(s). MDPI and/or the editor(s) disclaim responsibility for any injury to people or property resulting from any ideas, methods, instructions or products referred to in the content.

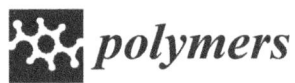 

Article

A Comparative Study on the Properties of Rosin-Based Epoxy Resins with Different Flexible Chains

Lianli Deng [1], Zehua Wang [2], Bailu Qu [3], Ying Liu [4,*], Wei Qiu [4] and Shaohe Qi [1]

[1] Hunan Provincial Key Laboratory of Xiangnan Rare-Precious Metals Compounds and Applications, School of Chemistry and Environmental Science, Xiangnan University, Chenzhou 423000, China; lianlideng@xnu.edu.cn (L.D.)
[2] School of Chemistry & Chemical Engineering, Guangxi University, Nanning 530004, China
[3] Changsha Ecological and Environmental Monitoring Centre of Hunan Province, Changsha 410001, China
[4] School of Chemical Engineering, Guizhou Institute of Technology, Guiyang 550003, China
* Correspondence: liuychem@git.edu.cn

Abstract: This study aims to reveal the effects of flexible chain lengths on rosin-based epoxy resin's properties. Two rosin-based epoxy monomers with varying chain lengths were synthesized: AR-EGDE (derived from ethylene glycol diglycidyl ether-modified acrylic acid rosin) and ARE (derived from acrylic acid rosin and epichlorohydrin). Diethylenetriamine (DETA), triethylenetetramine (TETA), and tetraethylenepentamine (TEPA) with different flexible chain lengths were used as curing agents. The adhesion, impact, pencil hardness, flexibility, water and heat resistance, and weatherability of the epoxy resins were systematically examined. It was found that when the flexible chains of rosin-based epoxy monomers were grown from ARE to AR-EGDE, due to the increased space of rosin-based fused rings, the toughness, adhesion, and water resistance of the rosin-based epoxy resins were enhanced, while the pencil hardness and heat resistance decreased. However, when the flexible chains of curing agents were lengthened, the resin's performance did not change significantly because the space between the fused rings changed little. This indicates that the properties of the rosin-based resins can only be altered when the introduced flexible chain increases the space between the fused rings. The study also compared rosin-based resins to E20, a commercial petroleum-based epoxy of the bisphenol A type. The rosin-based resins demonstrated superior adhesion, water resistance, and weatherability compared to the E20 resins, indicating the remarkable durability of the rosin-based resin.

Keywords: rosin; flexible chain; film property; bio-based epoxy; curing agent

Citation: Deng, L.; Wang, Z.; Qu, B.; Liu, Y.; Qiu, W.; Qi, S. A Comparative Study on the Properties of Rosin-Based Epoxy Resins with Different Flexible Chains. *Polymers* 2023, 15, 4246. https://doi.org/10.3390/polym15214246

Academic Editors: Jiangtao Xu and Sihang Zhang

Received: 27 September 2023
Revised: 21 October 2023
Accepted: 24 October 2023
Published: 28 October 2023

Copyright: © 2023 by the authors. Licensee MDPI, Basel, Switzerland. This article is an open access article distributed under the terms and conditions of the Creative Commons Attribution (CC BY) license (https://creativecommons.org/licenses/by/4.0/).

1. Introduction

Epoxy resins are three-dimensional network structure polymers formed by the reaction of epoxy monomer and a curing agent; thus, the properties of epoxy resins depend on the structure of the epoxy monomer and curing agent [1]. Consequently, epoxy resins containing polar groups such as hydroxyl and ether feature excellent mechanical strength, good dielectric properties, chemical resistance, strong adhesion, low shrinkage, and ease of processability [2,3], allowing for their widespread applications in different fields such as adhesives, coatings, composites, construction, and electronics. However, almost 70% of the world's production of epoxy resin is derived from bisphenol A (BPA), a petroleum-based compound [4,5]. Using renewable resources instead of BPA can protect petrochemicals, fix CO_2, accelerate the C-cycle process, and reduce greenhouse gases, all the while meeting society's requirements. A number of studies have demonstrated that humans could overcome the resource problem by utilizing just seven percent of the biomass that exists [6].

Rosin, a renewable resource that is abundantly available, is produced from pine and conifer trees, with a yearly global yield of approximately 1.27 million tons [7,8]. It is a mixture containing around 75% rosin acid, and the rosin acid molecules consist of a fused ring and active functional groups, such as double bonds and a carboxyl group, which are easily chemically modified. Diels–Alder type reactions produce rosin derivatives such as fumaropimaric acid (FPA), acrylicpimaricacid (APA), maleopimaric acid (MPA), and methyl maleopidate (MMP). Owing to their structural characteristics, low cost, biodegradability, biocompatibility, and corrosion resistance, rosin and its derivatives are valuable feedstocks for an array of polymers [9,10] and have found a broad range of applications [8], such as coatings [11–18], packaging [19,20], electrical equipment [6], antibacterial and antiviral polymers [19,21–24], surfactants, and adhesives [25–29]. Hence, the replacement of BPA with rosin and its derivatives as a fossil feedstock has gained increasing attention [4,5,30–33]. It has been demonstrated that the mechanical and thermal properties of rosin-based epoxy resin are comparable to those of bisphenol A epoxy resin, due to the analogous rigidity of the fused ring of rosin and the benzene ring of bisphenol A. It enables rosin to be a potential partial or full substitute for petroleum-based epoxy monomers and curing agents [2,30,34–42]. On the other hand, in order to improve the mechanical and thermal properties of bio-based epoxy resins and vitrimers such as vegetable oil-based epoxy resins that meet the application requirements of industry, rosin-based derivatives are often introduced as epoxy monomers or curing agents [43,44]. However, the fused ring of rosin-based epoxy resin is not only rigid but also brittle, leading to brittle fractures in reaction to external pressures and a decrease in its mechanical properties. For example, MPA (rosin-based curing agents) was employed by Liu et al. as a co-curing agent for DGEBA. As the MPA content grew, the material's mechanical parameters fell. The flexible chain was incorporated into the rosin-based epoxy resin to reduce the brittleness of the rosin-based fused ring to obtain a resin with improved performance. The effect of the quantity of flexible chains on the mechanical properties of rosin-based epoxy resin was studied in our previous research [45]. It was shown that the introduction of appropriate flexible chains into the rosin-based epoxy monomer can reduce brittleness and increase toughness, thereby improving its mechanical properties. Huang et al. combined the rigid rosin epoxy monomer with a flexible dimer acid epoxy monomer, producing resins with enhanced mechanical and thermal properties [46]. Li et al. combined curing agents derived from rigid rosin and flexible fatty acids to achieve balanced mechanical and thermal properties [47].

However, to our knowledge, the effects of epoxy monomers and curing agents with different flexible chain lengths on the properties of rosin-based epoxy resins have not been studied systematically, nor have the effects of flexible chains with different structures been compared. In this study, rosin-based epoxy monomers with various flexible chains were obtained: ARE, a diglycidyl ester derived from acrylic acid rosin and epichlorohydrin that lacked flexible chains, and AR-EGDE, an ethylene glycol diglycidyl ether-modified acrylic acid rosin with flexible chains. Diethylenetriamine (DETA), triethylenetetramine (TETA), and tetraethylenepentamine (TEPA) with different chain lengths were used concurrently to cure the aforementioned epoxy resins. The impacts of epoxy monomers and curing agents with various flexible chains on the adhesion, impact, pencil hardness, flexibility, water and heat resistance, and weatherability of rosin-based epoxy resins were studied. In the meantime, the effects of two distinct flexible chains on the properties of rosin-based epoxy resin were compared, and the structure of the flexible chain required to increase its toughness was determined. In addition, commercial epoxy resin was utilized for comparison purposes. According to the results, the bio-based resin with superior adhesion outperforms the commercial one (DGEBA).

2. Experimental

2.1. Materials

Wuzhou Sun Shine Forestry & Chemicals Co., Ltd. (Wuzhou, China) provided the acrylic acid rosin (ARA) (238 mg KOH·g^{-1}). Ethylene glycol diglycidyl ether (EGDE) with an epoxy value of 0.73 mol/100 g was acquired from Wuhan Yuancheng Create Technology Co. (Wuhan, China) E-20 (DGEBA with an epoxy value of 0.20 mol/100 g) was purchased from Wuxi Resin Factory of Blue Star New Chemical Material Co., Ltd. (Wuxi, China). From Sinopharm Group Chemical Reagent Co., Ltd. (Shanghai, China), diethylenetriamine (DETA) and triethylenetetramine (TETA) were purchased. Epichlorohydrin (EC), ethanol (EtOH), acetone, potassium hydroxide (KOH), and triethylamine (Et$_3$N) were of chemically pure quality and were utilized directly.

2.2. Preparation of Epoxy Monomers

2.2.1. ARE, Diglycidyl Ester from ARA (Figure 1)

ARA (100.00 g; 0.42 mol of carboxyl group), EC (156.67 g; 1.68 mol of epoxy group), and Et$_3$N (0.51 g; 0.2 wt% of the total weight of ARA and EC) were added to a 500 mL four-neck round-bottom flask equipped with a reflux condenser, a mechanical stirrer, a thermometer, and an N$_2$ inlet. The temperature of the system was raised to 110 °C and maintained until the acid number declined below 0.5 mg KOH/g. After cooling to 70 °C, 0.42 mol of solid NaOH (17.00 g) was added, and the temperature was maintained for 3 h. After the reaction, the precipitate was filtered out, and the filtrate was neutralized with water. After separation, the organic layer was evaporated at approximately 50 °C to recover excess epichlorohydrin, resulting in a yellow, transparent, viscous liquid.

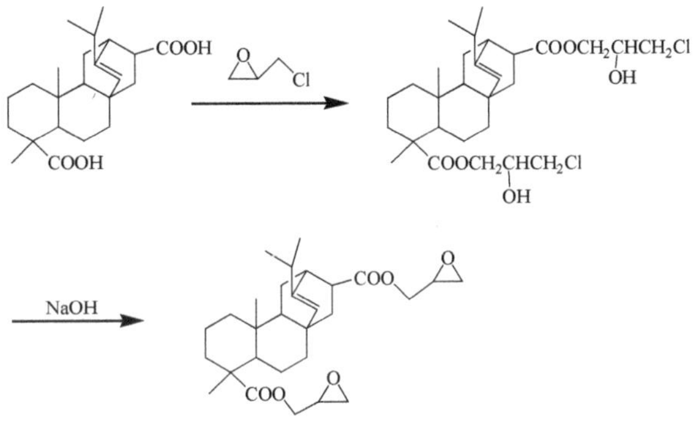

Figure 1. Synthesis route of ARE.

2.2.2. AR-EGDE, the Copolymers of Ethylene Glycol Diglycidyl Ether-Modified ARA (Figure 2)

EGDE (116.50 g, 0.85 mol of epoxy group) and ARA (50 g, 0.21 mol of carboxyl group) were added to a 500 mL four-necked flask equipped with a reflux condenser, agitator, thermometer, and nitrogen tubing. Et$_3$N (0.02% of the ARA mass) was added as a catalyst once the acrylic rosin had been completely dissolved. After reacting at 130 °C for 1 h, an additional 50.00 g of ARA (0.2117 mol carboxyl group) was added, and the reaction temperature was maintained until the acid value fell below 0.5 mg KOH/g, at which point the reaction was terminated. The product was a yellow, viscous, transparent liquid.

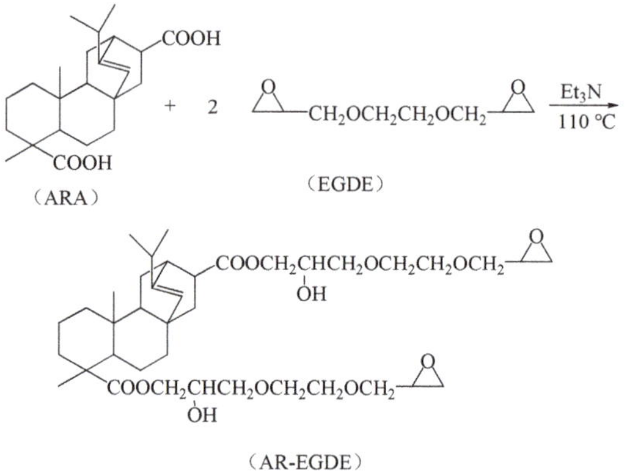

Figure 2. Synthesis route of AR-EGDE.

2.3. Preparation of Test Samples

Stoichiometrically, the epoxy monomers and curing agents were blended. Then, certain amounts of mixed solvents (xylene and butanol) were added to the above blends. The ratio of xylene to butanol by mass was 70:30. The method yielded a mixture that was aged for 30 min before being coated on 120 mm × 50 mm tinplates. This was carried out in accordance with GB/T 1727-2021 [48]. For each test, at least three tinplates were prepared. The formulations for the epoxy resins are listed in Table 1, and Figure 3 depicts the structure of epoxy resin based on rosin. Simultaneously, the epoxy monomer and curing agent were theoretically combined to produce 80 mm × 10 mm × 4 mm strip resin samples. The curing procedure required 12 h at room temperature, followed by 4 h at 100 °C in the oven.

Table 1. The formulas of epoxy resins.

Epoxy Resins	Epoxy Monomer	Curing Agent
ARED	ARE	DETA
ARET	ARE	TETA
AREP	ARE	TEPA
AR-EGDED	AR-EGDE	DETA
AR-EGDET	AR-EGDE	TETA
E20D	E20	DETA
E20T	E20	TETA

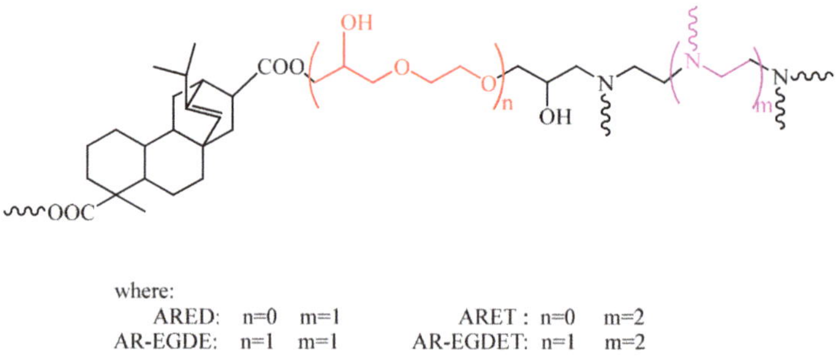

where:
ARED: n=0 m=1   ARET : n=0   m=2
AR-EGDE: n=1 m=1   AR-EGDET: n=1 m=2

Figure 3. Rosin-based epoxy resins with different flexible chains.

2.4. Characterization

Using the KBr method, an RFX-65A (Analect) FTIR instrument was used to acquire the infrared spectra of the prepared epoxy monomers.

The hardness was assessed in accordance with GB/T 6739-2006 [49].

Using the paint film scriber A2012058 (Elcometer), the adhesion was evaluated using a cross-cut test per GB/T 9286-2021 [50].

The impact resistance was evaluated in accordance with GB/T 1732-2020 [51], using a 20 cm impact height and a 1 kg mallet.

The flexibility was determined via GB/T 1731-2020 [52].

The heat resistance was measured as per GB/T 1735-2009 [53]. The tinplates were placed in an oven at 100 °C for 3 days. They were cooled to room temperature and examined for color changes or other signs of coating deterioration.

The tinplates were placed in the sun for 60 days to test weatherability.

For 49 days, painted tinplates were immersed vertically in 3000 mL water-filled glass beakers to study the water resistance behavior of the films. Regular inspections of the tinplates were conducted to assess the level of attack on the paint coatings and the substrate. At various intervals, the tinplates were removed, rinsed with water, and visually inspected for film integrity, overall appearance, and film failure.

The swelling behavior was investigated by immersing strip-shaped samples in water at room temperature. The percent change in mass was obtained using the following equation: percent change in mass = $(W_2 - W_1)/W_1 \times 100\%$, where W_2 and W_1 are the weights after and before absorption, respectively.

3. Results

3.1. Characterization of Epoxy Resin Monomers

By reacting the carboxyl groups of ARA with EC or EGDE, respectively, it is possible to generate ester-containing ARE and AR-EGDE. Figures 1 and 2 depict the formulations of the chemical reactions. The FTIR spectra of the EGDE, ARA, and epoxy monomers (AR-EGDE and ARE) are shown in Figure 4. For ARE and AR-EGDE, the broad absorption peaks attributed to the -COOH of ARA between 3000 and 3500 cm^{-1} and around 1698 cm^{-1} nearly disappeared. Simultaneously, the conspicuous absorption peaks at 1726 cm^{-1}, ascribed to the elongation of carbonyl groups in the ester groups are clearly visible. This indicates that the ARE and AR-EGDE formed. The epoxy groups were also confirmed by absorption bands at 1245, 910, and 852 cm^{-1}. The acid number of the reaction mixture decreased over the course of the reaction, reaching 0.5 mg KOH/g at the conclusion. This also demonstrated the esterification of the carboxyl and epoxy groups. In addition, the epoxy values for ARE and AR-EGDE were 0.35 mol·100 g^{-1} and 0.20 mol·100 g^{-1}, respectively.

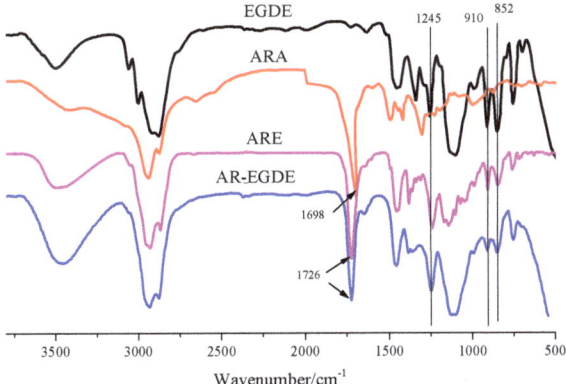

Figure 4. FTIR spectra of EGDE, ARA, ARE, and AR-EGDE.

3.2. Film Properties of the Rosin-Based Epoxy Resins with Different Flexible Chains

The rigidity and brittleness of a rosin-based fused ring structure will have distinct effects on the material properties [45]. Epoxy resin film properties, such as the pencil hardness, adhesion, flexibility, and impact resistance, are listed in Table 2.

Table 2. Results of film performance.

Coating Sample	Pencil Hardness	Grading of Cross-Cut Tests	Flexibility	Impact Resistance
ARED	3H	3	Destroyed	Cracks and spalling
ARET	3H	3	Destroyed	Cracks and spalling
AREP	-	3	-	Cracks and spalling
AR-EGDED	2H	0	Not affected	No cracks and spalling
AR-EGDET	2H	0	Not affected	No cracks and spalling
E20D	3H	5	Not affected	No cracks and spalling, slight wrinkles
E20T	3H	5	Not affected	and stress whitening phenomenon.

3.2.1. The Effects of Epoxy Monomer Flexible Chain on the Film Properties of Rosin-Based Epoxy Resin

As shown in Table 3, the hardness of the ARE coatings (ARED and ARET) was 3H. Upon replacing ARE with AR-EGDE, the hardness of AR-EGDED and AR-EGDET decreased to 2H when the epoxy monomer flexible chain length was increased. This is due to the presence of more rosin-based fused rings in the former. The rigidity of the rosin-based fused ring results in an increase in the hardness value as its content increases. However, the adhesion test results were the opposite of the hardness test results. The evaluation for adhesion, which refers to the capacity of the coating to adhere to the substrate, was conducted utilizing the cross-cut test. This test involves the classification of adhesion on a numerical scale ranging from 0 to 5. A smaller numerical value is indicative of superior adhesion. For the AR-EGDE coatings with flexible chains, the grade was 0, with no signs of damage, while for the ARE coatings, the grade was 2-3, with flaking of about 5–35 percent. The potential reason for this phenomenon can be attributed to the inherent characteristics of the rosin-based fused ring structure, which possesses materials of both rigidity and brittleness. The presence of brittleness has a negative impact on the level of adhesion. The incorporation of a flexible chain into rosin-based epoxy resin leads to a reduction in brittleness, thereby resulting in enhanced adhesion properties [45,54]. In order to find out the reason, the flexibility and impact resistance of the epoxy coatings were tested. The AR-EGDE coatings with flexible chains have better flexibility and impact resistance than the ARE coatings(see Table 3). It suggests that introducing a flexible chain of glycidyl ether into rosin epoxy monomers increases the toughness and impact resistance of rosin-based epoxy coatings, and thus improves adhesion.

Table 3. Water absorption rates of cured resins.

Epoxy Resins	16 h	24 h	48 h
ARED	0.24%	0.40%	0.57%
ARET	0.18%	0.39%	0.60%
AR-EGDED	5.78%	6.54%	9.35%
AR-EGDET	6.37%	7.30%	10.26%
E20D	0.26%	0.54%	1.34%
E20T	0.25%	1.14%	1.76%

3.2.2. The Effects of Curing Agent Flexible Chains on the Film Properties of Rosin-Based Epoxy Resin

The test results of pencil hardness, adhesion, flexibility, and impact resistance of rosin-based epoxy resins remained nearly identical when the epoxy monomers remained unchanged, but the curing agent was switched from DETA, which possesses a short flexible chain, to TETA, which possesses a long flexible chain. For instance, the ARE coatings exhibited a pencil hardness value of 2H, a cross-cut test grade of 3, and failed in the flexibility and impact resistance tests, regardless of the type of curing agent used. The above results show that the use of the amine curing agents, which possess flexible chains of longer lengths, does not improve the brittleness of the rosin-based fused ring. Consequently, it does not lead to enhancements in impact resistance, flexibility, and adhesion while reducing the hardness of the rosin-based epoxy resin.

3.2.3. A Comparison of Film Properties between the Rosin-Based Epoxy Resin and Bisphenol A Epoxy Resin

The film properties of bisphenol A epoxy (E20) resin were also found to be unaffected by the flexible chain length of its curing agent. The pencil hardness of the E20 coatings was 3H, which was equivalent to that of the ARE coatings. This indicates that the rosin-based fused ring structure is comparable to the rigidity of the benzene ring of the bisphenol A type. The results of the E20 coating tests for flexibility and impact resistance were superior to those of ARE, indicating that the brittleness of rosin-based fused rings is greater than that of benzene rings. However, the cross-test grades of the E20 coatings were 5, which were not only lower than the AR-EGDE coatings with flexible chains, but also lower than that of the ARE coating, which indicates that the rosin-based fused ring has better adhesion than the benzene ring of bisphenol A epoxy resin.

3.3. Water Resistance Tests of the Rosin-Based Epoxy Resin with Different Flexible Chain Lengths

Figure 5 shows test images of tinplates immersed in water. The ARE epoxy coating without flexible chains turned white, whereas the AR-EGDE epoxy coating with flexible chains became wrinkly. This indicates that the water resistance of the rosin-based epoxy coating decreases as the flexible chain length of the epoxy monomers increases [55]. The reason for this is that as the flexible chain increases, the rigidity of the rosin-based fused ring decreases, and the cross-linking density of AR-EGDE resins decreases. Moreover, the flexible chain of the epoxy monomers contains hydroxyl groups and ether bonds that can form hydrogen bonds with water (see Figure 3). This conclusion was also supported by the water absorption tests of cured resins. In Table 3 and Figure 6, the AR-EGDE resins had 10 times higher water absorption rates than the ARE resins. When the epoxy monomers were identical and the curing agents were different, such as ARED and ARET, AR-EGDED and AR-EGDET, there was little variation in the water resistance of the resins. This demonstrates that the length of fatty amine curing agents has little effect on the water resistance of epoxy coatings.

There was no whitening phenomenon observed in E20 epoxy coatings, but there was slight rusting. The water resistance of the E20 epoxy coatings was even worse than that of the AR-EGDE from a rust perspective. In most cases, a material's low water resistance results from its high water absorption. Nevertheless, the water absorption rates of the E20 coatings were significantly lower than those of AR-EGDE coatings (Figure 6). This suggests that the fused ring structure of the AR-EGDE resins is more resistant to water than the benzene ring of bisphenol A types, which should be related to their superior adhesion.

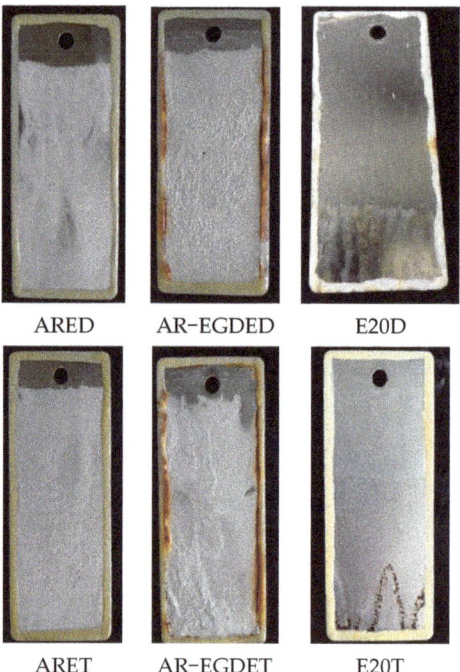

Figure 5. The images of immersion tests with water for 49 days.

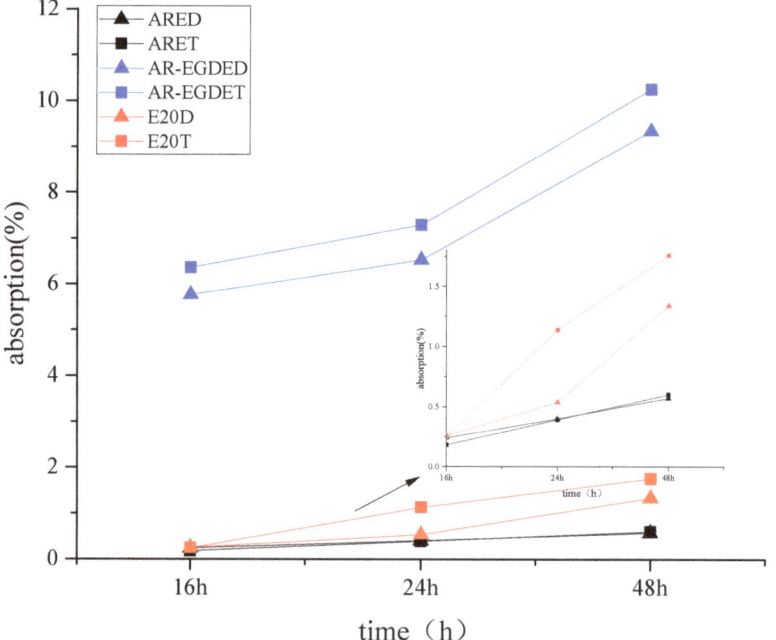

Figure 6. Water absorption rates of cured resins.

3.4. Heat Resistance of the Rosin-Based Epoxy Resin with Different Flexible Chain Lengths

Figure 7 depicts photographs of each epoxy coating heated in an oven at 100 °C for three days. Based on Figure 7, the AR-EGDE coatings were the darkest brown, followed by the ARE coatings in golden yellow and the E20 in pale yellow when the same curing agent was used. The lower the heat resistance, the darker the color. Accordingly, the heat resistance decreases with increasing flexible chains and decreasing rosin-based fused rings, indicating that the fused ring structure has some heat resistance, but its heat resistance at 100 °C is inferior to that of bisphenol A epoxy resin. When the epoxy monomers are identical, there is no significant color difference between the ARED and ARET, E20D, and E20T coatings. This indicates that changes in the flexible chain of the curing agent have no effect on heat resistance. However, AR-EGDED has a slightly darker hue than AR-EGDEDT, indicating that the longer the flexible chain length, the greater the heat resistance of AR-EGDEDT.

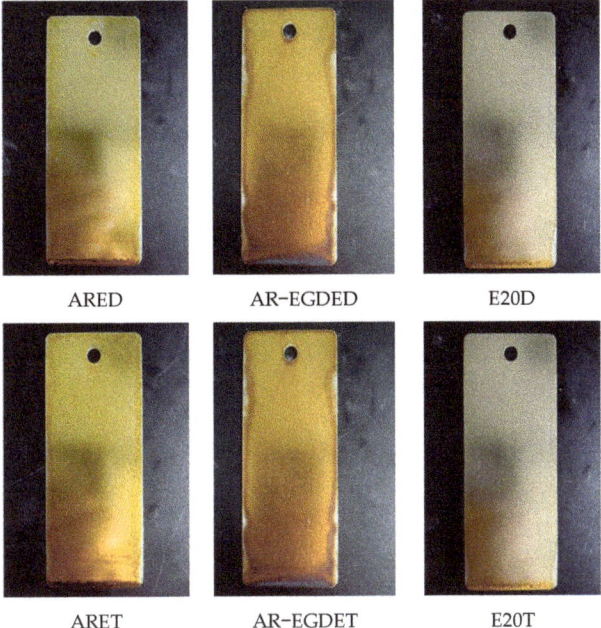

Figure 7. The images of the samples after heat resistance.

In addition, TGA was used to evaluate the thermal degradation of the material. Figure 8 depicts the TG and DTG as diagrams. From the TG curve, the temperature at which degradation begins ($T_{5\%}$) and the remaining solid residue at 700 °C were determined. From the DTG curve, the maximal weight loss rate temperatures (T_{max}) were determined. The information is shown in Table 4. Figure 8b,d show that the T_{max} of rosin-based epoxy resins has two distinct peaks, and their second peak height (approximately 420 °C) is nearly identical. It suggests that the first peaks correspond to the breakdown of the flexible chain segment, while the second peaks correspond to the breakdown of the rosin-based fused ring. The first T_{max} of the E20 resins exhibits a lower value compared to rosin-based epoxy resins, accompanied by a notably weaker peak intensity. Consequently, the $T_{5\%}$ of the E20 resins is lower than that of rosin-based epoxy resin. On the other hand, the second T_{max} which corresponds to the degradation of the benzene ring, is higher in the E20 resins compared to rosin-based epoxy resin. This demonstrates that, in comparison to rosin-based epoxy resin, the thermal degradation of bisphenol A epoxy resin is weaker at low temperatures and stronger at high temperatures.

Table 4. Thermal properties of the epoxy resins.

Cured Resins	$T_{5\%}$	T_{max}	Residue at 700 °C (%)
ARED	259.93	350.27, 423.73	16.883
ARET	249.9	340.69, 419.88	26.248
AR-EGDED	287.1	343.21, 418.98	8.549
AR-EGDET	268.41	346.11, 413.81	8.612
E20D	203.94	193.69, 434.1	16.254
E20T	208.8	188.83, 433.94	20.368

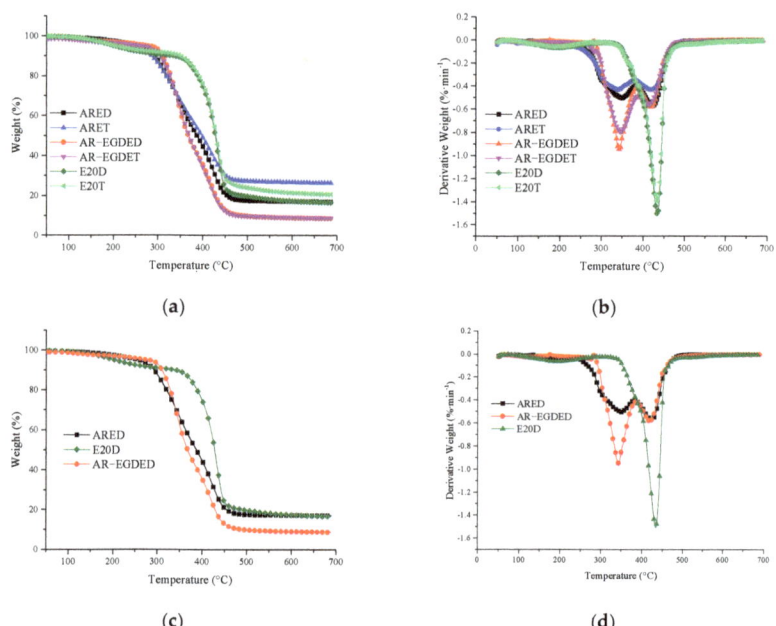

Figure 8. Thermal degradation curve of epoxy resins: (**a**) TG curves of epoxy resins; (**b**) D TG curves of cured resins; (**c**) TG curves of epoxy resins from DETA; (**d**) DTG curves of epoxy resins from DETA.

3.5. Weather Resistance of Rosin-Based Epoxy Resin Using Monomers and Curing Agents Containing Different Amounts of Flexible Chains

Figure 9 displays photographs depicting the coatings' conditions following the 60-day outdoor weather resistance tests. The specific outcomes of these tests are provided in Table 5. It is evident that none of the rosin-based coatings sustained any damage, whereas the E20 coatings exhibited wrinkling, flaking, and pulverization. The information suggests that the fused ring in rosin-based epoxy resin exhibits superior weather resistance properties compared to bisphenol A epoxy resin. This characteristic has the potential to compensate for the inadequate weather resistance of bisphenol A epoxy resin.

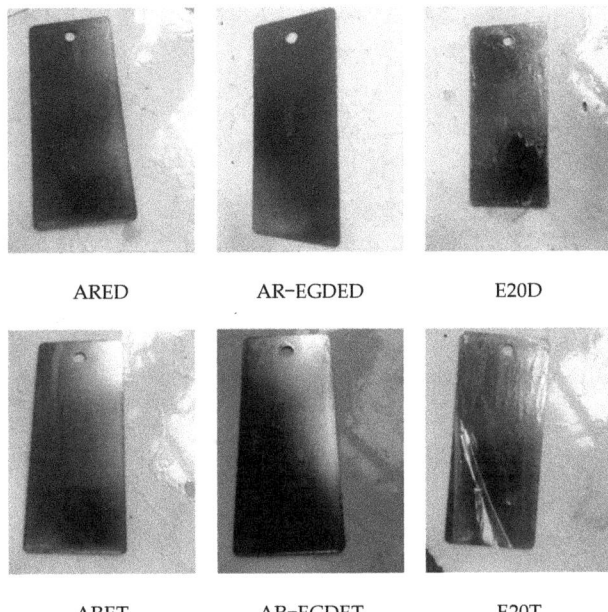

Figure 9. Photographs of the samples after the weatherability tests for 60 days.

Table 5. Results of the weatherability tests for 60 days.

Coating Sample	Observations
ARED	unchanged
ARET	unchanged
AR-EGDED	unchanged
AR-EGDET	unchanged
E20D	wrinkling, flaking, and pulverized
E20T	wrinkling, flaking, and pulverized

3.6. Effects of Flexible Chains on Properties of Rosin-Based Cured EPOXY Coating

When the curing agent remained the same, the adhesion, flexibility, and impact resistance of the rosin-based epoxy resins improved as the flexible chain of the epoxy monomer increased from ARE to AR-EGDE. This indicates that the introduction of flexible chains decreased the brittleness of rosin-based fused rings and enhanced the toughness of rosin-based epoxy resins. When the epoxy monomer remained unchanged and the curing agent's flexible chain grew from DETA to TETA, the performance of the resin, including its flexibility and impact resistance, remained relatively constant. This demonstrates that although the flexible chain of the curing agent was lengthened, the brittleness of the resin was not significantly reduced, so the toughness was not enhanced.

To investigate the reasons, structural fragments of three rosin-based epoxy resins were illustrated (see Figure 10). By comparing the spacing of the fused rings in ARED and AR-EGDED (rosin-based epoxy resins using the same curing agents and different monomers), it can be seen that the spacing increases with the growth of the flexible chain, whereas the spacing of ARED and ARET (rosin-based epoxy resins with the same epoxy monomer and different curing agents) changes very little with the growth of the flexible chain of the curing agent. This is because the flexible chain of the epoxy monomer did not cross-link in the resin, so the introduction of the flexible chain increased the spacing of the fused ring; the N atom in the flexible chain of the curing agent was cross-linked with the epoxy monomer containing the fused ring, which increases the number of fused rings while extending the flexible chain, resulting in a minimal change in the spacing of the fused rings. The spacing between fused rings is a significant factor in determining toughness. As the spacing widens, the molecular chain of motion increases, resulting in an improvement in toughness [35]. The spacing does not change significantly, and thus neither does the toughness.

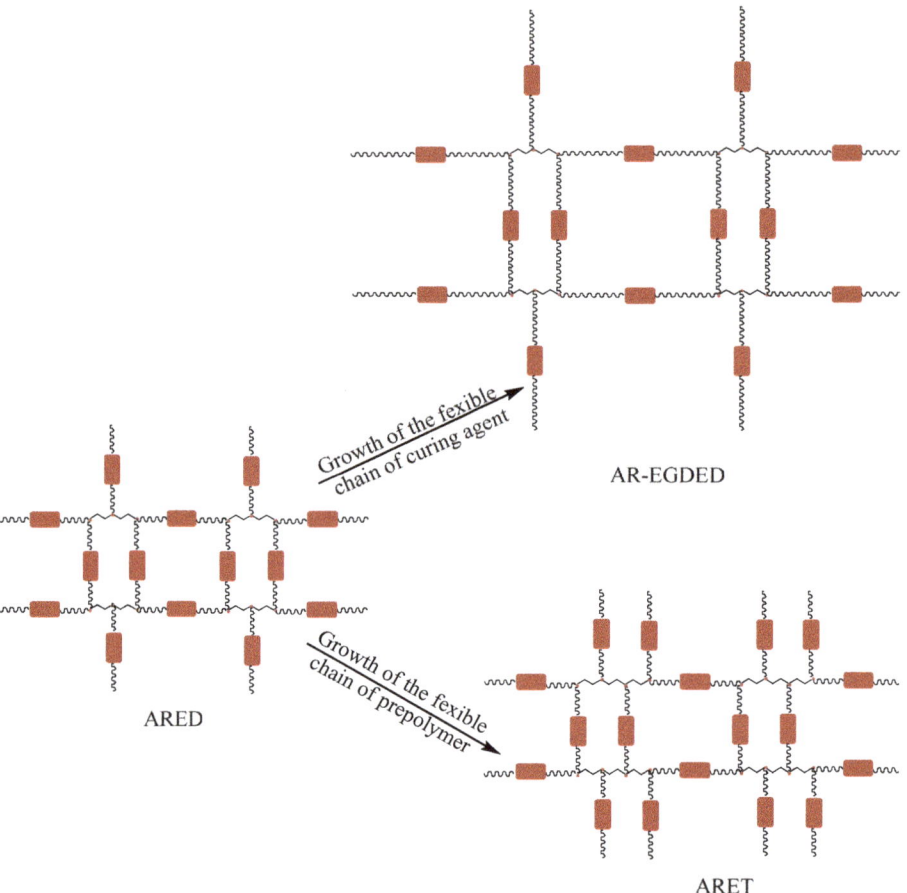

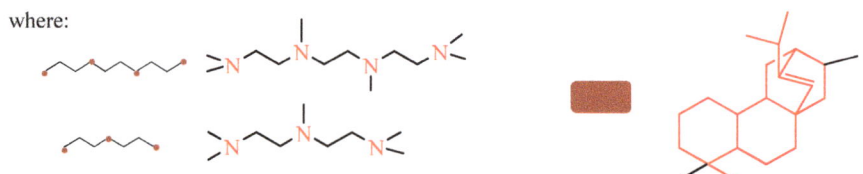

Figure 10. Structural diagram of acrylic rosin-based epoxy resin.

To reduce the negative effects of the brittleness of rosin-based epoxy resin and increase its toughness, it is not necessary to simply increase the flexible chains of the epoxy monomer and curing agent; rather, the structure of the cured resin must be designed so that the rigid rosin-based fused ring is separated by an appropriate distance.

4. Conclusions

The adhesion, flexibility, impact resistance, water resistance, and heat resistance of rosin-based epoxy resins containing epoxy monomers and curing agents with different flexible chains were compared. When the curing agent was left unchanged and the length of the flexible chain of the epoxy monomer was increased (from ARE to AR-EGDE), the toughness of the rosin-based epoxy resins increased, resulting in an increase in adhesion and a decrease in heat resistance. When the epoxy monomer was left unchanged and the length of the flexible chain of the curing agent was increased (from DETA to TETA), the performance of the rosin-based epoxy resin was not significantly altered. Analyzing the structure of rosin-based epoxy resin revealed that changing the epoxy monomer from ARE to AR-EGDE with a longer flexible chain increased the spacing between the rosin-based fused ring, thereby decreasing the brittleness and altering its properties. However, changing the curing agent from DETA to TETA, which has a longer flexible chain, had minimal effects on brittleness and resin performance. Therefore, when introducing a flexible chain to improve the brittleness of rosin-based resins, it is necessary to increase the spacing between the fused rings in the resins. In addition, rosin-based epoxy resin outperformed bisphenol A epoxy resin E20 in adhesion and weather resistance, allowing it to have a broader range of applications.

Author Contributions: Conceptualization, L.D. and Y.L.; data curation, L.D. and Z.W.; formal analysis, L.D., Z.W. and B.Q.; funding acquisition, Y.L., L.D. and W.Q.; investigation, B.Q.; methodology, Z.W., S.Q. and W.Q.; project administration, L.D.; software, Z.W. and S.Q.; validation, Z.W.; writing—original draft, L.D., B.Q. and Y.L.; writing—review and editing, L.D. and Y.L. All authors have read and agreed to the published version of the manuscript.

Funding: This research was funded by the National Natural Science Foundation of China (No. 52160010), the Doctoral Initiation Fund of Xiangnan University (2020 No. 11; QiankeheLH [2014]7368), the Science and Technology Plan Project of Guizhou Province (QiankeheJichu ZK[2023]yiban 132), the Excellent Talents Award Plan of Guizhou Education Department (Grant No. Qianjiaohe KY [2014]245), the Academic New Seedling Plan Project of Guizhou Institute of Technology (GZLGXM-16), the Guizhou Science and Technology Support Project (Granted No. Qiankehe Zhicheng [2017]2817; [2018]2835), and the Startup Foundation for Advanced Talents of Guizhou Institute of Technology (XJGC20190623; XJGC20190646).

Institutional Review Board Statement: Not applicable.

Data Availability Statement: The authors confirm that the data supporting the findings of this study are available within the article.

Conflicts of Interest: The authors declare no conflict of interest.

References

1. Thakur, T.; Jaswal, S.; Parihar, S.; Gaur, B.; Singha, A.S. Bio-based epoxy thermosets with rosin derived imidoamine curing agents and their structure-property relationships. *Express Polym. Lett.* **2020**, *14*, 512–529. [CrossRef]
2. Ding, C.; Matharu, A.S. Recent Developments on Biobased Curing Agents: A Review of Their Preparation and Use. *ACS Sustain. Chem. Eng.* **2014**, *2*, 2217–2236. [CrossRef]
3. Kumar, S.; Samal, S.K.; Mohanty, S.; Nayak, S.K. Recent Development of Biobased Epoxy Resins: A Review. *Polym. Technol. Eng.* **2016**, *57*, 133–155. [CrossRef]
4. Liu, J.; Zhang, L.; Shun, W.; Dai, J.; Peng, Y.; Liu, X. Recent development on bio-based thermosetting resins. *J. Polym. Sci.* **2021**, *59*, 1474–1490. [CrossRef]
5. Gonçalves, F.A.M.M.; Santos, M.; Cernadas, T.; Ferreira, P.; Alves, P. Advances in the development of biobased epoxy resins: Insight into more sustainable materials and future applications. *Int. Mater. Rev.* **2021**, *67*, 119–149. [CrossRef]
6. Liu, H.; Liu, X.; Liu, Y.; Guo, Z.; Guo, Q.; Sun, Z. The curing characteristics and properties of bisphenol A epoxy resin/maleopimaric acid curing system. *J. Mater. Res. Technol.* **2022**, *21*, 1655–1665. [CrossRef]
7. El-Ghazawy, R.A.; El-Saeed, A.M.; Al-Shafey, H.I.; Abdul-Raheim, A.R.M.; El-Sockary, M.A. Rosin based epoxy coating: Synthesis, identification and characterization. *Eur. Polym. J.* **2015**, *69*, 403–415. [CrossRef]
8. Kugler, S.; Ossowicz, P.; Malarczyk-Matusiak, K.; Wierzbicka, E. Advances in Rosin-Based Chemicals: The Latest Recipes, Applications and Future Trends. *Molecules* **2019**, *24*, 1651. [CrossRef]

9. Mandal, M.; Borgohain, P.; Begum, P.; Deka, R.C.; Maji, T.K. Property enhancement and DFT study of wood polymer composites using rosin derivatives as co-monomers. *New J. Chem.* **2018**, *42*, 2260–2269. [CrossRef]
10. Zhang, H.; Yang, Y.; Shen, M.; Shang, S.; Song, J.; Jiang, J.; Song, Z. Soybean oil-based thermoset reinforced with rosin-based monomer. *Iran. Polym. J.* **2018**, *27*, 405–411. [CrossRef]
11. Mirabedini, S.M.; Zareanshahraki, F.; Mannari, V. Enhancing thermoplastic road-marking paints performance using sus-tainable rosin ester. *Prog. Org. Coat.* **2020**, *139*, 105454. [CrossRef]
12. Zhou, W.; Wang, Y.; Ni, C.; Yu, L. Preparation and evaluation of natural rosin-based zinc resins for marine antifouling. *Prog. Org. Coat.* **2021**, *157*, 106270. [CrossRef]
13. Frances, M.; Gardere, Y.; Rubini, M.; Duret, E.; Leroyer, L.; Cabaret, T.; Athomo, A.B.B.; Charrier, B. Effect of heat treatment on Pinus pinaster rosin: A study of physico chemical changes and influence on the quality of rosin linseed oil varnish. *Ind. Crop. Prod.* **2020**, *155*, 112789. [CrossRef]
14. Dizman, C.; Ozman, E. Preparation of rapid (chain-stopped) alkyds by incorporation of gum rosin and investigation of coating properties. *Turk. J. Chem.* **2020**, *44*, 932–940. [CrossRef]
15. Nakanishi, E.Y.; Cabral, M.R.; Fiorelli, J.; Christoforo, A.L.; Gonçalves, P.d.S.; Junior, H.S. Latex and rosin films as alternative waterproofing coatings for 3-layer sugarcane-bamboo-based particleboards. *Polym. Test.* **2019**, *75*, 284–290. [CrossRef]
16. Li, Z.; Yang, X.; Liu, H.; Yang, X.; Shan, Y.; Xu, X.; Shang, S.; Song, Z. Dual-functional antimicrobial coating based on a qua-ternary ammonium salt from rosin acid with in vitro and in vivo antimicrobial and antifouling properties. *Chem. Eng. J.* **2019**, *374*, 564–575. [CrossRef]
17. Saat, A.M.; Yaakup, S.; Alaauldin, S.; Johor, H.; Azaim, F.Z.Z.; Isa, M.D.M.; Kamil, M.S.; Samsudin, S.; Lokman, M.I. Performance of rosin modified antifouling coated on mild steel surface at various immersion orientation. *Int. J. Innov. Technol. Explor. Eng.* **2019**, *8*, 5562–5565. [CrossRef]
18. Atta, A.M.; El-Saeed, A.M.; El-Mahdy, G.M.; Al-Lohedan, H.A. Application of magnetite nano-hybrid epoxy as protective marine coatings for steel. *RSC Adv.* **2015**, *5*, 101923–101931. [CrossRef]
19. Leiteabc, L.S.F.; Bilattob, S.; Paschoalinc, R.T.; Soaresb, A.C.; Moreirad, F.K.V.; Oliveirac, O.N.; Mattosoab, L.H.C.; Brasef, J. Eco-friendly gelatin films with rosin-grafted cellulose nanocrystals for antimicrobial packaging. *Int. J. Bio-Log. Macromol.* **2020**, *165*, 2974–2983. [CrossRef]
20. Majeed, Z.; Khurshid, K.; Ajab, Z.; Guan, Q.; Ahmad, B.; Mumtaza, I.; Ahmed, M.N.; Khan, R.A.W.; Andleeb, S. Agronomic evaluation of controlled release of micro urea encapsulated in rosin maleic anhydride adduct. *J. Plant Nutr.* **2020**, *43*, 1794–1812. [CrossRef]
21. Mao, S.; Wu, C.; Gao, Y.; Hao, J.; He, X.; Tao, P.; Li, J.; Shang, S.; Song, Z.; Song, J. Pine Rosin as a Valuable Natural Resource in the Synthesis of Fungicide Candidates for Controlling *Fusarium oxysporum* on Cucumber. *J. Agric. Food Chem.* **2021**, *69*, 6475–6484. [CrossRef] [PubMed]
22. Li, Z.; Wang, S.; Yang, X.; Liu, H.; Shan, Y.; Xu, X.; Shang, S.; Song, Z. Antimicrobial and antifouling coating constructed using rosin acid-based quaternary ammonium salt and N-vinylpyrrolidone via RAFT polymerization. *Appl. Surf. Sci.* **2020**, *530*, 147193. [CrossRef]
23. Yu, C.; Yan, C.; Shao, J.; Zhang, F. Preparation and properties of rosin-based cationic waterborne polyurethane dispersion. *Colloid Polym. Sci.* **2021**, *299*, 1489–1498. [CrossRef]
24. Rosu, L.; Mustata, F.; Rosu, D.; Varganici, C.D.; Rosca, I.; Rusu, T. Bio-based coatings from epoxy resins crosslinked with a rosin acid derivative for wood thermal and anti–fungal protection. *Prog. Org. Coat.* **2020**, *151*, 106008. [CrossRef]
25. Mahendra, V. Rosin Product Review. *Appl. Mech. Mater.* **2019**, *890*, 77–91. [CrossRef]
26. Popova, L.; Ivanchenko, O.; Pochkaeva, E.; Klotchenko, S.; Plotnikova, M.; Tsyrulnikova, A.; Aronova, E. Rosin Derivatives as a Platform for the Antiviral Drug Design. *Molecules* **2021**, *26*, 3836. [CrossRef] [PubMed]
27. Pan, Y.; Ge, B.; Zhang, Y.; Li, P.; Guo, B.; Zeng, X.; Pan, J.; Lin, S.; Yuan, P.; Hou, L. Surface activity and cleaning performance of Gemini surfactants with rosin groups. *J. Mol. Liq.* **2021**, *336*, 116222. [CrossRef]
28. Xu, C.A.; Qu, Z.; Lu, M.; Meng, H.; Zhan, Y.; Chen, B.; Wu, K.; Shi, J. Effect of rosin on the antibacterial activity against S.aureus and adhesion properties of UV-curable polyurethane/polysiloxane pressure-sensitive adhesive. *Colloids Surf. A Physicochem. Eng. Asp.* **2021**, *614*, 126146. [CrossRef]
29. Pavon, C.; Aldas, M.; Hernández-Fernández, J.; López-Martínez, J. Comparative characterization of gum rosins for their use as sustainable additives in polymeric matrices. *J. Appl. Polym. Sci.* **2021**, *139*, 51734. [CrossRef]
30. Mantzaridis, C.; Brocas, A.L.; Llevot, A.; Cendejas, G.; Auvergne, R.; Caillol, S.; Carlotti, S.; Cramail, H. Rosin acid oligomers as precursors of DGEBA-free epoxy resins. *Green Chem.* **2013**, *15*, 3091–3098. [CrossRef]
31. Gao, T.Y.; Wang, F.-D.; Xu, Y.; Wei, C.X.; Zhu, S.E.; Yang, W.; Lu, H.D. Luteolin-based epoxy resin with exceptional heat resistance, mechanical and flame retardant properties. *Chem. Eng. J.* **2021**, *428*, 131173. [CrossRef]
32. Zhang, Y.; Zhai, M.; Ma, F.; Li, Y.; Lyu, B.; Gao, Z.; Wang, L.; Vincent, D.; Kessler, M.R. Fully Eugenol-Based Epoxy Thermosets: Synthesis, Curing, and Properties. *Macromol. Mater. Eng.* **2021**, *307*, 2100833. [CrossRef]
33. Nabipour, H.; Niu, H.; Wang, X.; Batool, S.; Hu, Y. Fully bio-based epoxy resin derived from vanillin with flame retardancy and degradability. *React. Funct. Polym.* **2021**, *168*, 105034. [CrossRef]
34. Brocas, A.L.; Llevot, A.; Mantzaridis, C.; Cendejas, G.; Auvergne, R.; Caillol, S.; Carlotti, S.; Cramail, H. Epoxidized rosin acids as co-precursors for epoxy resins. *Des. Monomers Polym.* **2013**, *17*, 301–310. [CrossRef]

35. Li, C.; Liu, X.; Zhu, J.; Zhang, C.; Guo, J. Synthesis, Characterization of a Rosin-based Epoxy Monomer and its Comparison with a Petroleum-based Counterpart. *J. Macromol. Sci. Part A* **2013**, *50*, 321–329. [CrossRef]
36. Liu, X.Q.; Huang, W.; Jiang, Y.H.; Zhu, J.; Zhang, C.Z. Preparation of a bio-based epoxy with comparable properties to those of petroleum-based counterparts. *Express Polym. Lett.* **2012**, *6*, 293–298. [CrossRef]
37. Liu, X.Q.; Zhang, J.W. High-performance biobased epoxy derived from rosin. *Polym. Int.* **2010**, *59*, 607–609. [CrossRef]
38. Sun, J.; Zhang, Z.; Wang, L.; Liu, H.; Ban, X.; Ye, J. Investigation on the epoxy/polyurethane modified asphalt binder cured with bio-based curing agent: Properties and optimization. *Constr. Build. Mater.* **2021**, *320*, 126221. [CrossRef]
39. Zeng, Y.; Yang, B.; Luo, Z.; Pan, X.; Ning, Z. Fully rosin-based epoxy vitrimers with high mechanical and thermostability properties, thermo-healing and closed-loop recycling. *Eur. Polym. J.* **2022**, *181*, 111643. [CrossRef]
40. Zeng, Y.; Li, J.; Liu, S.; Yang, B. Rosin-Based Epoxy Vitrimers with Dynamic Boronic Ester Bonds. *Polymers* **2021**, *13*, 3386. [CrossRef]
41. Qin, J.; Liu, H.; Zhang, P.; Wolcott, M.; Zhang, J. Use of eugenol and rosin as feedstocks for biobased epoxy resins and study of curing and performance properties. *Polym. Int.* **2013**, *63*, 760–765. [CrossRef]
42. Zhang, H.; Li, W.; Xu, J.; Shang, S.; Song, Z.Q. Synthesis and characterization of bio-based epoxy thermosets using rosin-based epoxy monomer. *Iran. Polym. J.* **2021**, *30*, 643–654. [CrossRef]
43. Yang, X.; Guo, L.; Xu, X.; Shang, S.; Liu, H. A fully bio-based epoxy vitrimer: Self-healing, triple-shape memory and reprocessing triggered by dynamic covalent bond exchange. *Mater. Des.* **2019**, *186*, 108248. [CrossRef]
44. Huang, X.; Yang, X.; Liu, H.; Shang, S.; Cai, Z.; Wu, K. Bio-based thermosetting epoxy foams from epoxidized soybean oil and rosin with enhanced properties. *Ind. Crop. Prod.* **2019**, *139*, 111540. [CrossRef]
45. Deng, L.; Ha, C.; Sun, C.; Zhou, B.; Yu, J.; Shen, M.; Mo, J. Properties of Bio-based Epoxy Resins from Rosin with Different Flexible Chains. *Ind. Eng. Chem. Res.* **2013**, *52*, 13233–13240. [CrossRef]
46. Huang, K.; Zhang, J.; Li, M.; Xia, J.; Zhou, Y. Exploration of the complementary properties of biobased epoxies derived from rosin diacid and dimer fatty acid for balanced performance. *Ind. Crops Prod.* **2013**, *49*, 497–506. [CrossRef]
47. Li, R.; Zhang, P.; Liu, T.; Muhunthan, B.; Xin, J.N.; Zhang, J.W. Use of Hempseed-Oil-Derived Polyacid and Rosin-Derived Anhydride Acid as Cocuring Agents for Epoxy Materials. *ACS Sustain. Chem. Eng.* **2018**, *6*, 4016–4025. [CrossRef]
48. GB/T 1727-2021; General Methods for Preparation of Coating Films. Standards Press of China: Beijing, China, 2021.
49. GB/T 6739−2006; Paints and Varnishes—Determination of Film Hardness by Pencil Test. Standards Press of China: Beijing, China, 2007.
50. GB/T 9286−2021; Paints and Varnishes—Cross-Cut Test. Standards Press of China: Beijing, China, 2007.
51. GB/T 1732−2020; Determination of Impact Resistance of Coating Films. Standards Press of China: Beijing, China, 2020.
52. GB/T 1731−2020; Determination of Flexibility of Coating and Putty Films. Standards Press of China: Beijing, China, 2020.
53. GB/T 1735−2009; Paints and Varnishes—Determination of Heat Resistance. Standards Press of China: Beijing, China, 2009.
54. Atta, A.M.; El-Saeed, S.M.; Farag, R.K. New vinyl ester resins based on rosin for coating applications. *React. Funct. Polym.* **2006**, *66*, 1596–1608. [CrossRef]
55. Liu, G.; Wu, G.; Chen, J.; Kong, Z. Synthesis, modification and properties of rosin-based non-isocyanate polyurethanes coatings. *Prog. Org. Coat.* **2016**, *101*, 461–467. [CrossRef]

Disclaimer/Publisher's Note: The statements, opinions and data contained in all publications are solely those of the individual author(s) and contributor(s) and not of MDPI and/or the editor(s). MDPI and/or the editor(s) disclaim responsibility for any injury to people or property resulting from any ideas, methods, instructions or products referred to in the content.

Article

Effects of Electron Irradiation and Temperature on Mechanical Properties of Polyimide Film

Jian Qiu [1,2], Jusha Ma [3], Wenjia Han [3], Xiao Wang [1], Xunchun Wang [3], Maliya Heini [4], Bingyang Li [5,6,*], Dongyang Sun [7], Ruifeng Zhang [8], Yan Shi [1,8,*] and Cunfa Gao [1]

[1] State Key Laboratory of Mechanics and Control for Aerospace Structures, Nanjing University of Aeronautics and Astronautics, Nanjing 210016, China; swordqiu@163.com (J.Q.); xiaowang@nuaa.edu.cn (X.W.); cfgao@nuaa.edu.cn (C.G.)
[2] School of Chemistry and Civil Engineering, Shaoguan University, Shaoguan 512005, China
[3] Shanghai Institute of Space Power-Sources, Shanghai 200245, China; jushama@163.com (J.M.); hanbeihang@163.com (W.H.); runze0916@126.com (X.W.)
[4] Xinjiang Key Laboratory of Electronic Information Materials and Devices, Chinese Academy of Sciences, Urumqi 830011, China; maliya@ms.xjb.ac.cn
[5] China Academy of Aerospace Science and Innovation, Beijing 100871, China
[6] College of Engineering, Peking University, Beijing 100871, China
[7] School of Naval Architecture and Ocean Engineering, Jiangsu University of Science and Technology, Zhenjiang 212003, China; sundongyang@cqu.edu.cn
[8] School of Electrical Information Engineering, Ningxia Institute of Science and Technology, Shizuishan 753000, China; zhangruifeng@nuaa.edu.cn
* Correspondence: libingyang@stu.pku.edu.cn (B.L.); yshi@nuaa.edu.cn (Y.S.)

Citation: Qiu, J.; Ma, J.; Han, W.; Wang, X.; Wang, X.; Heini, M.; Li, B.; Sun, D.; Zhang, R.; Shi, Y.; et al. Effects of Electron Irradiation and Temperature on Mechanical Properties of Polyimide Film. *Polymers* **2023**, *15*, 3805. https://doi.org/10.3390/polym15183805

Academic Editor: Dan Rosu

Received: 10 August 2023
Revised: 1 September 2023
Accepted: 9 September 2023
Published: 18 September 2023

Copyright: © 2023 by the authors. Licensee MDPI, Basel, Switzerland. This article is an open access article distributed under the terms and conditions of the Creative Commons Attribution (CC BY) license (https://creativecommons.org/licenses/by/4.0/).

Abstract: Polyimide (PI) is widely deployed in space missions due to its good radiation resistance and durability. The influences from radiation and harsh temperatures should be carefully evaluated during the long-term service life. In the current work, the coupled thermal and radiation effects on the mechanical properties of PI samples were quantitatively investigated via experiments. At first, various PI specimens were prepared, and electron irradiation tests were conducted with different fluences. Then, both uniaxial tensile tests at room temperature and the dynamic mechanical analysis at varied temperatures of PI specimens with and without electron irradiation were performed. After that, uniaxial tensile tests at low and high temperatures were performed. The fracture surface of the PI film was observed using a scanning electron microscope, and its surface topography was measured using atomic force microscopy. In the meantime, the Fourier-transform infrared spectrum tests were conducted to check for chemical changes. In conclusion, the tensile tests showed that electron irradiation has a negligible effect during the linear stretching period but significantly impacts the hardening stage and elongation at break. Moreover, electron irradiation slightly influences the thermal properties of PI according to the differential scanning calorimetry results. However, both high and low temperatures dramatically affect the elastic modulus and elongation at break of PI.

Keywords: electron irradiation; high and low temperature; polyimide film; mechanical property

1. Introduction

Due to its good mechanical properties and lightweight, a polyimide (PI) film is usually applied as a substrate material of space solar cells and even structural material in spacecraft. Recently, a PI-based flexible solar wing was successfully deployed in the core module assembly of China Space Station for the first time [1]. The materials and structures employed in space missions will endure unfavorable irradiation and alternate temperatures, which usually affect their mechanical properties [2,3]. Considering the radiation effect, the National Aeronautics and Space Administration (NASA) has taken all kinds of PI films into the International Space Station (ISS) to assess their durability via the Materials International Space Station Experiment (MISSE). Many achievements have been reported [4–7]. After one to two years of space exposure in a low Earth orbit (LEO), the PI specimens show degradation in surface characteristics and mechanical properties. However, the results from ground tests show that electron beam (EB) and ultraviolet (UV) irradiation had an insignificant impact on its mechanical properties. It is reported that the erosion from atomic oxygen (AO) irradiation in a LEO causes the degradation of the mechanical properties of PI films [8,9]. Ground tests rather than in-orbit experiments are more applicable for cost and time-saving. Recent research reveals that an irradiation-load coupling effect appeared by applying high tensile stress to PI during electron irradiation on the ground. The radiation damage effect of polyimide fibers under electron irradiation and gamma-ray radiation and the coupling effect of irradiation with preload are presented [10–12]. Electron irradiations were reported to have considerable influence on the mechanical properties of other components, such as polymers [13–15] and semiconductor materials [16]. In summary, high-energy electron irradiation in space has a noticeable effect on PI's morphology and mechanical properties [17].

Moreover, PI will experience harsh high- and low-temperature cycles in space. The risk of thermal stress concentration may occur after a thermal cycling test to solar array with flexible printed circuit (FPC), whose main structure is made of PI, with temperatures ranging from $-100\ °C$ to $130\ °C$ [18]. Similar fracture behaviors have been observed for thin-film solar cells after thermal cycling and thermo-vacuum tests [19]. Furthermore, the tensile strength, stretching modulus, and elongation variations in polyimide films after temperature alternating from $-190\ °C$ to $+200\ °C$ have also been reported [20]. The stability of PI under electron irradiation can be attributed to the aromatic imide ring structure [21]. From the DMA and DSC analyses [22], there exists a decrease in the thermal stability of irradiated PI. The tensile properties of PI are hardly affected by electron irradiation, and the formation of crosslinking results in a change in glass transition temperature observed in irradiated polyimides [23,24]. Under the joint effect of 2 MeV electron irradiation and high temperature at 373 K, mechanical tests showed an increase in plasticity, and long-term hardening was observed [25]. The effects of low temperature and mechanical loading were seldom investigated.

Systematic research on the mechanical behaviors of a PI film under coupling effects of electron irradiation and harsh temperature is presented in this work. The PI specimens were first irradiated by high-energy electrons. Then, quasi-static, dynamic stretching, and high- and low-temperature tests were carried out. At last, various characterization experiments were performed.

2. Materials and Methods

The workflow of this study is summarized in Figure 1. At first, all the PI specimens including dumbbell-shaped, rectangular, and square tensile test specimens were cut by a silhouette CAMEO desktop cutting machine (see Figure 1a). Moreover, specimens for mechanical tests were fabricated along the same axial direction. Then, some PI specimens were subjected to an electron irradiation test (see Figure 1b). After that, uniaxial tensile and DMA tests were performed to evaluate the mechanical properties of the PI specimens with or without electron irradiation, and variable temperature factors were also considered (see Figure 1c).

More specific information about the test type and quantity of the PI specimens is listed in Table 1. All the PI specimens were double-checked under an optical microscope to eliminate samples with unexpected flaws.

Electron irradiation experiments were conducted at Xinjiang Technical Institute of Physics & Chemistry, Chinese Academy of Sciences. The ELV-8 II electron accelerator can provide 1 MeV electron beam with an electron flux of 1.012×10^{12} e/(cm^2·s). According to ISO standard 23038 [26] and China national standard GB 38190 [27], the PI specimens were exposed to a maximum total fluence of 1×10^{16} e/cm^2. Details are shown in Table 2.

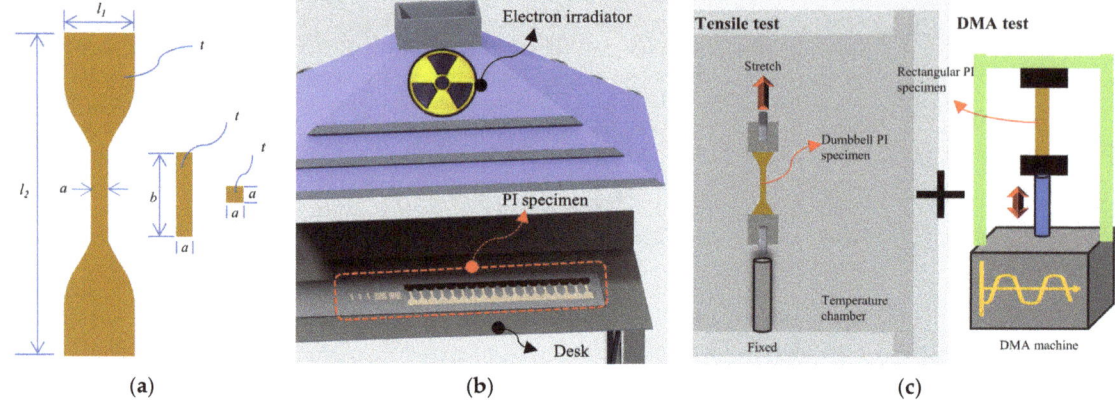

Figure 1. Workflow of this study. (**a**) PI specimens of different shapes. (**b**) Schematic illustration of electron irradiation. (**c**) Uniaxial tensile test and DMA test.

Table 1. List of PI samples.

Shape/Type	Test Type	Dimension (mm)	Quantity/Pieces
Dumbbell	Uniaxial tensile test	$l_1 = 25$; $l_2 = 115$; $a = 6$; $t = 0.05$	15
Dumbbell	Tensile test with temperature	$l_1 = 25$; $l_2 = 115$; $a = 6$; $t = 0.05$	35
Rectangular	Dynamic mechanical analysis	$a = 6$; $b = 30$; $t = 0.05$	10
Square	AFM surface topography	$a = 6$; $t = 0.05$	5

Table 2. Electron irradiation arrangements.

Flux	Electron Fluence (e/cm^2)	Quantity/Pieces		
		Dumbbell	Rectangular	Square
1×10^{12} e/(cm$^2 \cdot$s)	5×10^{14}	3	2	1
	1×10^{15}	3	2	1
	5×10^{15}	3	2	1
	1×10^{16}	3	2	1

Uniaxial tensile tests were carried out by a universal material testing machine (Range 500 N, Instron 5943, Boston, MA, USA). PI specimens both with and without electron irradiation were elongated to breakage. The loading speed was set as 100 mm/min, as suggested by China national standard GB/T 1040.3 [28], at ambient temperature and humidity. The whole loading process was recorded by a high-resolution CCD camera.

To investigate the high and low temperature effects, tensile tests with different temperatures were carried out in a temperature chamber. The testing temperature was increased or decreased from room temperature at a constant rate of 10 °C/min to a specific value and then was maintained for about 30 min. Based on the working temperature in space, high temperatures were chosen from 60 °C to 200 °C at intervals of 20 °C, and low temperatures were chosen from 0 °C to −90 °C at intervals of 30 °C.

The dynamic mechanical analysis (DMA) test of PI was carried out by TA DMA850 to investigate the coupling effect of temperature and electron irradiation. Rectangular PI specimens and the thin-film tensile kit were selected to perform the tensile test in the atmosphere. At first, the temperature was increased from room temperature to about 390 °C at a speed of 5 °C/min. Then, a new set of specimens fabricated from the same slice was tested as temperature decreased from room temperature to about −90 °C at a speed of 5 °C/min. The frequency was 1 Hz, and the amplitude was 20 microns.

The influence of electron irradiation on the glass transition temperature (Tg) was measured by a differential scanning calorimetry (DSC) instrument (TA Q2000, New Castle, DE, USA). The temperature ranged from room temperature to 500 °C in a nitrogen environment, at a rate of 5 °C/min. Moreover, the comparative group was cooled down to −120 °C rapidly, then heated up to 20 °C at a rate of 5 °C/min by TA DSC250. The fracture surface of PI after the tensile test was observed using scanning electron microscopy (SEM, FEI Quanta 650, Columbus, OH, USA). The surface topography was measured using atomic force microscopy (AFM, Oxford Instruments Asylum Research Cypher ES, San Diego, CA, USA) in a tapping mode. Firstly, an area of 5 μm × 5 μm on a PI specimen was randomly selected to perform the coarse scan, and then, an area of 1 μm × 1 μm within was selected to perform the fine scan. Fourier-transform infrared spectroscopy (FTIR) test was carried out using ThermoFisher Nicolet iS20 (Waltham, MA, USA). The wavenumber was chosen from 400 to 4000, and a transmittance mode was selected to perform the test. All the above specimens were cleaned with absolute alcohol and deionized water before the test.

3. Results
3.1. Uniaxial Tensile Test
3.1.1. Effect of Electron Irradiation

Based on the force–displacement curves of the uniaxial tensile test (Figure 2a), electron irradiation has negligible influence on the elastic modulus of PI in the linear elastic stage. However, in the hardening stage, PI becomes "stronger" after electron irradiation and is gradually reinforced by the radiation fluence. Moreover, the secant modulus of PI in the hardening stage shows an increasing trend with electron irradiation.

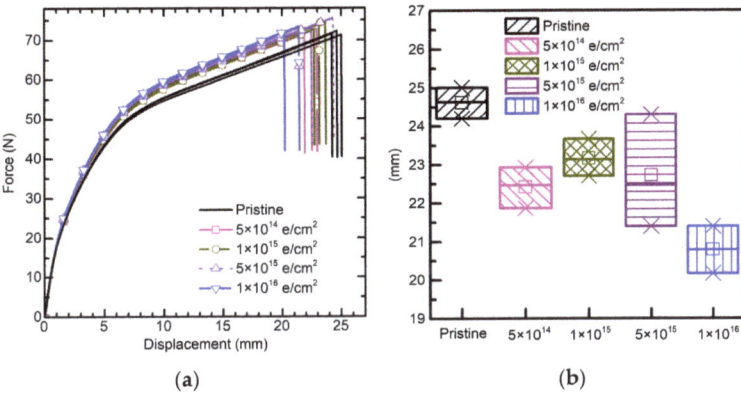

Figure 2. (a) Force–displacement curve from uniaxial tensile test for polyimide specimens with none, 5×10^{14} e/cm^2, 1×10^{15} e/cm^2, 5×10^{15} e/cm^2, and 1×10^{16} e/cm^2 electron irradiation fluence. (b) Maximum elongation at break.

Furthermore, electron irradiation significantly affects elongation at break of PI. When the total electron fluence reaches 5×10^{14} e/cm^2, the mean elongation at break decreases by 8.9%. When electron fluence reaches a maximum of 1×10^{16} e/cm^2, the mean elongation at break decreases by 15.5% (Figure 2b). On the contrary, the ultimate tensile strength tends to increase after electron irradiation.

Amorphous regions and semi-crystalline regions coexist and overlap each other in PI material. The latter is related to mechanical properties. The elastic stage is dominated by the elongation of semi-crystalline and amorphous regions, which cross each other and are insensitive to electron irradiation. However, the hardening stage is dominated by the alignment and sliding of the crystalline region, where electron irradiation may cause chemical changes and affect the crystalline phase. Finally, the electron irradiation may bring reinforcement and embrittlement to the PI specimen. As a result, the ultimate tensile strength of PI film increases, but elongation at break decreases after irradiation. Finally, the tensile process of PI was further evaluated by digital image correlation (DIC) algorithm, more details can be seen in Figure A1 of Appendix A.

3.1.2. Effect of High Temperature

As shown in Figure 3a, the high temperature of the tensile test chamber was set from 60 °C to 200 °C, with increments of 20 °C for each tensile set. Compared to the tensile results at room temperature, the mean break elongation shows an increasing trend with the rise in temperature. In contrast, elastic modulus and tensile strength present an opposite trend. Compared to the elongation at break at room temperature, an increase of 12.1% and 18.7% were observed at the temperatures of 160 °C and 200 °C, respectively (see Figure 3b). Regarding the ultimate tensile strength, a decrease of 34% and 39.9% appeared at the temperatures of 160 °C and 200 °C, respectively. The mean elastic modulus at room temperature was about 2.45 GPa, which decreased to 1.58 GPa and 1.43 GPa at the temperatures of 160 °C and 200 °C, respectively.

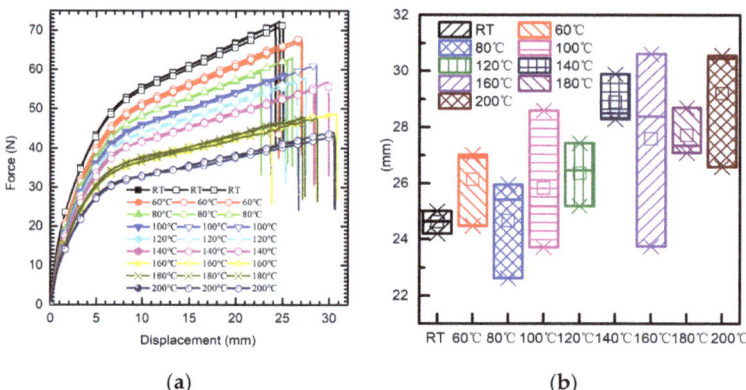

Figure 3. (**a**) Force–displacement curve from uniaxial tensile test with high temperature. (**b**) Maximum elongation at break.

3.1.3. Effect of Low Temperature

The low temperature range was set from 0 °C to −90 °C, with decrements of 30 °C for each tensile set as shown in Figure 4a. Similar to the cases in high temperatures, the mechanical properties of PI show an enhancement trend with the decrease in temperature. Compared to the ultimate tensile strength at room temperature, a maximum increase of 23% was achieved at the temperature of −90 °C. Moreover, the secant modulus of PI in the hardening stage presents an increase of 77.8%. However, a decrease in mean elongation at break comes to 35.7% at the temperature of −90 °C compared to that at room temperature (see Figure 4b).

3.2. DMA Test

3.2.1. Effect of Electron Irradiation and High Temperature

The storage modulus of PI with or without electron irradiation presents a kind of decrease as the temperature increases from 30 °C to 390 °C (Figure 5a), which exhibits a "softening" process. However, this trend shows a bifurcation when the temperature comes to about 350 °C. The storage modulus for the samples without electron irradiation appears to drop rapidly, while for those with irradiation, it appears to be smaller. That means electron irradiation may influence the glass transition temperature (Tg), which was validated in the following DSC test.

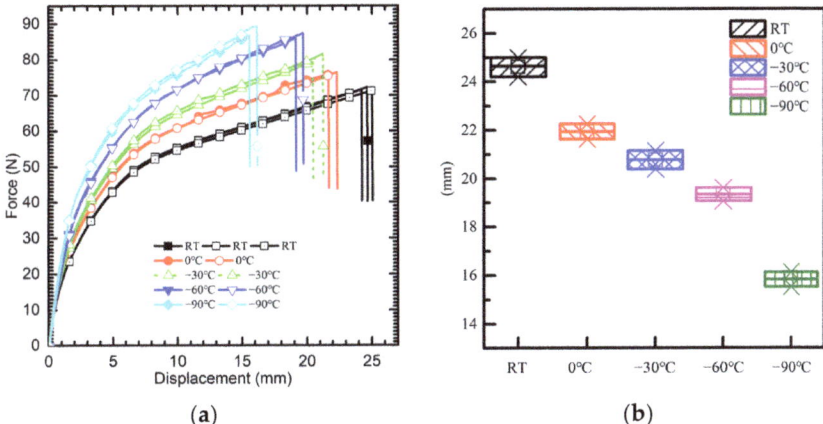

Figure 4. (**a**) Force–displacement curve from uniaxial tensile test with low temperature. (**b**) Maximum elongation at break.

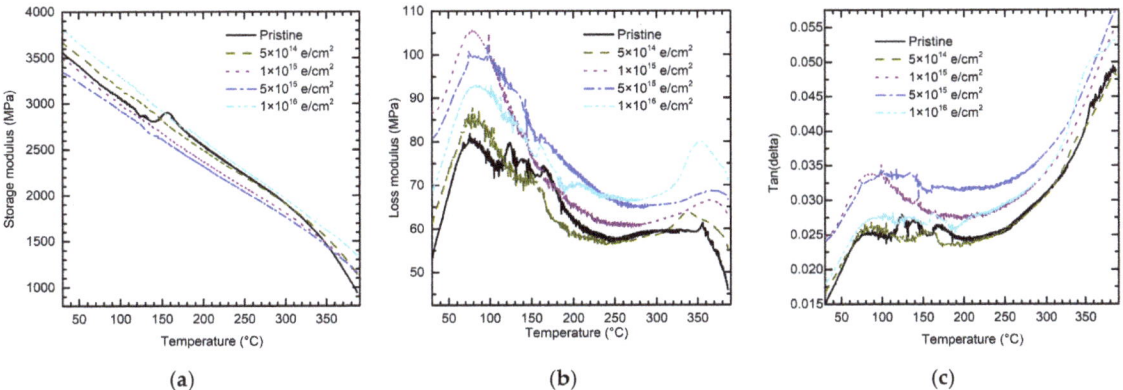

Figure 5. DMA test from 30 °C to 390 °C. (**a**) Storage modulus. (**b**) Loss modulus. (**c**) Phase angle (damping factor).

From the loss modulus curve (Figure 5b), PI with or without electron irradiation can be divided into three to four stages. When the temperature increases from 30 °C to about 90 °C, the loss modulus of PI presents an increasing trend. However, when the temperature is over 90 °C, the loss modulus slope shows a kind of oscillation and a decreasing trend. When the temperature is more than 250 °C, the loss modulus slope shows a kind of smoothness. Finally, when the temperature is over 350 °C, there is a clear bifurcation in loss modulus curve of PI before and after electron irradiation.

Similarly, the damping factor δ (see Figure 5c) also undergoes three to four dissimilar stages as the temperature changes, but in a different trend. When the temperature is more than 300 °C, the damping curve presents an opposite trend compared to the loss modulus (Figure 5b), that is, because the storage modulus shows a decline after the temperature is higher than 300 °C, as shown in Figure 5a.

According to previous research [11,12], the loss modulus at around 250 °C depends on the transitions due to the rotation of aromatic rings (β-transition). The loss modulus slope around 250 °C shows no significant difference with or without electron irradiation, which may indicate that the structure of aromatic rings was not influenced by electron irradiation. When the temperature is over 350 °C, the loss modulus is related to the segmental motion of the backbone (α-transition). Combined with the storage modulus results, it can be concluded that electron irradiation brings the scission of chemical bonds in PI and forms radicals. This may cause molecular chain crosslinking and therefore change the PI's crystalline state.

3.2.2. Effect of Electron Irradiation and Low Temperature

As the temperature decreases from 20 °C to −90 °C, the storage moduli of PI with or without electron irradiation all present a kind of increase, which exhibits a "hardening" process (see Figure 6a). After receiving a total electron radiation fluence of 1×10^{16} e/cm^2, the storage modulus of PI presents a most significant increase as the temperature approaches −90 °C. Meantime, a similar phenomenon can be observed in the case of loss modulus (see Figure 6b) and damping factor (see Figure 6c). Unlike the high temperature situation, storage modulus, loss modulus, and damping factor all present an ascending trend as the temperature descends. Referring to the results of Section 3.1.1, electron irradiation had an insignificant effect on the elastic stage of PI, when quasi-static tests were carried out at ambient temperature, while the DMA test focused on small displacement in a cyclic loading process. Therefore, under the dynamic test condition, there is a coupling effect of electron irradiation and low temperature on the mechanical behavior of PI.

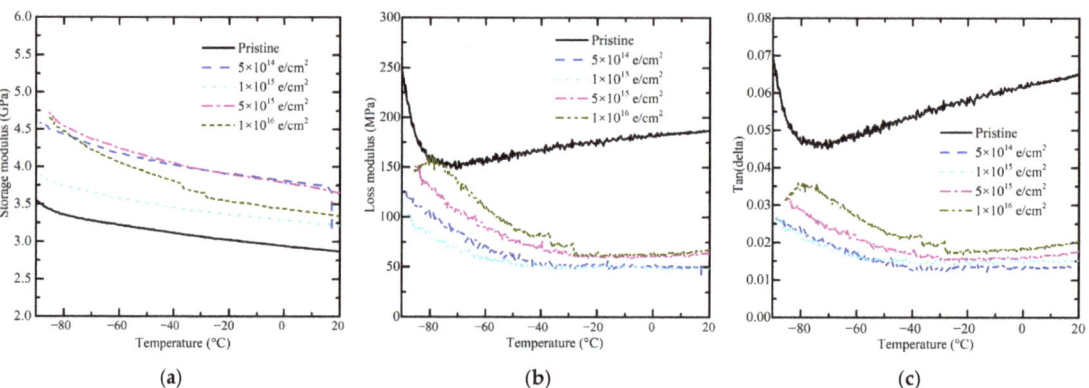

Figure 6. DMA test from 20 °C to −90 °C. (**a**) Storage modulus. (**b**) Loss modulus. (**c**) Phase angle (damping factor).

3.3. Changes in Thermal Properties

DSC tests were carried out to investigate whether electron irradiation impacts thermal properties of PI, such as glass transition temperature (Tg). Firstly, a small bump around 100 °C is observed, and the glass transition temperature fluctuates between 380 °C and 390 °C. After receiving a total fluence of 5×10^{15} e/cm^2, the Tg temperature reaches a maximum of 388.8 °C (Figure 7a). Then, as the temperature rises from −90 °C to 20 °C, the DSC curve presents a slowly ascending slope both with and without electron irradiation (Figure 7b). From the above DSC results, the thermal properties of PI are almost unaffected by electron irradiation. From the reference, electron irradiation leads to crosslinking of the polyimide chains and may induce a higher glass transition temperature [29]. As a result, crosslinking of polyimide chains is not likely to happen when receiving a low electron fluence such as 5×10^{14} e/cm^2.

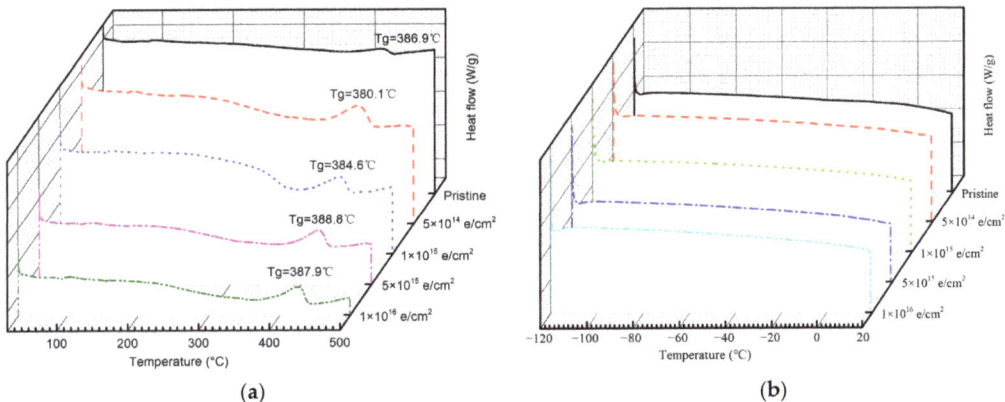

Figure 7. DSC curve of pristine PI samples and those after receiving radiation fluence of 5×10^{14} e/cm^2, 1×10^{15} e/cm^2, 5×10^{15} e/cm^2, and 1×10^{16} e/cm^2. (**a**) Room temperature to 500 °C; (**b**) −120 °C to room temperature.

3.4. Changes in Morphology

3.4.1. Morphology Changes on Fracture Surface

After the tensile test, the fracture surfaces of the PI specimens were studied under SEM. As shown in Figure 8, when down to the scale of microns, the inner space of PI is not entirely intact and spaceless. The morphology of fracture section presents a layered structure full of microvoids because some gas was produced during electron radiation [30]. The inner structure may change topography during electron irradiation and other "defects" are generated [31]. Then, the aggregation and entanglement of the "defects" chains changed the matrix topology. Another noticeable topology change is that more fiber-like structures appear after a different fluence of electron irradiation, as shown in Figure 8b–e.

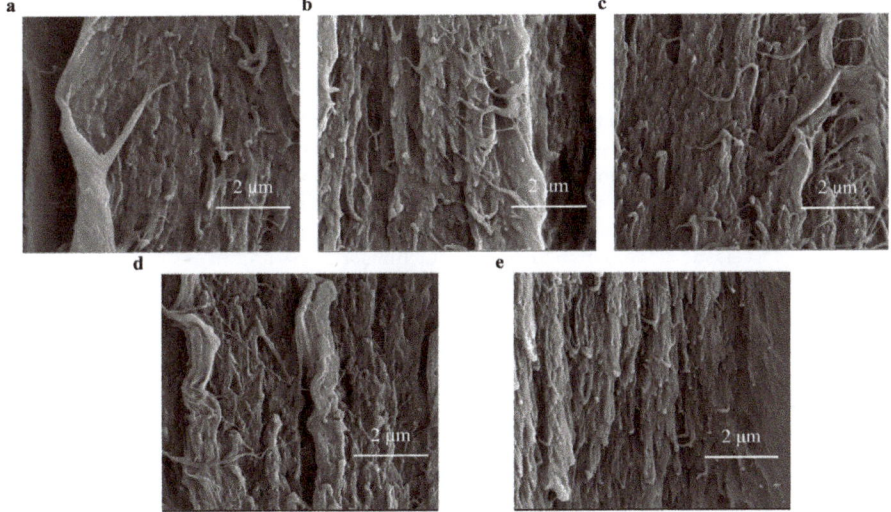

Figure 8. SEM images. (**a**) Fracture surface of pristine polyimide film after tensile test. (**b**) Fracture surface of polyimide film after receiving radiation fluence of 5×10^{14} e/cm^2, (**c**) 1×10^{15} e/cm^2, (**d**) 5×10^{15} e/cm^2, and (**e**) 1×10^{16} e/cm^2.

3.4.2. Morphology Changes on the Surface

Furthermore, the surface topography of the specimens was measured by AFM. From previous research, the surface roughness of PI may either increase significantly [11] or decrease after electron irradiation [32]. According to our research, the variations of the surface roughness are nonmonotonic. As shown in Figure 9, the average roughness of pristine PI is 2.603 nm and drops to 2.318 nm after receiving electron irradiation of 5×10^{14} e/cm^2. Then, it reaches a maximum of 2.737 nm as irradiation fluence increases to 1×10^{15} e/cm^2. Further, it comes to a minimum of 2.118 nm in the case of 5×10^{15} e/cm^2 and increases to 2.716 nm after irradiation fluence of 1×10^{16} e/cm^2. More details about measured roughness are listed in Table A1 of Appendix A, and other influencing factors are discussed. Due to previous experimental data [33], 1 MeV electron has sufficient energy to penetrate through the 50-micron thickness of the PI specimens employed in this study. Apparent morphology changes can be observed both on the interior and exterior surfaces.

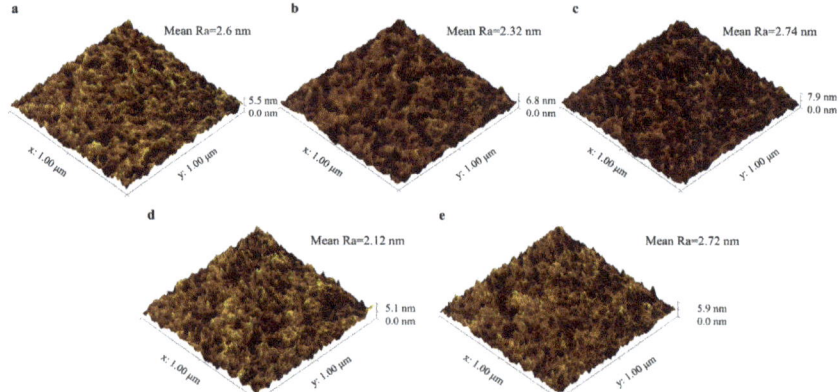

Figure 9. AFM testing. Three-dimensional surface topography of (**a**) pristine polyimide film and those after receiving radiation fluence of (**b**) 5×10^{14} e/cm^2, (**c**) 1×10^{15} e/cm^2, (**d**) 5×10^{15} e/cm^2, and (**e**) 1×10^{16} e/cm^2.

4. Discussion and Conclusions

Based on the FTIR test results, no new substance is created after electron irradiation. However, significant differences on each peak of the pristine and irradiated PI specimens indicate a change in the quantity of each chemical composition. More details can be seen in Figure A2 and Table A2 of Appendix A. From the tensile test results after electron irradiation, electron irradiation has little influence on mechanical properties in the linear elastic stage. An increase in stretching stiffness and decrease in elongation at break can be observed during the nonlinear stretching stage, which is believed to be induced by electron irradiation. Regarding the temperature effect on the mechanical properties of PI, high temperature causes a decline in elastic modulus but an increase in elongation at break. The opposite trend is observed in the case of low temperatures. Changes in thermal properties can be attributed to molecular crosslinking and chain scission. In this study, electron irradiation had negligible influence on the thermal stability of the PI samples. The rotation of aromatic rings (β-transition) of PI is not likely influenced by electron irradiation, but the segmental motion of the backbone (α-transition) of PI is slightly changed. Both surface and inner topography of the PI samples changes with electron irradiation in a complex way. The aggregation and entanglement of molecular chains as well as generation of internal gas may be the reasons; both bring adverse influences on PI-based flexible space devices such as thin-film solar cells (see Figure A3 in Appendix A). Electron irradiation and low temperature have a coupling effect on PI under dynamic test conditions, which should be thoroughly investigated in the future.

Author Contributions: Conceptualization, J.Q., J.M., W.H., B.L., D.S. and R.Z; methodology, J.Q., J.M., W.H., B.L., D.S. and R.Z.; Validation, X.W. (Xunchun Wang); investigation, X.W. (Xiao Wang) and M.H.; resources, X.W. (Xunchun Wang); writing—original draft preparation, J.Q.; writing—review and editing, Y.S. and C.G. All authors have read and agreed to the published version of the manuscript.

Funding: This study was co-supported by the Joint Fund of the Advanced Aerospace Manufacturing Technology Research (U1937601), the Research Fund of State Key Laboratory of Mechanics and Control of Mechanical Structures (Nanjing University of Aeronautics and astronautics, No. MCMS-I-0221Y01), the Research Funding from Xi'an Jiaotong University (U22B2013), National Natural Science Foundation of China for Creative Research Groups (No. 51921003).

Institutional Review Board Statement: Not applicable.

Data Availability Statement: The data presented in this study are available on reasonable request from the corresponding author.

Acknowledgments: We acknowledge the facilities in the Center for Microscopy and Analysis at Nanjing University of Aeronautics and Astronautics.

Conflicts of Interest: The authors declare no conflict of interest.

Appendix A

After recording the uniaxial tensile test process of the PI film in real time, a DIC algorithm such as Ncorr [34] was taken to analyze the tensile process. Meantime, commercial software ABAQUS (version 6.14) was used to simulate the uniaxial tensile process of a PI specimen. The PI specimen was set as an isotropic elastic material with Poisson's ratio of 0.34 and Young's modulus of 2.45 GPa, measured from the experiment. The geometry dimensions of the dumbbell specimen are the same as those in Figure 1a and Table 1. A number of 18,150 8-node linear brick elements with reduced integration and hourglass control (C3D8R) were employed for the PI specimen in the simulation. While one end of the dumbbell specimen was fixed, the other end was subjected to the measured maximum displacement just before elongation at break. Finally, a comparison of the DIC results and FEM simulation is shown in Figure A1. The width of the PI film becomes narrow after elongation, this is also confirmed by the simulation, see Figure A1a. According to Figure A1b, the maximum strain along the stretch direction happened in the middle section of the PI specimen. However, from the DIC results on the left, there is an area with the darkest color, at which point, finally, the PI film breaks.

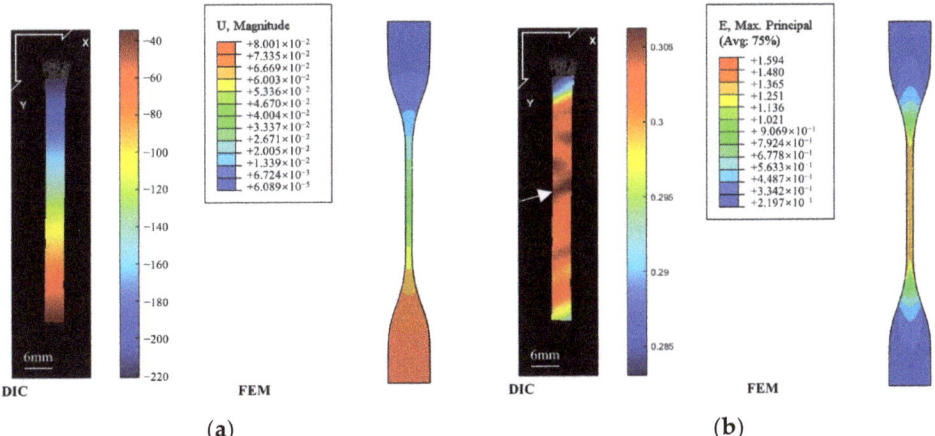

Figure A1. DIC results and FEM simulation. (**a**) Displacement. (**b**) Strain.

Three different square areas of 1 μm × 1 μm from the pristine and irradiated PI surfaces were randomly selected for an AFM scan. The measured roughness and calculated average roughness (mean Ra) are listed in Table A1. The standard deviations σ of AFM roughness can be calculated by the following formula:

$$\sigma = \sqrt{\frac{1}{V\_npnts - 1}\Sigma(Y_i - V\_avg)^2} \quad (A1)$$

where V_npnts represents the number of points, V_avg denotes the mean Ra, and Y_i is Ra of each area.

Table A1. Measured average roughness of pristine and irradiated PI.

	Pristine	5×10^{14} e/cm²	1×10^{15} e/cm²	5×10^{15} e/cm²	1×10^{16} e/cm²
Area 1	2.606 nm	2.375 nm	2.758 nm	2.14 nm	3.05 nm
Area 2	2.646 nm	2.132 nm	2.931 nm	1.88 nm	2.559 nm
Area 3	2.556 nm	2.447 nm	2.521 nm	2.334 nm	2.538 nm
Mean Ra	2.603 nm	2.318 nm	2.737 nm	2.118 nm	2.716 nm
σ	0.045094346	0.165054536	0.205831242	0.227798156	0.289731773

After the FTIR test of PI with or without electron irradiation, a noticeable difference exists on each peak of different specimens in the transmittance mode, as shown in Figure A2. Based on a previous study [15], the peak observed at 2932 cm^{-1} represents the C-H bonds in tert-butyl and cyclohexyl groups. The peaks around 1775 cm^{-1} and 1718 cm^{-1} represent coupled stretching vibration of carbonyl group C=O. The wavenumber of 1500 cm^{-1} represents stretching vibration band of phenyl ring C=C. Here, 1375 cm^{-1} represents stretching vibration band of amide group =C-N, and 1242 cm^{-1} represents stretching vibration band of aromatic oxide C-O-C. More details are presented in Table A2. However, the test results reveal that fluctuation does not simply increase with irradiation fluence. Finally, the wavenumber presents a most noticeable increase after receiving electron irradiation of 1×10^{16} e/cm². Each band mentioned above shows a remarkable difference. When the radiation fluence reaches 1×10^{16} e/cm², the chemical bond is more likely to break and cannot recombine, which in consequence causes a fluctuation of the vibration peak in the FTIR spectra. No new characterizing peak is created, meaning no new substance is created after electron irradiation. Chemical bond breakage and recombining of radicals are the most potential reasons for the degradation of mechanical properties and a change in glass transition temperature.

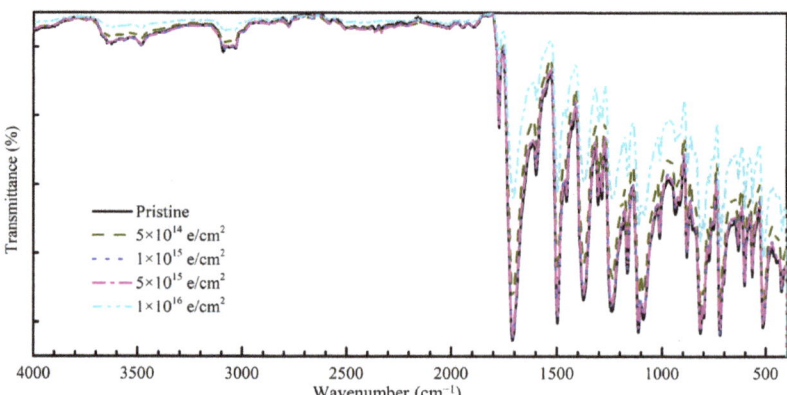

Figure A2. FTIR curve of pristine PI samples and those after receiving radiation fluence of 5×10^{14} e/cm², 1×10^{15} e/cm², 5×10^{15} e/cm², and 1×10^{16} e/cm².

Table A2. Spectrum bands at different wavenumbers of pristine and irradiated PI.

Wavenumbers	Pristine	5×10^{14} e/cm^2	1×10^{15} e/cm^2	5×10^{15} e/cm^2	1×10^{16} e/cm^2
2932 cm^{-1}	97.2	97.3	97.4	97.2	98.3
1775 cm^{-1}	67.5	73.3	68.4	68.4	83.3
1718 cm^{-1}	15.0	26.5	15.7	16.6	52.0
1500 cm^{-1}	11.8	23.6	13.2	14.0	49.5
1375 cm^{-1}	16.7	27.5	18.1	18.7	50.9
1242 cm^{-1}	13.5	23.3	14.9	15.5	46.8

Figure A3 shows a thin-film solar cell based on a PI film from Shanghai Institute of Space Power-Sources. The thin-film GaAs solar cell is an inverted metamorphic triple junction (3J IMM) solar cell, with InGaAs, GaAs, and GaInP as the bottom, middle, and top junctions. A bonding metal layer connects the flexible PI substrate and epitaxial layer of the battery. The thickness of the PI substrate is about 50 microns, and the total thickness of the solar cell is about 70 to 80 microns. The influences of radiation and temperature will be investigated in future work.

Figure A3. Thin-film solar cell based on PI film.

References

1. Wang, K.; Wang, X.; Qian, B.; Ma, J.; Lu, J. Recent Development OnSpace Application for High-Efficiency Solar Cells and Array Technology. *Kuei Suan Jen Hsueh Pao/J. Chin. Ceram. Soc.* **2022**, *50*, 1436–1446.
2. Zhang, F.; Han, W.; Chen, H.; He, H.; Wang, X. The Design of Flexible Printed Circuit in Solar Cell Array and Experimental Study of Its Thermal Adaptability. *Spacecr. Environ. Eng.* **2021**, *38*, 670–676. [CrossRef]
3. Yin, M.S.; Yang, G.; Wang, X.C.; Fan, B.; Jiang, D.P.; Yang, H.D. Strain-Testing Research of Space Solar Cell Array. *Wuli Xuebao/Acta Phys. Sin.* **2021**, *70*, 198801. [CrossRef]
4. Miller, S.K.R.; Dever, J.A. Space Environment Exposure Results from the MISSE 5 Polymer Film Thermal Control Experiment on the International Space Station. In Proceedings of the 11th International Symposium on Materials in the Space Environment, Aix-en-Provence, France, 15–18 September 2009; p. 7.
5. Miller, S.K.R.; Dever, J.A.; Banks, B.A.; Kline, S. MISSE 6 Polymer Film Tensile Experiment. In Proceedings of the 2010 National Space and Missile Materials Symposium (NSMMS), Scottsdale, AZ, USA, 28 June–1 July 2010; p. 13.
6. Miller, S.K.R.; Banks, B.A.; Sechkar, E. An Investigation of Stress Dependent Atomic Oxygen Erosion of Black Kapton Observed on MISSE 6. In Proceedings of the 2010 National Space and Missile Materials Symposium (NSMMS), Scottsdale, AZ, USA, 28 June–1 July 2010; p. 8.
7. Miller, S.K.R.; Sechkar, E.A. An Examination of Radiation Induced Tensile Failure of Stressed and Unstressed Polymer Films Flown on MISSE 6. 2013. Available online: https://ntrs.nasa.gov/api/citations/20130001603/downloads/20130001603.pdf (accessed on 8 September 2023).
8. Shimamura, H.; Nakamura, T. Investigation of Degradation Mechanisms in Mechanical Properties of Polyimide Films Exposed to a Low Earth Orbit Environment. *Polym. Degrad. Stab.* **2010**, *95*, 21–33. [CrossRef]
9. Shimamura, H.; Yamagata, I. Degradation of Mechanical Properties of Polyimide Film Exposed to Space Environment. *J. Spacecr. Rocket.* **2012**, *46*, 15–21. [CrossRef]
10. Shen, Z.; Liang, G.; Ziliang, M.; Yu, B.; Yigang, D.; Yenan, L.; Zhihao, W. Mechanical Property Degradation of Polyimide Film under Gamma Ray Radiation. *Spacecr. Environ. Eng.* **2016**, *33*, 100–104.

11. Dong, S.S.; Shao, W.Z.; Yang, L.; Ye, H.J.; Zhen, L. Surface Characterization and Degradation Behavior of Polyimide Films Induced by Coupling Irradiation Treatment. *RSC Adv.* **2018**, *8*, 28152–28160. [CrossRef]
12. Dong, S.S.; Shao, W.Z.; Yang, L.; Ye, H.J.; Zhen, L. Microstructure Evolution of Polyimide Films Induced by Electron Beam Irradiation-Load Coupling Treatment. *Polym. Degrad. Stab.* **2018**, *155*, 230–237. [CrossRef]
13. Manas, D.; Mizera, A.; Manas, M.; Ovsik, M.; Hylova, L.; Sehnalek, S.; Stoklasek, P. Mechanical Properties Changes of Irradiated Thermoplastic Elastomer. *Polymers* **2018**, *10*, 87. [CrossRef]
14. Manas, D.; Ovsik, M.; Mizera, A.; Manas, M.; Hylova, L.; Bednarik, M.; Stanek, M. The Effect of Irradiation on Mechanical and Thermal Properties of Selected Types of Polymers. *Polymers* **2018**, *10*, 158. [CrossRef]
15. Stelescu, M.D.; Airinei, A.; Manaila, E.; Craciun, G.; Fifere, N.; Varganici, C.; Pamfil, D.; Doroftei, F. Effects of Electron Beam Irradiation on the Mechanical, Thermal, and Surface Properties of Some EPDM/Butyl Rubber Composites. *Polymers* **2018**, *10*, 1206. [CrossRef]
16. Qiu, J.; Heini, M.; Ma, J.; Han, W.; Wang, X.; Yin, J.; Shi, Y.; Gao, C. Mechanical Properties of Multi-Scale Germanium Specimens from Space Solar Cells under Electron Irradiation. *Chin. J. Aeronaut.* **2023**; *in press*. [CrossRef]
17. Plis, E.A.; Engelhart, D.P.; Cooper, R.; Johnston, W.R.; Ferguson, D.; Hoffmann, R. Review of Radiation-Induced Effects in Polyimide. *Appl. Sci.* **2019**, *9*, 1999. [CrossRef]
18. Zhang, H.; Liu, H.W.; Hu, Z.Y.; Zhang, W. On-Orbit Load Analysis of Solar Wing with Flexible Characteristics. *Yuhang Xuebao/J. Astronaut.* **2019**, *40*, 139–147. [CrossRef]
19. Sibin, K.P.; Mary Esther, A.C.; Shashikala, H.D.; Dey, A.; Sridhara, N.; Sharma, A.K.; Barshilia, H.C. Environmental Stability of Transparent and Conducting ITO Thin Films Coated on Flexible FEP and Kapton® Substrates for Spacecraft Applications. *Sol. Energy Mater. Sol. Cells* **2018**, *176*, 134–141. [CrossRef]
20. Cherkashina, N.I.; Pavlenko, V.I.; Noskov, A.V. Synthesis and Property Evaluations of Highly Filled Polyimide Composites under Thermal Cycling Conditions from -190 °C to $+200$ °C. *Cryogenics* **2019**, *104*, 102995. [CrossRef]
21. Kang, P.H.; Jeon, Y.K.; Jeun, J.P.; Shin, J.W.; Nho, Y.C. Effect of Electron Beam Irradiation on Polyimide Film. *J. Ind. Eng. Chem.* **2008**, *14*, 672–675. [CrossRef]
22. Mishra, R.; Tripathy, S.P.; Dwivedi, K.K.; Khathing, D.T.; Ghosh, S.; Müller, M.; Fink, D. Spectroscopic and Thermal Studies of Electron Irradiated Polyimide. *Radiat. Meas.* **2003**, *36*, 621–624. [CrossRef]
23. Sasuga, T. Electron Irradiation Effects on Dynamic Viscoelastic Properties and Crystallization Behaviour of Aromatic Polyimides. *Polymer* **1991**, *32*, 1539–1544. [CrossRef]
24. Hirade, T.; Hama, Y.; Sasuga, T.; Seguchi, T. Radiation Effect of Aromatic Thermoplastic Polyimide (New-TPI). *Polymer* **1991**, *32*, 2499–2504. [CrossRef]
25. Kupchichin, A.I.; Muradov, A.D.; Omarbekova, Z.A.; Taipova, B.G. Mechanooptical Investigations of Polyimide Films Exposed to Electrons, Mechanical Loads, and Temperatures. *Russ. Phys. J.* **2007**, *50*, 153–160. [CrossRef]
26. Space Systems—Space Solar Cells—Electron and Proton Irradiation Test Methods. 2018, Volume 9. Available online: https://www.iso.org/standard/69495.html (accessed on 8 September 2023).
27. China, S.A. of Test Method of Electron Irradiation Aerospace Solar Cells. 2019. Available online: http://c.gb688.cn/bzgk/gb/showGb?type=online&hcno=B3F21AA1A502A459C4E7208AD380B3B8 (accessed on 8 September 2023).
28. China, S.A. of Plastics—Determination of Tensile Properties—Part 3: Test Conditions for Films and Sheets. 2006. Available online: https://openstd.samr.gov.cn/bzgk/gb/newGbInfo?hcno=39C9E88D2852DFAF876475C8AB7A97E2 (accessed on 8 September 2023).
29. Rahnamoun, A.; Engelhart, D.P.; Humagain, S.; Koerner, H.; Plis, E.; Kennedy, W.J.; Cooper, R.; Greenbaum, S.G.; Hoffmann, R.; van Duin, A.C.T. Chemical Dynamics Characteristics of Kapton Polyimide Damaged by Electron Beam Irradiation. *Polymer* **2019**, *176*, 135–145. [CrossRef]
30. Ennis, C.P.; Kaiser, R.I. Mechanistical Studies on the Electron-Induced Degradation of Polymethylmethacrylate and Kapton. *Phys. Chem. Chem. Phys.* **2010**, *12*, 14902–14915. [CrossRef] [PubMed]
31. Zhao, F.; Zhang, H.; Zhang, D.; Wang, X.; Wang, D.; Zhang, J.; Cheng, J.; Gao, F. Molecular Insights into the 'Defects' Network in the Thermosets and the Influence on the Mechanical Performance. *RSC Adv.* **2022**, *12*, 22342–22350. [CrossRef] [PubMed]
32. Plis, E.A.; Bengtson, M.T.; Engelhart, D.P.; Badura, G.P.; Cowardin, H.M.; Reyes, J.A.; Hoffmann, R.C.; Sokolovskiy, A.; Ferguson, D.C.; Shah, J.R.; et al. Characterization of Novel Spacecraft Materials under High Energy Electron and Atomic Oxygen Exposure. In Proceedings of the AIAA Science and Technology Forum and Exposition, AIAA SciTech Forum 2022, Virtual, 3–7 January 2022; AIAA: San Diego, CA, USA, 2022; pp. 1–12.
33. Cherkashina, N.I.; Pavlenko, V.I.; Noskov, A.V.; Romanyuk, D.S.; Sidelnikov, R.V.; Kashibadze, N.V. Effect of Electron Irradiation on Polyimide Composites Based on Track Membranes for Space Systems. *Adv. Sp. Res.* **2022**, *70*, 3249–3256. [CrossRef]
34. Blaber, J.; Adair, B.; Antoniou, A. Ncorr: Open-Source 2D Digital Image Correlation Matlab Software. *Exp. Mech.* **2015**, *55*, 1105–1122. [CrossRef]

Disclaimer/Publisher's Note: The statements, opinions and data contained in all publications are solely those of the individual author(s) and contributor(s) and not of MDPI and/or the editor(s). MDPI and/or the editor(s) disclaim responsibility for any injury to people or property resulting from any ideas, methods, instructions or products referred to in the content.

Review

Research Status of and Prospects for 3D Printing for Continuous Fiber-Reinforced Thermoplastic Composites

Yuan Yang [1,*], Bo Yang [1], Zhengping Chang [2], Jihao Duan [1] and Weihua Chen [1]

[1] Key Laboratory of Manufacturing Equipment of Shaanxi Province, Xi'an University of Technology, Xi'an 710048, China
[2] School of Mechanical Engineering, Northwestern Polytechnical University, Xi'an 710072, China
* Correspondence: yuanyang@xaut.edu.cn

Abstract: Continuous fiber-reinforced thermoplastic composites (CFRTPCs) have advantages such as high specific strength, high specific modulus, corrosion resistance, and recyclability and are widely used in the fields of aerospace, rail transit, new energy, and so on. However, traditional methods for preparing CFRTPCs, such as placement and molding, rely more on forming molds, resulting in high manufacturing costs and a slow response speed, which limits the promotion and application of the new generation of CFRTPCs with complex configurations and designable performance. Three-dimensional printing can efficiently create products with multiple materials, complex structures, and integrated functions, introducing new ways and opportunities for the manufacturing of CFRTPCs. However, poor mechanical properties are the bottleneck problem in achieving 3D printing of CFRTPCs. This paper summarizes the research status of the fused deposition modeling (FDM) 3D printing process and the corresponding mechanical properties of CFRTPCs. The focus is on analyzing the influences of the FDM process parameters, such as the material type, printing temperature, speed parameters, layer thickness, scanning space, stacking direction, and fiber volume content, on the mechanical properties of CFRTPCs. Finally, the main problems and future prospects of current CFRTPCs-FDM are analyzed and forecasted, providing new references and ideas for 3D printing of high-performance CFRTPCs.

Keywords: three-dimensional printing; CFRTPCs; FDM; process parameters; mechanical properties

Citation: Yang, Y.; Yang, B.; Chang, Z.; Duan, J.; Chen, W. Research Status of and Prospects for 3D Printing for Continuous Fiber-Reinforced Thermoplastic Composites. *Polymers* **2023**, *15*, 3653. https://doi.org/10.3390/polym15173653

Academic Editors: Jiangtao Xu and Sihang Zhang

Received: 6 June 2023
Revised: 8 August 2023
Accepted: 11 August 2023
Published: 4 September 2023

Copyright: © 2023 by the authors. Licensee MDPI, Basel, Switzerland. This article is an open access article distributed under the terms and conditions of the Creative Commons Attribution (CC BY) license (https://creativecommons.org/licenses/by/4.0/).

1. Introduction

Continuous Fiber-Reinforced Thermoplastic Composites (CFRTPCs), composed of reinforcing fiber and matrix resin through certain forming processes, have advantages such as high specific strength, high specific modulus, corrosion resistance, fatigue resistance, recyclability, good damping and shock absorption, designable performance, and multifunctional integration [1–3]. As shown in Figure 1, they have been widely used in various fields such as aviation, aerospace, rail transit, new energy vehicles, nuclear power, wind energy, and so on [4,5].

The traditional process methods for preparing CFRTPCs (such as compression molding [6], resin transfer molding [6], automated winding/placement [7], autoclave molding [8], etc., as shown in Figure 2) are excessively dependent on the forming die, with high manufacturing costs and slow response speeds. However, with the continuous promotion and application of composite materials, the design of the appearance of products is becoming increasingly complex, the accuracy requirement is becoming higher, and the market rapid response and lightweight requirement are becoming more stringent. Relying on traditional process methods, it is difficult to meet the above technical and efficiency requirements of complex high-performance CFRTPCs.

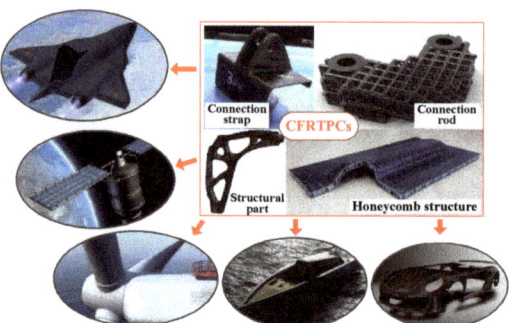

Figure 1. CFRTPCs products and their applications in various fields.

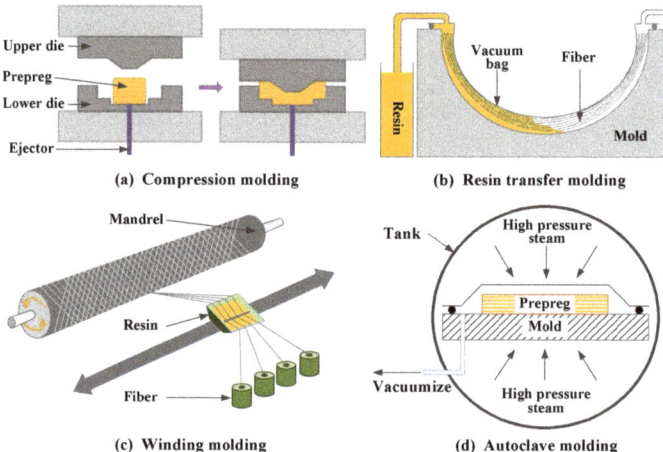

Figure 2. Traditional preparation processes for CFRTPCs.

Three-dimensional printing can directly build products with multiple materials, complex structures, and integrated functions, according to the CAD model without the need for traditional mold development and mechanical processing, which makes the forming not limited by the shape of the parts and the mold, and greatly shortens the product development cycle. Therefore, 3D printing of CFRTPCs has become a new generation of composite material forming methods, namely composites 2.0 [4,5,9,10].

Applying 3D printing technology to the field of forming CFRTPCs can fully leverage the manufacturing advantages of 3D printing and the performance advantages of composite materials [11], which can truly achieve the transformation of the design and manufacturing philosophy of CFRTPCs from "design for manufacturing" to "manufacturing for design". Fused deposition modeling (FDM) is a widely used 3D printing method due to its advantages of low cost, easy operation, and relatively simple process. This paper analyzes and summarizes the current research status of CFRTPCs-FDM. Firstly, the process principles, classification, and corresponding mainstream printing equipment of CFRTPCs-FDM are analyzed and introduced. Secondly, a detailed discussion is conducted on the impact mechanism and research status of the FDM process parameters on the mechanical properties of CFRTPCs. Finally, based on the current research status, the main problems, future research directions, and development trends of CFRTPCs-FDM are considered and prospected.

2. CFRTPCs-FDM Principles and Equipment

According to the different ways of embedding continuous fibers into the resin matrix, the process principles of CFRTPCs-FDM can be mainly divided into two categories: online infiltration co-extrusion and offline prepreg double nozzles extrusion.

Online infiltration co-extrusion: As shown in Figure 3a, continuous fiber dry bundles and pure resin wires are used as raw materials, and both are fed into the same printing head. The fibers are embedded in the resin and soaked inside the printing head and then extruded and deposited on the current printing layer together. The infiltrating time and pressure inside the printing head directly affect the infiltrating effect of fibers/resins. By controlling the feeding speed of resin wires, the fiber volume content of the part can be adjusted.

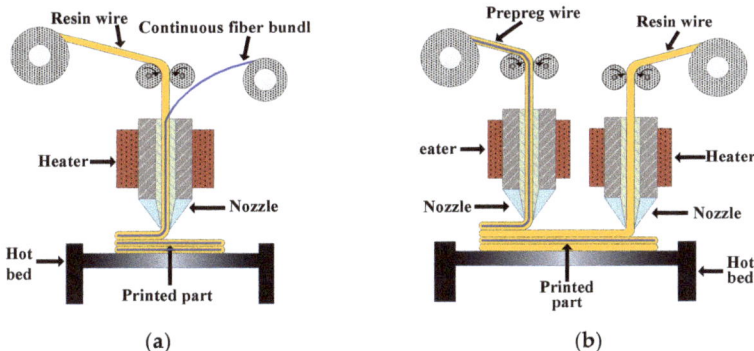

Figure 3. CFRTPCs-FDM principle. (**a**) Online infiltration co-extrusion; (**b**) Offline prepreg double nozzles extrusion.

Offline prepreg double nozzles extrusion: As shown in Figure 3b, a prepreg wire premade of continuous fibers and resin is used as the raw material, and the prepreg is melted, extruded, and deposited in one of the printing nozzles. The other printing head uses pure resin wire as the raw material. On the one hand, the pure matrix resin is melted and deposited to form a good bonding interface between the prepreg layers required by the design, and on the other hand, the pure matrix resin can be used to print the outer frame of the product or fill the internal space to obtain a better appearance and dimensional accuracy. By controlling the printing amount of the pure matrix resin, the fiber volume content can be indirectly adjusted.

The representative printer of online infiltration co-extrusion is the Combot series printing equipment developed by Xi'an Jiaotong University, as shown in Figure 4a; the representative printer of offline prepreg double nozzles extrusion is the Mark series printing equipment developed by the MarkForged Company (Waltham, MA, USA), as shown in Figure 4b.

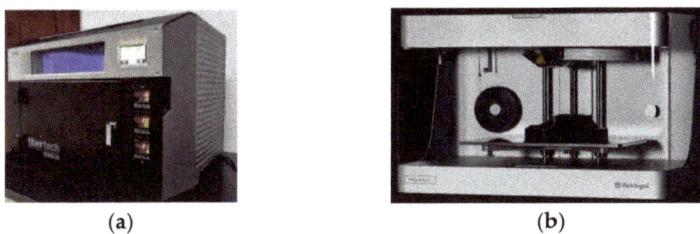

Figure 4. CFRTPCs-FDM printing equipment. (**a**) Combot-1^{TM}; (**b**) MarkTwo^{TM}.

The principle of CFRTPCs-FDM determines the structural form of the 3D printers. For CFRTPCs with the same component, using different process principles and printer structures results in significant differences in the degree of infiltration between the fibers and resins and significant differences in the mechanical properties of the parts. Therefore, when selecting the process principle and corresponding printing equipment, factors such as printing cost, printing efficiency, material performance, and molding quality should be comprehensively considered [12,13].

At present, CFRTPCs-FDM is still in its early development stages, with problems such as low forming accuracy and poor mechanical properties caused by high porosity and multiple interfaces [14], as shown in Figure 5. Figure 6 shows the fracture surface SEM micrographs of the CF/ABS composites printed by FDM and by compression molding, respectively, indicating that pore enlargement is evident around the fibers in the FDM sample, while no significant enlargement is seen in the compression molding sample. How to efficiently integrate CFRTPCs with AM and achieve rapid and accurate additive manufacturing of CFRTPCs is of great significance.

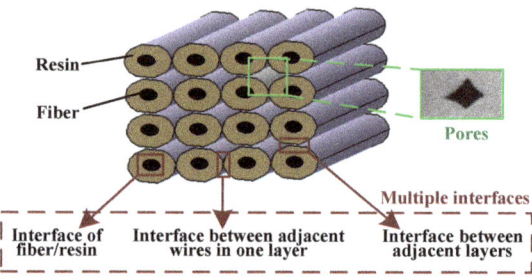

Figure 5. CFRTPCs-FDM printing equipment.

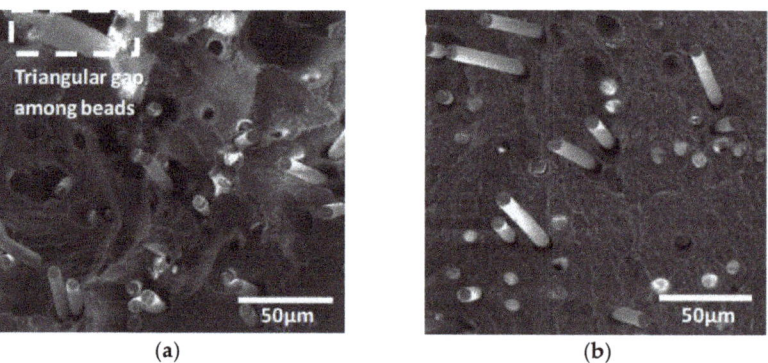

Figure 6. Fracture surface SEM micrographs [15]. (**a**) CF/ABS composites printed by FDM; (**b**) CF/ABS composites formed by compression molding.

3. Influence of the Process Parameters on the Mechanical Properties

The schematic diagram of CFRTPCs-FDM and its process parameters based on online infiltration co-extrusion is shown in Figure 7. The mapping relationship between the process parameters and the mechanical properties is the basis for parameter optimization. Table 1 summarizes the relationship that has been studied in the existing literature, where the mechanical properties include tensile, bending, shear, and compression, and the process parameters include the printing temperature, speed parameters, printing layer thickness, scanning space, etc.

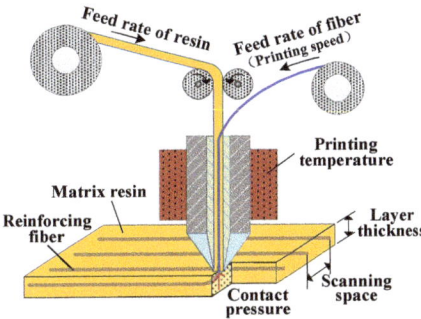

Figure 7. CFRTPCs-FDM based on online infiltration co-extrusion.

Table 1. A list of the literature on the parameters and performance relationship of CFRTPCs.

	Tensile/Elastic	Bending/Flexural	Shear	Compressive	Impact	Fracture	Porosity	Fiber Content
Material type	Refs. [16–20]	Refs. [17,18,20]	Ref. [21]		Ref. [22]		Ref. [17]	Ref. [17]
Printing temperature	Refs. [23–26]	Refs. [26–30]	Refs. [25,30]		Ref. [30]	Ref. [31]		
Printing speed	Refs. [23,24,26,32]	Refs. [26,28,30,33]	Ref. [4]		Ref. [4]	Ref. [9]		
Layer thickness	Refs. [18,23,24,26,34,35]	Refs. [18,26,27,29,30,33]	Refs. [21,30,36]	Ref. [37]	Refs. [22,30]			Refs. [26,27,35–37]
Scanning space	Ref. [23]	Refs. [27,28,33,38]	Ref. [36]					Refs. [27,36]
Stacking orientation	Ref. [18]	Ref. [18]			Ref. [22]			
Fiber orientation	Refs. [17,19,39,40]	Refs. [17,28,41]	Ref. [40]	Ref. [41]			Ref. [17]	Ref. [17]
Fiber content	Refs. [17,39,42]	Refs. [17,41,43]	Ref. [21]	Refs. [41,43]	Ref. [22]		Ref. [17]	

3.1. Material Type

CFRTPCs are composed of two types of materials: a matrix phase thermoplastic resin and a reinforcing phase continuous fiber. The continuous fiber is used to enhance the load-bearing capacity, while the resin is used to support and protect the fibers and evenly transmit and distribute loads. Currently, for CFRTPCs-FDM, commonly used fibers include carbon fiber (CF), glass fiber (GF), and Kevlar fiber (KF); commonly used thermoplastic resins include polylactic acid (PLA), acrylonitrile butadiene styrene copolymers (ABS), polyamide (commonly known as nylon) (PA), and polyether ether ketone (PEEK). Table 2 shows the mechanical parameters of the materials.

Different fiber/resin combinations result in significant differences in the mechanical properties of the 3D printed CFRTPCs due to the different interphase interface properties. Oztan et al. [16] used CF and KF to enhance the nylon matrix and found that the tensile strength of the 3D printed part increased by 2 to 11 times, reaching the strength level of aviation aluminum alloys. In the literature [17–19,22], CFRTPCs were 3D printed with continuous CF, GF, and KF as reinforcing phases and nylon as the matrix phase, respectively. It was found that the mechanical properties of the CF-reinforced composites were the best, followed by GF, and KF was the worst. In addition to the differences in the properties of the fibers themselves, Dickson et al. [17], Caminero et al. [21], and Goh et al. [20,31] also found in the study of the interlayer mechanical properties that the interfacial bonding performance between the reinforced fibers and the matrix is an important factor affecting the mechanical properties of the workpiece.

Table 2. Parameters of commonly used fibers and resins in CFRTPCs-FDM.

Material Parameters	Reinforcing Fibers			Matrix Resins			
	CF	GF	KF	PLA	ABS	PA	PEEK
Density (g/cm^3)	1.27–1.76	1.5	1.2	1.25	1.04	1.1	1.32
Tensile Strength (MPa)	700	590	610	15.5–72.2	36–71.6	54	97
Tensile Modulus (GPa)	54	21	27	2.02–3.55	0.1–2.413	0.94	2.8
Tensile Strain at Break (%)	1.5	3.8	2.7	0.5–9.2	3–20	260	-
Flexural Strength (MPa)	470	210	190	52–115.1	48–110	32	142
Flexural Modulus (GPa)	51	22	26	2.392–4.93	1.917–2.507	0.84	3.7
Flexural Strain at Break (%)	1.2	1.1	2.1	-	-	-	-
Compressive Strength (MPa)	320	140	97	-	-	-	-
Compressive Modulus (GPa)	54	21	28	-	-	-	-
Compressive Strain at Break (%)	0.7	-	1.5	-	-	-	-

The above research indicates that the material properties of fibers and resins, as well as their interfacial properties, are important factors affecting the mechanical properties of CFRTPCs. To improve the mechanical properties, on the one hand, efforts can be made to develop high-performance fibers and matrices and expand the material system; on the other hand, fiber modification technology can be combined to make the fiber surface rougher or undergo chemical reactions to generate new polar groups to promote matrix impregnation of the fibers and improve the interfacial bonding performance between the fibers and the matrix.

3.2. Printing Temperature

The printing temperature directly affects the melt flow performance of the matrix resin after heating, affecting the mechanical properties of the CFRTPCs from two aspects:

(1) Influence on the infiltration between fibers and resins

The printing temperature is too low, resulting in high viscosity and poor fluidity during resin melting, making it difficult to complete the printing work. The increase in printing temperature enhances the resin melt fluidity, making it easier to enter the interior of the fiber bundle, improving the degree of infiltration between the fibers and resin, and thus improving the forming quality of the workpiece. However, excessive temperature can easily lead to strong resin fluidity and even decomposition and vaporization, which is not conducive to 3D forming. In general, the reasonable range for 3D printing temperature selection is above the glass transition temperature of the resin material and below the thermal decomposition temperature. Tian et al. [27] confirmed this through experiments. When the printing temperature is below 180 °C, the fluidity of the PLA is poor, and the nozzle is prone to blockage. When the temperature is above 240 °C, PLA is in liquid form and naturally flows out from the nozzle, which cannot guarantee printing accuracy. Based on comprehensive analysis, it can be concluded that the reasonable printing temperature for PLA is 180–230 °C. Shan et al. [25] found that when the printing temperature increased from 180 °C to 220 °C, the tensile strength of the specimen increased by about 20% (188 MPa vs. 225 MPa), and the bending strength increased by about 8% (274 MPa vs. 296 MPa).

(2) Influence on the multi-interface bonding performance of CFRTPCs

Three-dimensional printed CFRTPCs have multiple interfaces, as shown in Figure 6, mainly including the interphase interface of fiber/resin, the interface between adjacent layers, and the interface between adjacent wires in the layer, among which the latter two belong to the resin wire interface.

(1) For the fiber/resin interface, a reasonable printing temperature is beneficial for the resin matrix to remain above the glass transition temperature for a relatively long time, which is conducive to the full infiltration between the resin and the fiber bundle, thereby obtaining a phase interface with good performance.

(2) For the interface between resin wires, as shown in Figure 8, the formation process of the wire interface includes three stages: the contact between the wire surfaces, the radial growth between adjacent wires, and the diffusion and fusion of molecular chains. A reasonable printing temperature is conducive to the full diffusion and fusion of molecular chains near the contact surface of the printed wire, promoting the radial growth of the wire contact surface, and finally forming a good wire interface.

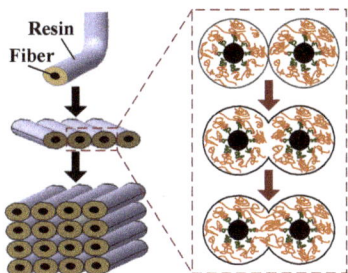

Figure 8. Multi-interface formation process of CFRTPCs-FDM.

3.3. Speed Parameters

The speed parameters include the printing speed and substrate feeding speed. The printing speed refers to the movement speed of the printing head, which is also the wire output speed of continuous fibers at the printer nozzle. The printing speed determines the forming efficiency of the workpiece and the soaking time of the continuous fibers and molten resin in the printing head. For online infiltration co-extrusion, the substrate feeding speed refers to the volume of resin material entering the printing head per unit time. Since the nozzle inner diameter remains unchanged, the substrate feeding speed determines the impregnation pressure between the fiber and the molten matrix, and to some extent determines the fiber volume content of the product.

When determining the speed parameters, it is necessary to comprehensively consider the nozzle diameter and the type and performance of the fiber and resin. In general, a lower printing speed and a higher substrate feeding speed can improve the impregnation pressure and degree, which is beneficial for improving the interfaces' bonding performance. However, too low a printing speed can prolong the heating time of the resin matrix inside the nozzle, causing thermal decomposition of the resin and preventing normal printing. Excessive substrate feeding speed may cause the molten resin to overflow from the nozzle, affecting the normal printing work. Zeng et al. [28] also found that the bending strength and modulus of the specimen decreased with the increasing printing speed (1 mm–6 mm/s). Dou et al. [23] concluded that the tensile strength and modulus of the CF-reinforced PLA specimens decreased by 7.70% (200.43 MPa vs. 185 MPa) and 17.07% (23.31 GPa vs. 19.33 GPa), respectively, by increasing the printing speed from 50 mm/min to 400 mm/min.

3.4. Layer Thickness

As shown in Figure 5, the printing layer thickness refers to the spatial distance between the nozzle and the previous printing layer. A smaller layer thickness is beneficial for enhancing the compaction effect of the nozzle on the material during the process [37,44,45]. On the one hand, it can promote the resin in the molten state to better infiltrate the reinforced fibers and improve the fiber/matrix interface performance. On the other hand, it is also beneficial for improving the interface performance between adjacent printing wires within and between layers. At the same time, the layer thickness also affects the fiber volume content and porosity in the material. As the thickness decreases, the fiber volume content of the product shows an upward trend, and the corresponding porosity decreases.

For 3D printed CFRTPCs, Shan et al. [25] found that with the increase in the layer thickness (0.8–1.2 mm), the tensile strength and bending strength of the workpiece decreased by 33.82% (253.28 MPa vs. 167.63 MPa) and 37.17% (224.78 MPa vs. 141.22 MPa), respectively. Tian et al. [27] found that with the increase in the thickness (0.3–0.8 mm), the bending strength and modulus decreased by 58.9% and 66.3%, respectively. Hu et al. [29] found in their study that the thickness has the most significant impact on the bending performance of composite materials. As the thickness decreases, the bending strength and modulus of the 3D printed continuous CF-reinforced PLA composite can reach 610.1 MPa and 40.1 GPa, respectively. The reduction in the pore defects and the increase in the fiber volume content in materials are the main reasons for the significant improvement in the bending performance. Ning et al. [24] also found that a decrease in the thickness can reduce the porosity of the composite, thereby improving the mechanical properties. Chacón et al. [18,21,22,46] found in 3D printing research that an increase in the layer thickness leads to a decrease in the printing efficiency, and the impact of the thickness on the material mechanical properties is also related to the specific material stacking direction and load action form. Ming et al. [33] found that when the printing layer thickness is large, weak compaction leads to poor interface bonding performance between adjacent layers. But when the thickness is too small, the strong compaction effect can lead to fiber breakage and printing nozzle blockage. Only when the thickness is reasonably selected, can the surface of composite materials be relatively flat, the material can have good fiber continuity and low porosity, and the mechanical properties of the composite are at their highest.

Therefore, when selecting the thickness of the printing layer, it is necessary to comprehensively consider the relationship between the material mechanical properties, printing accuracy, and printing efficiency and make a compromise.

3.5. Scanning Space

As shown in Figure 5, the scanning space refers to the center distance between adjacent printing wires within the same printing layer. Usually, to ensure sufficient contact between printed wires and reduce porosity, a certain overlap area is required. Different scanning spaces can lead to differences in overlap and contact pressure, which in turn affect the degree of fiber/resin infiltration and the multi-interface bonding performance. When the scanning space is too small and the overlapping proportion is too high, fiber wear and fracture occur in the printed structure. If the scanning spacing is too large, there is no overlap between the adjacent wires, and there are obvious pore defects. And the different scanning space also affects the fiber volume content and mechanical properties of the composites.

In the study of the 3D printing of CFRTPCs, Shan. et al. [25] found that with the increase in the scanning space (0.5–1.1 mm), the tensile strength and bending strength of the composites increased first and then decreased. When the scanning space was 0.65 mm, the mechanical property was the highest. When the scanning space was less than 0.65 mm, fiber wear, fracture and warpage occurred during printing, resulting in poor mechanical properties. When the space was greater than 0.65 mm, the increase in the scanning space reduced the impregnation degree and also caused the decrease in the fiber volume content, which led to the decrease in the mechanical strength. Tian et. al. [27,38] discussed the effect of the scanning space on the flexural properties in detail. It was found that with the increase in the scanning space (0.4–1.8 mm), the flexural strength and flexural modulus decreased by 60.7% and 79.3%, respectively. In summary, the scanning space has a significant effect on the mechanical properties of the composites.

3.6. Stacking Direction

According to the different geometric shapes and performance requirements of the parts, different stacking directions can be used to stack and accumulate materials layer by layer. The stacking direction is also an important factor affecting the mechanical properties of CFRTPCs. As shown in Figure 9, in the x-y-z coordinate system, the z-axis represents the stacking direction. It can be seen that the standard test pieces with the same geometric shape have various stacking directions such as a flat normal direction, a side legislative direction, and a straight legislative direction. Three-dimensional prints with different stacking directions have different microstructures and mechanical properties.

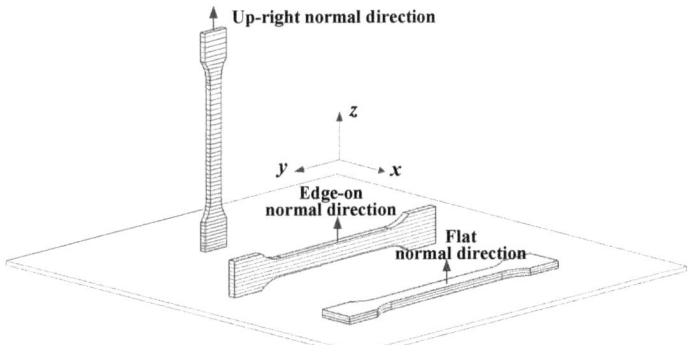

Figure 9. Schematic diagram of stacking direction.

Currently, most research on the stacking direction is mainly based on experiments. Chacón et. al. [18,22,46] systematically studied the influence of the three stacking directions shown in Figure 7 on the mechanical properties of CFRTPCs. In terms of the tensile tests, the performance of stacking along the flat normal and side legislative directions was close to and good, while the performance of stacking along the straight legislative direction was poor. In terms of the bending and impact experiments, the performance of stacking along the lateral direction was the best, followed by the horizontal normal direction and the vertical normal direction. It is worth noting that the impact of the stacking methods on the mechanical properties varies under different loads [41]. The reasons can be summarized: (a) the microstructure of composites with different stacking directions is different, resulting in different bearing characteristics under different loads; (b) different stacking directions can also affect the fiber volume content; (c) different stacking directions may also lead to differences in the porosity of the composite materials.

3.7. Fiber Volume Content

Reinforcing fibers are the main load-bearing phase of CFRTPCs, and their volume content directly determines the part's mechanical properties. Normally, an increase in fiber volume content results in a significant improvement in the mechanical properties. The commonly used methods for calculating fiber content include thermogravimetric analysis [16], geometric calculation [23,37], and image analysis [20,39]. At present, the fiber volume content of 3D printed CFRTPCs does not exceed 50% [11,17,19,47,48], which is lower than that of composite materials prepared by traditional processes. The low fiber volume content is one of the important reasons for the low mechanical properties of 3D printed CFRTPCs. Since in the 3D printing process, with the increase in the fiber volume content, the full infiltration of fibers becomes more difficult, this results in a decrease in the interfacial bonding strength between the fiber/resin phases of 3D printed composites. At the same time, it also introduces more pore defects in the material, ultimately having an adverse impact on the mechanical properties of the composite. Dickson et. al. [17] found that the tensile properties of 3D printed composite materials significantly improved with the increase in the fiber volume content within a certain range. However, when the fiber volume content exceeded a certain level, the improvement in the mechanical properties of the composite material decreased. Chacón et. al. [18,22] believe that an increase in fiber volume content will have two distinct effects on 3D printed composite materials: on the one hand, an increase in the fiber content will hinder the damage evolution in composite materials and play a positive role in improving the mechanical properties of the material; on the other hand, an increase in fiber content will make it more difficult for the fibers to fully infiltrate, leading to a decrease in the strength of the fiber/matrix interface and the introduction of more pore defects in the material, ultimately affecting the mechanical properties. It should be noted that the fiber volume content of 3D printed composite materials is still relatively low now, and with the increase in the fiber volume content, the sufficient infiltration of fibers cannot be effectively guaranteed, and the number of pore defects in the material also increases, which has adverse effects on the improvement of the mechanical properties of composite materials.

Therefore, in future research, on the one hand, effective methods to increase the fiber volume content of 3D printed composite materials should be further explored; on the other hand, in-depth research should also be conducted on how to improve the sufficient infiltration and high porosity of fibers under high fiber volume content, in order to ultimately achieve effective printing of high fiber volume content composite materials with good mechanical properties, meeting the application requirements of complex engineering structures for high-performance composite materials.

4. Improvement and Perfection for CFRTPCs-FDM
4.1. Structural Topology Optimization and Fiber Path Planning for CFRTPCs-FDM

Structural topology optimization is an important means to achieve lightweight in CFRTPCs-FDM products, and fiber path planning aims to determine the distribution of continuous fibers within the product. The main challenge of CFRTPCs-FDM structural topology optimization and fiber path planning is the coupling between structural design and the fiber distribution. Due to the anisotropic mechanical properties of fibers, the change in the fiber distribution caused by the shape change of composite materials should be fully considered in the topology optimization process. There are two main ideas for topology optimization and path planning of a CFRTPCs-FDM structure. One is sequential optimization, which first considers the overall topological shape and then considers the fibers distribution. For example, Li et al. [49] and Fedulov et al. [50] proposed a method of additive manufacturing of continuous carbon-fiber-reinforced nylon composites based on path planning. The topological optimization method is used to analyze the transmission path of the load in isotropic materials, and then the fiber trajectory design is carried out. This method considers the load transmission path of continuous fibers and the anisotropic mechanical properties. Papapetrou et al. [51] proposed a shape optimization method based on the density shape level set (Figure 10a) and gave three fiber filling methods: the offset method (Figure 10b), the equally spaced method (Figure 10c), and the streamline method (Figure 10d). These three methods can ensure the continuity of the fiber in the filling area.

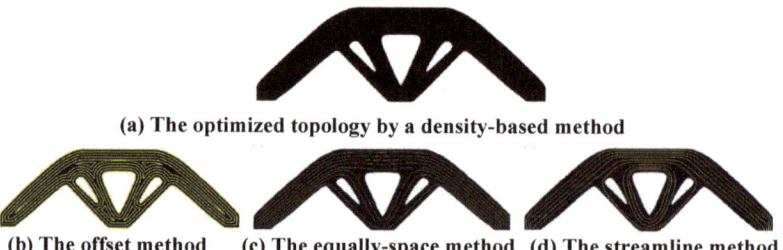

Figure 10. Stiffness-based optimization for the topology and fiber paths by Papapetrou.

The other is a parallel optimization that considers both the structural design and fiber distribution. For example, Lee et al. [52,53] proposed a topology optimization method for functionally graded composite structures, which can simultaneously design the optimal composite topology and spatially variable fiber distribution. A three-dimensional topology optimization method of continuous fibers based on the natural evolution method proposed by Alexander [54] can dynamically find the local optimal distribution of the material density and obtain a lighter structure. Huang et al. [55] proposed a design and manufacturing strategy integrating concurrent optimization of fiber orientation and structural topology for CFRTPCs, realized by ingenious path planning for the 3D printing process.

4.2. Assisted Processes and Devices for CFRTPCs FDM

(1) Microwave-heating-assisted CFRTPCs-FDM

At present, the 3D printing of composite materials mainly adopts the traditional electric heating method to melt and then solidify and deposit, which has a long curing time and a high energy consumption. Li et al. [56] applied microwave technology to the 3D printing of composite materials and proposed a microwave-assisted heating CFRTPCs-FDM method, which uses microwaves to instantaneously volume heat CFRTPCs filaments. As shown in Figure 11, microwaves can quickly heat the fiber and quickly transfer heat to the resin, reducing the temperature change caused by the change in the printing speed, thereby improving the mechanical properties of the parts. At the same time, compared with the traditional heating method, the maximum printable wire diameter (1.75 mm vs. 5.48 mm) and the maximum printable speed (10 mm/s vs. 35 mm/s) are increased. Microwave-assisted CFRTPCs-FDM has the advantages of a fast heating speed, high thermal energy utilization, easy control, energy savings, and environmental protection.

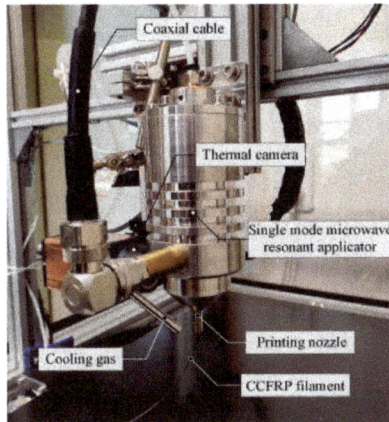

Figure 11. Illustration of the microwave printing head.

(2) Ultrasonic-assisted CFRTPCs-FDM

Qiao et al. [57] proposed an ultrasonic-assisted fiber interface modification method to improve the wettability of fiber resin and the mechanical properties of CFRTPCs parts. In the 3D printing process of carbon fiber composites based on online infiltration co-extrusion, the carbon fiber is first guided to infiltrate once in the resin liquid with ultrasonic action, which reduces the porosity and improves the interfacial properties of the composites, as shown in Figure 12.

In order to improve the bonding between the resin and the fiber and reduce the porosity to improve the printing quality of CFRTPCs, in addition to the above process improvements, scholars have also proposed assisted measures such as substrate ultrasonic vibration assistance, laser melting assistance, roller pressure followup assistance, and hot pressing post treatment to improve the mechanical properties of CFRTPCs–FDM.

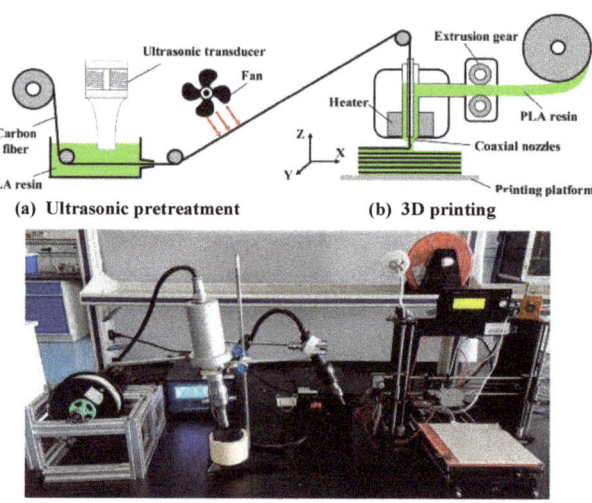

Figure 12. A schematic diagram and a photograph of the ultrasonic-assisted CFRTPCs-FDM.

4.3. Recycling and Remanufacturing of CFRTPCs–FDM

With the wide application of CFRTPCs-FDM, the recycling and remanufacturing of waste composite parts has become a problem that must be faced. The recovery technologies of carbon fiber composite materials are divided into three categories: the physical recovery method, the energy recovery method, and the chemical recovery method. The physical recovery method is to mill, cut, or crush the waste composite materials to obtain short fibers, particles, powders, and other substances. This method has the advantages of being low cost, a simple process, and having no pollution, which is only suitable for uncontaminated composite waste parts, and the strength of the fiber is seriously reduced after treatment, and the reuse value is low. So, it can only be used as some fillers and cannot become a material with high purity requirements. The energy recovery method is to incinerate waste composite materials to obtain heat energy. However, this method causes secondary pollution, which should be avoided as much as possible. The technology that can really further recycle and reuse fibers is the chemical recovery method. The method is to degrade the resin matrix into small molecular compounds or oligomers, so as to achieve separation from fibers and fillers. It can not only recover the reinforced fibers but also recover the resin as raw material. It is the most promising recovery method at present [58].

The carbon fiber obtained by the chemical recovery method is usually in a messy and fluffy state, which is difficult to make into prepreg products with long tow lengths. Therefore, the recovered fiber is more suitable for nonwoven materials, cut into short fibers, or even ground into smaller particles or powders, added to the resin matrix for remanufacturing. At present, chopped-fiber-reinforced thermoplastic materials are widely used in the 3D printing of composite materials. Therefore, the recycled carbon fiber is particularly suitable for the 3D printing of chopped-fiber-reinforced composites. Figure 13 shows the CFRTPCs-FDM sustainable manufacturing and circular economy [59]. Compared with original fibers, using recycled fibers for 3D printing composite materials not only avoids the environmental pollution caused by waste materials but also helps to reduce the cost of the 3D printing composite materials.

Figure 13. CFRTPCs-FDM sustainable manufacturing and circular economy.

5. Summary and Prospects

Compared with traditional preparation processes, the mechanical properties of 3D printed CFRTPCs are still relatively poor. This paper analyzes and reviews the current research status of the 3D printing of CFRTPCs from the perspective of mechanical properties, and draws the following conclusions:

(1) The mechanical properties of 3D printed CFRTPCs are closely related to the properties of reinforcing fibers, matrix resins, and multiple interfaces. The influencing factors can be summarized into three aspects: the material type, the process principle, and the process parameters. The factors are coupled with each other, jointly determining the forming quality of CFRTPC products;

(2) The impact of the above factors on the mechanical properties of 3D printed CFRTPCs is ultimately reflected in three aspects: the fiber volume content, the interface bonding strength, and the porosity. Compared with traditional forming processes, current 3D printed CFRTPCs still face prominent problems such as a low fiber volume content, a weak interface bonding strength, and a high porosity.

(3) The fiber volume content is determined by a combination of speed parameters, layer thickness, scanning space, stacking direction, and so on. The pores of workpieces can be divided into macroscopic pores caused by the minimum bending radius or scanning overlap of the fibers and microscopic pores existing in continuous fibers, matrix resin, and interphase interface connections. Three-dimensional printed CFRTPCs have multiple interfaces: a fiber/resin interface and a resin interface between and within layers. Good interface bonding characteristics are conducive to reducing the porosity, improving the stress transfer efficiency, and interlaminar shear strength at the interface.

However, the existing research mainly focuses on analyzing the relevant factors through experimental methods along the lines of equipment development, experimental design, and result comparison, and exploring the optimal process methods and parameters under the relevant material types. In the future, research on CFRTPCs-FDM may focus on the following aspects:

(1) It is necessary to explore the rheological and time-dependent behavior of composite materials in the multiple states of FDM from a more microscopic perspective, so as to construct a mechanical model that accurately describes the complex rheological properties of materials and reveal the melting deposition mechanism of CFRTPCs at the molecular/atomic level, to realize scientific prediction of the mechanical properties of 3D printed CFRTPCs.

(2) The data show that 50% to 60% of the structural failures in composite materials are closely related to interlayer damage. The bonding ability of the CFRTPCs' 3D printing multi-interface (interphase interface, interlayer, and inner layer wire interface) determines the interlaminar mechanical properties of the CFRTPCs to a large extent. Therefore, it is necessary to further study the interface cross scale coupling model of CFRTPCs-FDM, in order to reveal the regulatory mechanism of the interface from the microstructure to the macroscopic performance.

(3) Researchers have established a relatively effective numerical simulation model for the traditional forming process of composite materials, achieving prediction and simulation of their mechanical behavior under typical load conditions. However, numerical simulations for the 3D printing of CFRTPCs are relatively lacking. Effective finite element and molecular dynamic models should be established based on the rheological mechanism and interface model of CFRTPCs-FDM to achieve effective simulation and prediction of 3D printed CFRTPCs properties.

(4) The 3D printing of CFRTPCs has anisotropic characteristics, and the stacking direction and fiber orientation seriously affect the optimal load-bearing condition of the product. Therefore, it is necessary to carry out configuration design in combination with topology optimization, develop five-axis 3D printing equipment, adjust the stacking direction in real time, and use a path planning algorithm to realize the controllable layout of the fiber direction, so as to realize the integrated design and manufacturing of CFRTPCs in terms of "performance–configuration–process".

(5) It is necessary to explore and develop new process principles and improvement methods for the 3D printing of CFRTPCs, innovate CFRTPCs' 3D printing equipment, and further improve the mechanical properties. Also needed is to develop new materials and improve the material system and, on the basis of the tensile, bending, and compression properties, further enrich the quality evaluation methods of the 3D printing of CFRTPCs, such as the impact properties, wear properties, creep properties, fatigue properties, and damage evolution laws.

Author Contributions: Conceptualization, Y.Y. and B.Y.; methodology, Y.Y. and Z.C.; validation, J.D. and W.C.; formal analysis, Z.C.; investigation, Y.Y. and B.Y.; resources, Y.Y.; data curation, Y.Y. and Z.C.; writing—original draft preparation, Y.Y.; writing—review and editing, Y.Y. and W.C.; visualization, Y.Y.; supervision, Z.C. and J.D.; project administration, Y.Y.; funding acquisition, Y.Y. All authors have read and agreed to the published version of the manuscript.

Funding: This work was supported by the Special Scientific Research Project of Shaanxi Education Department (19JK0594).

Institutional Review Board Statement: Not applicable.

Data Availability Statement: No new data were created or analyzed in this study. Data sharing is not applicable to this article.

Conflicts of Interest: The authors declare no conflict of interest.

References

1. Minchenkov, K.; Vedernikov, A.; Safonov, A.; Akhatov, I. Thermoplastic Pultrusion: A Review. *Polymers* **2021**, *13*, 180. [CrossRef] [PubMed]
2. Vedernikov, A.; Minchenkov, K.; Gusev, S.; Sulimov, A.; Zhou, P.; Li, C.; Xian, G.; Akhatov, I.; Safonov, A. Effects of the Pre-Consolidated Materials Manufacturing Method on the Mechanical Properties of Pultruded Thermoplastic Composites. *Polymers* **2022**, *14*, 2246. [CrossRef] [PubMed]
3. Tucci, F.; Rubino, F.; Pasquino, G.; Carlone, P. Thermoplastic Pultrusion Process of Polypropylene/Glass Tapes. *Polymers* **2023**, *15*, 2374. [CrossRef] [PubMed]
4. Tian, X.; Todoroki, A.; Liu, T.; Wu, L.; Hou, Z.; Ueda, M.; Hirano, Y.; Matsuzaki, R.; Mizukami, K.; Iizuka, K.; et al. 3D Printing of continuous fiber reinforced polymer composites: Development, application, and prospective. *Chin. J. Mech. Eng. Addit. Manuf. Front.* **2022**, *1*, 100016. [CrossRef]
5. Liu, G.; Xiong, Y.; Zhou, L. Additive manufacturing of continuous fiber reinforced polymer composites: Design opportunities and novel applications. *Compos. Commun.* **2021**, *27*, 100907. [CrossRef]

6. Santos, A.C.M.Q.S.; Monticeli, F.M.; Ornaghi, H.; Santos, L.F.D.P.; Cioffi, M.O.H. Porosity characterization and respective influence on short-beam strength of advanced composite processed by resin transfer molding and compression molding. *Polym. Polym. Compos.* **2021**, *29*, 1353–1362. [CrossRef]
7. Smith, R.P.; Qureshi, Z.; Scaife, R.J.; El-Dessouky, H.M. Limitations of processing carbon fibre reinforced plastic/polymer material using automated fibre placement technology. *J. Reinf. Plast. Compos.* **2016**, *35*, 1527–1542. [CrossRef]
8. Dong, C.; Zhou, J.; Ji, X.; Yin, Y.; Shen, X. Study of the curing process of carbon fiber reinforced resin matrix composites in autoclave processing. *Procedia Manuf.* **2019**, *37*, 450–458. [CrossRef]
9. Ryosuke, M.; Masahito, U.; Masaki, N.; Tae-kun, J.; Hirosuke, A.; Keisuke, H.; Taishi, N.; Akira, T.; Yoshiyasu, H. Three-dimensional printing of continuous-fiber composites by in-nozzle impregnation. *Sci. Rep.* **2016**, *6*, 23058.
10. Yang, C.; Tian, X.; Liu, T.; Cao, Y.; Li, D. 3D printing for continuous fiber reinforced thermoplastic composites: Mechanism and performance. *Rapid Prototyp. J.* **2017**, *23*, 209–215. [CrossRef]
11. Justo, J.; Távara, L.; García-Guzmán, L.; París, F. Characterization of 3D printed long fibre reinforced composites. *Compos. Struct.* **2018**, *185*, 537–548. [CrossRef]
12. Azarov, A.V.; Antonov, F.K.; Vasil'Ev, V.V.; Golubev, M.V.; Krasovskii, D.S.; Razin, A.F.; Salov, V.A.; Stupnikov, V.V.; Khaziev, A.R. Development of a two-matrix composite material fabricated by 3D printing. *Polym. Sci.* **2017**, *10*, 87–90. [CrossRef]
13. Azarov, A.V.; Antonov, F.K.; Golubev, M.V.; Khaziev, A.R.; Ushanov, S.A. Composite 3D printing for the small size unmanned aerial vehicle structure. *Compos. Part B Eng.* **2019**, *169*, 157–163. [CrossRef]
14. Heidari-Rarani, M.; Rafiee-Afarani, M.; Zahedi, A.M. Mechanical characterization of FDM 3D printing of continuous carbon fiber reinforced PLA composites. *Composites* **2019**, *175*, 107147. [CrossRef]
15. Tekinalp, H.L.; Kunc, V.; Velez-Garcia, G.M.; Duty, C.E.; Love, L.J.; Naskar, A.K.; Blue, C.A.; Ozcan, S. Highly oriented carbon fiber–polymer composites via additive manufacturing. *Compos. Sci. Technol.* **2014**, *105*, 144–150. [CrossRef]
16. Oztan, C.; Karkkainen, R.; Fittipaldi, M.; Nygren, G.; Roberson, L.; Lane, M.; Celik, E. Microstructure and mechanical properties of three dimensional-printed continuous fiber composites. *J. Compos. Mater.* **2019**, *53*, 271–280. [CrossRef]
17. Dickson, A.N.; Barry, J.N.; McDonnell, K.A.; Dowling, D.P. Fabrication of Continuous Carbon, Glass and Kevlar fibre reinforced polymer composites using Additive Manufacturing. *Addit. Manuf.* **2017**, *16*, 146–152. [CrossRef]
18. Chacón, J.M.; Caminero, M.A.; Núñez, P.J.; García-Plaza, E.; García-Moreno, I.; Reverte, J.M. Additive manufacturing of continuous fibre reinforced thermoplastic composites using fused deposition modelling: Effect of process parameters on mechanical properties. *Compos. Sci. Technol.* **2019**, *181*, 107688. [CrossRef]
19. Al Abadi, H.; Thai, H.T.; Paton-Cole, V.; Patel, V.I. Elastic properties of 3D printed fibre-reinforced structures. *Compos. Struct.* **2018**, *193*, 8–18. [CrossRef]
20. Goh, G.D.; Dikshit, V.; Nagalingam, A.P.; Goh, G.L.; Agarwala, S.; Sing, S.L.; Wei, J.; Yeong, W.Y. Characterization of mechanical properties and fracture mode of additively manufactured carbon fiber and glass fiber reinforced thermoplastics. *Mater. Des.* **2018**, *137*, 79–89. [CrossRef]
21. Caminero, M.A.; Chacón, J.M.; García-Moreno, I.; Reverte, J.M. Interlaminar bonding performance of 3D printed continuous fibre reinforced thermoplastic composites using fused deposition modelling. *Polym. Test.* **2018**, *68*, 415–423. [CrossRef]
22. Caminero, M.A.; Chacón, J.M.; García-Moreno, I.; Rodríguez, G.P. Impact damage resistance of 3D printed continuous fibre reinforced thermoplastic composites using fused deposition modelling. *Compos. Part B Eng.* **2018**, *148*, 93–103. [CrossRef]
23. Dou, H.; Cheng, Y.; Ye, W.; Zhang, D.; Li, J.; Miao, Z.; Rudykh, S. Effect of Process Parameters on Tensile Mechanical Properties of 3D Printing Continuous Carbon Fiber-Reinforced PLA Composites. *Materials* **2020**, *13*, 3850. [CrossRef]
24. Ning, F.; Cong, W.; Hu, Y.; Wang, H. Additive manufacturing of carbon fiber-reinforced plastic composites using fused deposition modeling: Effects of process parameters on tensile properties. *J. Compos. Mater.* **2017**, *51*, 451–462. [CrossRef]
25. Fan, C.; Shan, Z.; Zou, G.; Li, Z. Interfacial Bonding Mechanism and Mechanical Performance of Continuous Fiber Reinforced Composites in Additive Manufacturing. *Chin. J. Mech. Eng.* **2021**, *34*, 1–11. [CrossRef]
26. Wang, G.; Jia, Z.; Wang, F.; Dong, C.; Wu, B. Additive Manufacturing of Continuous Carbon Fiber Reinforced Thermoplastic Composites: An Investigation on Process-Impregnation-Property Relationship. In Proceedings of the ASME 2020 15th International Manufacturing Science and Engineering Conference, Online, 3 September 2020.
27. Tian, X.; Liu, T.; Yang, C.; Wang, Q.; Li, D. Interface and performance of 3D printed continuous carbon fiber reinforced PLA composites. *Compos. Part A Appl. Sci. Manuf.* **2016**, *88*, 198–205. [CrossRef]
28. Zeng, C.; Liu, L.; Bian, W.; Liu, Y.; Leng, J. 4D printed electro-induced continuous carbon fiber reinforced shape memory polymer composites with excellent bending resistance. *Compos. Part B Eng.* **2020**, *194*, 108034. [CrossRef]
29. Hu, Q.; Duan, Y.; Zhang, H.; Liu, D.; Yan, B.; Peng, F. Manufacturing and 3D printing of continuous carbon fiber prepreg filament. *J. Mater. Sci.* **2018**, *53*, 1887–1898. [CrossRef]
30. Chen, K.; Yu, L.; Cui, Y.; Jia, M.; Pan, K. Optimization of printing parameters of 3D-printed continuous glass fiber reinforced polylactic acid composites. *Thin-Walled Struct.* **2021**, *164*, 107717. [CrossRef]
31. Goh, G.D.; Dikshit, V.; An, J.; Yeong, W.Y. Process-structure-property of Additively Manufactured Continuous Carbon Fiber Reinforced Thermoplastic: An Investigation of Mode I Interlaminar Fracture Toughness. *Mech. Adv. Mater. Struct.* **2020**, *29*, 1418–1430. [CrossRef]

32. Mohammadizadeh, M.; Imeri, A.; Fidan, I.; Elkelany, M. 3D printed fiber reinforced polymer composites-Structural analysis. *Compos. Part B Eng.* **2019**, *175*, 107112. [CrossRef]
33. Ming, Y.; Zhang, S.; Han, W.; Wang, B.; Duan, Y.; Xiao, H. Investigation on process parameters of 3D printed continuous carbon fiber-reinforced thermosetting epoxy composites. *Addit. Manuf.* **2020**, *33*, 101184. [CrossRef]
34. Akhoundi, B.; Behravesh, A.H.; Saed, A.B. Mproving mechanical properties of continuous fiber-reinforced thermoplastic composites produced by FDM 3D printer. *J. Reinf. Plast. Compos.* **2019**, *38*, 99–116. [CrossRef]
35. Dong, K.; Liu, L.; Huang, X.; Xiao, X. 3D printing of continuous fiber reinforced diamond cellular structural composites and tensile properties. *Compos. Struct.* **2020**, *15*, 112610. [CrossRef]
36. Luo, M.; Tian, X.; Shang, J.; Zhu, W.; Li, D.; Qin, Y. Impregnation and Interlayer Bonding Behaviours of 3D-Printed Continuous Carbon-Fiber-Reinforced Poly-ether-ether-ketone Composites. *Compos. Part A Appl. Sci. Manuf.* **2019**, *121*, 130–138. [CrossRef]
37. Hou, Z.; Tian, X.; Zhang, J.; Li, D. 3D printed continuous fibre reinforced composite corrugated structure. *Compos. Struct.* **2018**, *184*, 1005–1010. [CrossRef]
38. Liu, T.; Tian, X.; Zhang, M.; Abliz, D.; Li, D.; Ziegmann, G. Interfacial performance and fracture patterns of 3D printed continuous carbon fiber with sizing reinforced PA6 composites. *Compos. Part A Appl. Sci. Manuf.* **2018**, *114*, 368–376. [CrossRef]
39. Chabaud, G.; Castro, M.; Denoual, C.; Duigou, A.L. Hygromechanical properties of 3D printed continuous carbon and glass fibre reinforced polyamide composite for outdoor structural applications. *Addit. Manuf.* **2019**, *26*, 94–105. [CrossRef]
40. Pyl, L.; Kalteremidou, K.A.; Hemelrijck, D.V. Exploration of specimen geometry and tab configuration for tensile testing exploiting the potential of 3D printing freeform shape continuous carbon fibre-reinforced nylon matrix composites. *Polym. Test.* **2018**, *71*, 318–328. [CrossRef]
41. Araya-Calvo, M.; López-Gómez, I.; Chamberlain-Simon, N.; León-Salazar, J.L.; Guillén-Girón, T.; Corrales-Cordero, J.S.; Sánchez-Brenes, O. Evaluation of compressive and flexural properties of continuous fiber fabrication additive manufacturing technology. *Addit. Manuf.* **2018**, *22*, 157–164. [CrossRef]
42. Pappas, J.M.; Thakur, A.R.; Leu, M.C.; Dong, X. A parametric study and characterization of additively manufactured continuous carbon fiber reinforced composites for high-speed 3D printing. *Int. J. Adv. Manuf. Technol.* **2021**, *113*, 2137–2151. [CrossRef]
43. Li, H.; Wang, T.; Joshi, S.; Yu, Z. The quantitative analysis of tensile strength of additively manufactured continuous carbon fiber reinforced polylactic acid (PLA). *Rapid Prototyp. J.* **2019**, *10*, 1624–1636. [CrossRef]
44. Yin, L.; Tian, X.; Shang, Z.; Wang, X.; Hou, Z. Characterizations of continuous carbon fiber-reinforced composites for electromagnetic interference shielding fabricated by 3D printing. *Appl. Phys. A Mater. Sci. Process.* **2019**, *125*, 266. [CrossRef]
45. Chang, B.; Li, X.; Parandoush, P.; Ruan, S.; Shen, C.; Lin, D. Additive manufacturing of continuous carbon fiber reinforced poly-ether-ether-ketone with ultrahigh mechanical properties. *Polym. Test.* **2020**, *88*, 106563. [CrossRef]
46. Chacón, J.M.; Caminero, M.A.; García-Plaza, E.; Núñez, P.J. Additive manufacturing of PLA structures using fused deposition modelling: Effect of process parameters on mechanical properties andtheir optimal selection. *Mater. Des.* **2017**, *124*, 143–157. [CrossRef]
47. Mei, H.; Ali, Z.; Yan, Y.; Ali, I.; Cheng, L. Influence of mixed isotropic fiber angles and hot press on the mechanical properties of 3D printed composites. *Addit. Manuf.* **2019**, *27*, 150–158. [CrossRef]
48. Werken, N.; Hurley, J.; Khanbolouki, P.; Sarvestani, A.N.; Tamijani, A.Y.; Tehrani, M. Design considerations and modeling of fiber reinforced 3D printed parts. *Compos. Part B Eng.* **2019**, *160*, 684–692. [CrossRef]
49. Li, N.; Link, G.; Wang, T.; Ramopoulos, V.; Neumaier, D.; Hofele, J.; Walter, M.; Jelonnek, J. Path-designed 3D printing for topological optimized continuous carbon fibre reinforced composite structures. *Compos. Part B Eng.* **2020**, *182*, 107612. [CrossRef]
50. Fedulov, B.; Fedorenko, A.; Khaziev, A.; Antonov, F. Optimization of parts manufactured using continuous fiber three-dimensional printing technology. *Compos. Part B Eng.* **2021**, *227*, 109406. [CrossRef]
51. Papapetrou, V.S.; Patel, C.; Tamijani, A.Y. Stiffness-based optimization framework for the topology and fiber paths of continuous fiber composites. *Compos. Part B Eng.* **2020**, *183*, 107681. [CrossRef]
52. Lee, J.; Kim, D.; Nomura, T.; Dede, E.M.; Yoo, J. Topology optimization for continuous and discrete orientation design of functionally graded fiber-reinforced composite structures. *Compos. Struct.* **2018**, *201*, 217–233. [CrossRef]
53. Kim, D.; Lee, J.; Nomura, T.; Dede, E.M.; Yoo, J.; Min, S. Topology optimization of functionally graded anisotropic composite structures using homogenization design method. *Comput. Methods Appl. Mech. Eng.* **2020**, *369*, 113220. [CrossRef]
54. Alexander, A. Safonov.3D topology optimization of continuous fiber-reinforced structures via natural evolution method. *Compos. Struct.* **2019**, *215*, 289–297.
55. Huang, Y.; Tian, X.; Zheng, Z.; Li, D.; Malakhov, A.V.; Polilov, A.N. Multiscale concurrent design and 3D printing of continuous fiber reinforced thermoplastic composites with optimized fiber trajectory and topological structure. *Compos. Struct.* **2022**, *285*, 115241. [CrossRef]
56. Li, N.; Guido, L.; John, J.; Morais Manuel, V.C.; Frank, H. Microwave additive manufacturing of continuous carbon fibers reinforced thermoplastic composites: Characterization, analysis, and properties. *Addit. Manuf.* **2021**, *44*, 102035. [CrossRef]
57. Qiao, J.; Li, Y.; Li, L. Ultrasound-assisted 3D printing of continuous fiber-reinforced thermoplastic (FRTP) composites. *Addit. Manuf.* **2019**, *30*, 100926. [CrossRef]

58. Scaffaro, R.; Di Bartolo, A.; Dintcheva, N.T. Matrix and Filler Recycling of Carbon and Glass Fiber-Reinforced Polymer Composites: A Review. *Polymers* **2021**, *13*, 3817. [CrossRef]
59. Liu, W.; Huang, H.; Zhu, L.; Liu, Z. Integrating carbon fiber reclamation and additive manufacturing for recycling CFRP waste. *Compos. Part B Eng.* **2021**, *215*, 108808. [CrossRef]

Disclaimer/Publisher's Note: The statements, opinions and data contained in all publications are solely those of the individual author(s) and contributor(s) and not of MDPI and/or the editor(s). MDPI and/or the editor(s) disclaim responsibility for any injury to people or property resulting from any ideas, methods, instructions or products referred to in the content.

Article

Fabrication of Highly Conductive Porous Fe$_3$O$_4$@RGO/PEDOT:PSS Composite Films via Acid Post-Treatment and Their Applications as Electrochemical Supercapacitor and Thermoelectric Material

Luyao Gao [1,2,3], Fuwei Liu [1,2,3,*], Qinru Wei [1], Zhiwei Cai [1], Jiajia Duan [1], Fuqun Li [1], Huiying Li [1], Ruotong Lv [1], Mengke Wang [1], Jingxian Li [1] and Letian Wang [1]

1 College of Physics and Electronic Engineering, Xinyang Normal University, Xinyang 464000, China
2 Key Laboratory of Advanced Micro/Nano Functional Materials of Henan Province, Xinyang Normal University, Xinyang 464000, China
3 Energy-Saving Building Materials Innovative Collaboration Center of Henan Province, Xinyang Normal University, Xinyang 464000, China
* Correspondence: liufuwei168@163.com

Citation: Gao, L.; Liu, F.; Wei, Q.; Cai, Z.; Duan, J.; Li, F.; Li, H.; Lv, R.; Wang, M.; Li, J.; et al. Fabrication of Highly Conductive Porous Fe$_3$O$_4$@RGO/PEDOT:PSS Composite Films via Acid Post-Treatment and Their Applications as Electrochemical Supercapacitor and Thermoelectric Material. *Polymers* 2023, 15, 3453. https://doi.org/10.3390/polym15163453

Academic Editors: Jeong In Han, Jiangtao Xu and Sihang Zhang

Received: 4 July 2023
Revised: 16 August 2023
Accepted: 16 August 2023
Published: 18 August 2023

Copyright: © 2023 by the authors. Licensee MDPI, Basel, Switzerland. This article is an open access article distributed under the terms and conditions of the Creative Commons Attribution (CC BY) license (https://creativecommons.org/licenses/by/4.0/).

Abstract: As a remarkable multifunctional material, ferroferric oxide (Fe$_3$O$_4$) exhibits considerable potential for applications in many fields, such as energy storage and conversion technologies. However, the poor electronic and ionic conductivities of classical Fe$_3$O$_4$ restricts its application. To address this challenge, Fe$_3$O$_4$ nanoparticles are combined with graphene oxide (GO) via a typical hydrothermal method, followed by a conductive wrapping using poly(3,4-ethylenedioxythiophene):poly(styrene sulfonic sulfonate) (PEDOT:PSS) for the fabrication of composite films. Upon acid treatment, a highly conductive porous Fe$_3$O$_4$@RGO/PEDOT:PSS hybrid is successfully constructed, and each component exerts its action that effectively facilitates the electron transfer and subsequent performance improvement. Specifically, the Fe$_3$O$_4$@RGO/PEDOT:PSS porous film achieves a high specific capacitance of 244.7 F g^{-1} at a current of 1 A g^{-1}. Furthermore, due to the facial fabrication of the highly conductive networks, the free-standing film exhibits potential advantages in flexible thermoelectric (TE) materials. Notably, such a hybrid film shows a high electric conductivity (σ) of 507.56 S cm^{-1}, a three times greater value than the Fe$_3$O$_4$@RGO component, and achieves an optimized Seebeck coefficient (S) of 13.29 µV K^{-1} at room temperature. This work provides a novel route for the synthesis of Fe$_3$O$_4$@RGO/PEDOT:PSS multifunctional films that possess promising applications in energy storage and conversion.

Keywords: PEDOT:PSS; Fe$_3$O$_4$; GO; composite film; supercapacitor; thermoelectric

1. Introduction

Considering the rapid growth of the world's economy and the continual consumption of fossil fuels, green energy (such as wind, hydropower, thermoelectric, etc.) and electrical energy storage devices are in urgent need for many applications such as portable electronic devices and electric vehicles [1–6]. Among the variety of energy storage devices, supercapacitors (SCs) have attracted widespread attention for their high power density, reliable safety, outstanding cycling stability, and low cost [2,7–10]. Generally, there are two kinds of supercapacitors: electrochemical double-layer capacitors (EDLCs) and pseudocapacitors. While EDLCs store electricity through the double-layer effect, the pseudocapacitor works through a fast redox reaction, which is essential for harvesting outstanding capacitive ability [11–13].

To explore desired electrode materials for supercapacitors, many efforts have been put into researching transition metal oxides, such as Fe$_3$O$_4$, Fe$_2$O$_3$, Co$_3$O$_4$, RuO$_2$, MnO$_2$, etc. [10,14–18]. Among them, iron ferrite of Fe_3O_4 has been proposed as a potential supercapacitor material because of its high specific capacitance, easy redox reaction, rich natural storage, and environmental friendliness [10]. Nevertheless, Fe_3O_4 has a low conductivity in nature, which limits its electrochemical performance. Furthermore, it remains a challenge to avoid nanoparticle agglomeration during the preparation of electrode materials. In order to solve the above problems, Fe_3O_4 is commonly combined with carbon-based materials, especially graphene and CNTs. For instance, through a layer-by-layer method, the obtained Fe_3O_4/RGO multilayer electrodes exhibited a specific capacitance of 151 F g^{-1} when a current density of 0.9 A g^{-1} was used, and after 1000 cycles, the capacitance retained 85% of its original value, indicating good cycling stability [19]. A sandwich-like Fe_3O_4/MnO_2/RGO nanocomposite was explored and the value of specific capacitance reached 77.5 F g^{-1} at 0.5 A g^{-1} and kept 35 F g^{-1} at 20 A g^{-1} in 1 M Na_2SO_4 [20]. CNT/Fe_3O_4 nanocomposites synthesized through the hydrothermal method also achieved a specific capacitance of 117.2 F g^{-1} at 10 mA cm^{-2} in a 6 M KOH electrolyte [21]. A novel BRGO/Fe_3O_4-MWCNT hybrid nanocomposite was successfully fabricated and possessed good supercapacitance performance (165 F g^{-1} at a current density of 2 A g^{-1}) [22]. Not only that, the obtained BRGO/Fe_3O_4-MWCNT composites also possessed a high photo degradation efficiency. As conductive frameworks, carbon materials in these strategies effectively avoid the collapse of the nano-Fe_3O_4 particles and thus improve their electrochemical properties. However, to realize large capacitance and practical applications requires high mass loading of active Fe_3O_4, which in turn increases the electrode resistance and thus limits the performance characteristics of the composite electrodes. In addition, binder materials like polytetrafluoroethylene (PTFE) and polyvinylidene fluoride (PVDF) are frequently used during the preparation of metal-oxide-based nanocomposite films for the preparation of flexible composite materials [23,24]. However, these binders are nonconductive and decrease the electrical conductivity of the electrodes.

To address these issues, one promising strategy is to incorporate Fe_3O_4 nanoparticles onto carbon-based frameworks and coat them with conducting polymers to form highly interconnected networks for charge transformation. Thus, there is an urgent demand for highly conductive binders that can further disperse the packed Fe_3O_4/carbon nanostructures. Among various conducting polymers, PEDOT:PSS is water soluble and can be used as a binder that is capable of dispersing carbon-based materials and/or other kinds of nanomaterials in water. Additionally, PEDOT:PSS can achieve high conductivities via incorporating additives (such as organic solvents [25,26], ionic liquids [27], and inorganic salts [28], etc.) and post-treatment through polar solvents (e.g., DMSO [29,30], EG [31,32], etc.) or acids [33,34]. It is believed that the Fe_3O_4/carbon/PEDOT:PSS composite can be employed as an excellent capacity electrode material that possesses considerable potential application in energy storage devices. Furthermore, such a unique structure can effectively increase the conductivity of the composite material, and its potential application in energy conversion technologies, such as thermoelectric, photoelectric, and thermal sensor, can also be expected.

Herein, we construct a ternary system based on Fe_3O_4@RGO/PEDOT:PSS; the graphene oxide (GO) in the compound acts as a base supporting material, while the PEDOT:PSS serves as a highly conductive wrapping material. Notably, after acid treatment, the as-prepared hybrid composite is easily stripped from the glass substrate and forms porous, highly conductive, and flexible electrode films. Benefiting from the large-scale construction of porous structures and the highly connected conducting networks, the Fe_3O_4@RGO/PEDOT:PSS electrodes exhibit a high specific capacitance of 244.7 F g^{-1} at 1 A g^{-1}, and a good rate capability that remains 146.0 F g^{-1} at 10 A g^{-1}. Except for energy storage, the constructed hybrid films can also be used as thermoelectric (TE) materials, which are capable of converting low-grade and/or waste heat into electricity, making them an important source of green energy. A dimensionless figure of merit, $ZT = S^2\sigma T/\kappa$, is usually applied to evaluate the TE materials' conversion efficiency, in which σ is the electrical conductivity of the material, S stands for the Seebeck coefficient, and T and κ

represent absolute temperature and thermal conductivity, respectively. For polymers and their composites, the thermal conductivity is relatively low (lower than that of the inorganic TE materials by almost one to three orders of magnitude). Therefore, the power factor $S^2\sigma$ is a good approximation for comparing organic and hybrid thermoelectric materials. The experiment results reveal that the Fe_3O_4@RGO/PEDOT:PSS hybrid films possess better TE properties than those of their single components. The related mechanism is also discussed in detail.

2. Materials and Methods

2.1. Materials

PEDOT:PSS (PH1000, Mw = 326.388) was purchased from Heraeus Company (Hanau, Germany); both ferroferric oxide nanoparticles (Fe_3O_4, Mw = 231.54 g/mol, around 20 nm particle size) and lithium sulfate (Li_2SO_4, Mw = 109.94 g/mol) from Beijing InnoChem Science & Technology Co., Ltd. (Beijing, China); GO aqueous solution (5 mg/mL) from Suzhou Tanfeng Graphene Technology Co., Ltd. (Suzhou, China); perchloric acid ($HClO_4$, 70~72%, ~1.76 g/mL) and hydroiodic acid (HI, 57%, 5.23 g/mL) were obtained from Tianjin DaMao Chemical Reagent Factory (Tianjin, China) and Shanghai Mclean Biochemical Technology Co., Ltd. (Shanghai, China), respectively. Water used in this work was all deionized (DI) water (its resistance is around 18.2 MΩ cm). All the agents employed in the experiments were utilized directly.

2.2. Preparation of Fe_3O_4@GO and Fe_3O_4@RGO

The Fe_3O_4@GO composites were prepared according to a typical hydrothermal method. Firstly, Fe_3O_4 nanoparticles (12.5 mg) were added to 12.5 mL of GO aqueous solution (2 mg mL^{-1}). Then, 12.5 mL of deionized water was added and the mixture was shaken well. After 30 min of sonication, the homogenous solution obtained was transferred to a 50 mL Teflon-lined steel autoclave and heated to 180 °C for 12 h. Subsequently, the steel autoclave was taken out and cooled to room temperature. The black sediment was collected and allowed to freeze-dry for 24 h. Finally, ~20.3 mg of black-brown fluffy product was obtained. For comparison, some Fe_3O_4@GO powder was also treated using HI for preparation of Fe_3O_4@RGO.

2.3. Preparation of Fe_3O_4@GO/PEDOT:PSS Composite Films

The Fe_3O_4@GO/PEDOT:PSS composite films were prepared via a drop-coating method; 25 mg of Fe_3O_4@GO lyophilized powder was dispersed in 10 mL PEDOT:PSS solution (1 mg mL^{-1}). The obtained suspension was sonicated for about 1 h at room temperature. Thereafter, the as-prepared solution was allowed to drop onto a precleaned glass substrate, followed by a drying process at room temperature.

2.4. Preparation of Fe_3O_4@RGO/PEDOT:PSS Free-Standing Films

For fabrication of the free-standing Fe_3O_4@RGO/PEDOT:PSS films, an acid post-treatment was applied. The as-prepared Fe_3O_4@GO/PEDOT:PSS hybrid films were first immersed in $HClO_4$ for 24 h. After that, the films were washed using DI water, followed by an air-drying process. To ensure good conductivities of the electrodes, the samples were further treated using HI with a similar procedure as described above. During immersion using HI, a chemical reductant, GO, can be effectively reduced to RGO. Finally, the free-standing Fe_3O_4@RGO/PEDOT:PSS hybrid films were successfully fabricated.

2.5. Characterization and Measurements

A Hitachi 4800 field emission scanning electron microscope (FE-SEM) was employed to analyze the morphologies of the films (Hitachi Limited, Tokyo, Japan). A Thermo K-Alpha X-ray photoelectron spectroscope (XPS) was used to determine the electron-binding energies of the samples (Thermo Fisher Scientific, Shanghai, China). Raman spectroscopy was performed using a LabRAM HR Evolution instrument with a 532 nm laser (HORIBA,

Shanghai, China). An SDT Q600 (TA Instruments, New Castle, DE, USA) was applied for thermogravimetric analysis (TGA). An electrochemical performance analysis of the samples was conducted using a CHI 660E electrochemical workstation (Shanghai CH instruments Co., Shanghai, China). The thermoelectric properties were measured by employing a thin-film thermoelectric test system (MRS-3 M, Wuhan Joule Yacht Science &Technology Co., Ltd., Wuhan, China).

In this study, cyclic voltammogram (CV) curves, galvanostatic charge–discharge (GCD) curves, and electrochemical impedance spectroscopy (EIS) were collected via a three-electrode system in which a platinum mesh, saturated calomel electrode, and the as-prepared sample were used as the counter electrode, reference electrode, and working electrode, respectively. A 1 M Li_2SO_4 aqueous solution was used as the electrolyte. The specific capacitance (C_g, F g^{-1}) of the samples was calculated with the formula $C_g = I\Delta t/m\Delta V$, where I stands for constant discharge current (A), Δt for discharge time (s), and m and ΔV for the mass of the film sample and the potential window, respectively.

3. Results and Discussion

3.1. Fabrication of Fe_3O_4@RGO/PEDOT:PSS Free-Standing Film

Figure 1 provides a schematic illustration of how Fe_3O_4@RGO/PEDOT:PSS free-standing films are fabricated. First, Fe_3O_4 nanoparticles are combined with GO nanosheets for preparation of the Fe_3O_4@GO nanocomposites. After washing and freeze-drying, the Fe_3O_4@GO framework is successfully fabricated. For further improvement of the electric conductivity, a PEDOT:PSS conducting polymer is incorporated for conductive wrapping. With a subsequent $HClO_4$ treatment, the conductivity of the PEDOT:PSS is dramatically improved, owing to the removal of nonconductive PSS and the formation of ordered molecular packing. Furthermore, a secondary acid treatment is conducted using HI, an efficient reducing agent that can eliminate the attached oxygen-containing functional groups and enhance the electrical conductivity of GO. Moreover, the Fe_3O_4 component is partly etched during the acid treatment, which contributes to the construction of interconnected porous nanostructures. Based on the above analysis, we infer that the obtained flexible, free-standing, and porous film should possess good electrochemical properties, which will be discussed latterly.

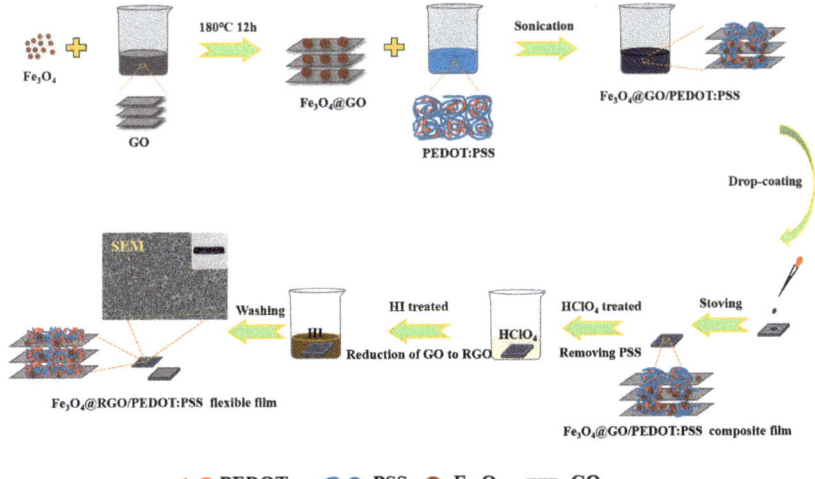

Figure 1. Schematic illustration of the fabrication process of the free-standing Fe_3O_4@RGO/PEDOT:PSS film.

3.2. Structure Characterization and Analysis

First, SEM was applied to examine the surface morphology of the obtained precursors and the final products. The as-prepared Fe_3O_4@GO exhibits an interconnected, highly porous microstructure, as shown in Figure 2a,b. First, the ultrasonic process results in homogeneous dispersion of the GO aqueous solution and Fe_3O_4 nanoparticles. And the subsequent liquid interfacial polymerization, under high pressure and temperature, allows the Fe_3O_4 structures to grow uniformly on the surface of the GO nanosheets. Finally, the freeze-drying procedure maintains the lamellar structure of the GO, and a homogeneous porous nanostructure is successfully constructed. With the introduction of the conductive PEDOT:PSS polymer, the gained composite films no longer possess a porous architecture. As is well known, PEDOT:PSS is water soluble and can wrap around the surface of the Fe_3O_4@GO precursor, forming an electrically conductive polymer shell. As depicted in Figure 2c,d, it can be observed visually that the Fe_3O_4@GO/PEDOT:PSS film shows a highly crumpled surface, which should be derived from the Fe_3O_4@GO core and PEDOT:PSS shell.

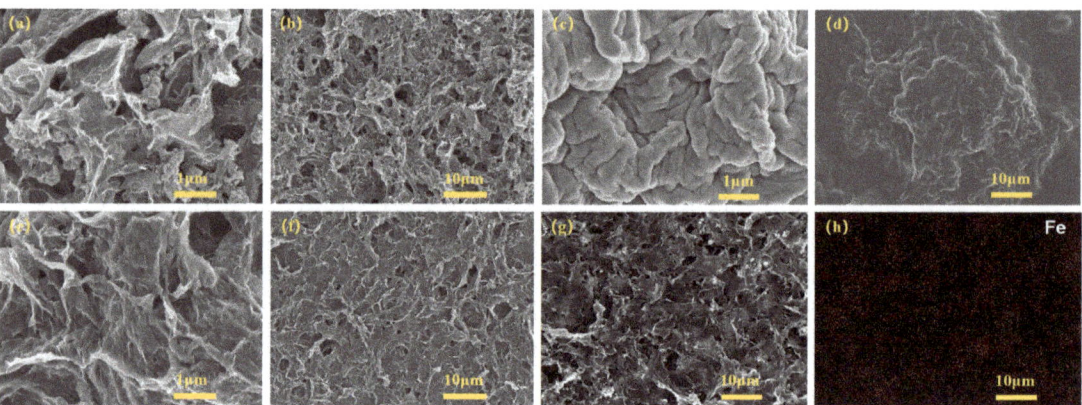

Figure 2. SEM images of (**a,b**) the Fe_3O_4@GO composites; (**c,d**) Fe_3O_4@GO/PEDOT:PSS hybrid films as prepared; (**e–g**) Fe_3O_4@RGO/PEDOT:PSS flexible free-standing films after acid treatment; (**h**) the corresponding EDS mapping image of (**g**) Fe_3O_4@RGO/PEDOT:PSS.

Notably, the superficial structure of the hybrid has changed significantly after the acid treatment. As exhibited in Figure 2e,f, the porous nanostructure is rediscovered in the complex architecture of Fe_3O_4@RGO/PEDOT:PSS, which can be ascribed to the acid treatment. As those meso-/macropores would provide a large surface area, electrolyte transport and access to active sites could be enhanced during the charging/discharging process. EDS element mapping was also conducted, and the results (Figure 2g,h) clearly reveal that the Fe element is evenly distributed in the Fe_3O_4@RGO/PEDOT:PSS sample. All these features would be beneficial for enhancing the supercapacitive performance of the hybrid films. On the other hand, when it is used as a thermoelectric material, the porous nanostructure is of importance in suppressing the thermal conductivity, whereas the large number of holes is not conducive to the formation of conductive networks and thus leads to a reduction in electrical conductivity and a decrease in mechanical properties. And thus, the presence of a conductive binder becomes very important. Furthermore, the introduction of conductive polymers is conducive to achieving relatively low thermal conductivity (≤ 1 W·m^{-1}·K^{-1} in general), and therefore exhibits enormous potential in TE applications.

In order to account for the structural variation in Fe_3O_4@RGO/PEDOT:PSS, a Raman spectroscopy experiment was performed. As is depicted in Figure 3a, the pristine Fe_3O_4@GO/PEDOT:PSS presents only two main characteristic peaks at 1341 and 1591 cm^{-1}, which are associated with the D and G bands of GO, respectively [8,33,35]. Furthermore, two small peaks (centered at 218 and 284 cm^{-1}), corresponding to Fe$_3$O$_4$, also appear in the Raman spectrum [21]. After acid treatment, the most obvious difference is that some novel peaks associated with PEDOT appear. As exhibited in Figure S1, for acid-treated PEDOT:PSS, peaks at 1561 cm^{-1} and 1504 cm^{-1} are assigned to the asymmetric C$_\alpha$=C$_\beta$ stretching, while the peak position 1430 cm^{-1} corresponds to the symmetric C$_\alpha$=C$_\beta$(–O) stretching in the five-membered ring, 1366 cm^{-1} to the C$_\beta$–C$_\beta$ stretching, 1254 cm^{-1} to the inter-ring C$_\alpha$–C$_\alpha$ stretching, 1095 cm^{-1} to the C–O–C deformation, 990 cm^{-1} and 576 cm^{-1} to oxyethylene ring deformation, 857 cm^{-1} and 699 cm^{-1} to C–S bonds, and 437 cm^{-1} to SO$_2$ bending [36]. With the incorporation of Fe$_3$O$_4$, a shift of some characteristic peaks is noticeable (see Figure 3b and Figure S1), indicating the interaction between the PEDOT:PSS and Fe$_3$O$_4$ filler. Furthermore, it is of importance to note that the peak at 1430 cm^{-1} (symmetric C$_\alpha$=C$_\beta$(–O) stretching) shifts to 1429 cm^{-1} in the composite sample. In general, the shift of symmetric C$_\alpha$=C$_\beta$(–O) stretching vibration is mainly related to the ratio between the benzoid and quinoid conformations. Due to the lack of conjugated π-electrons in C$_\alpha$–C$_\beta$, the band red-shift of the symmetric C$_\alpha$=C$_\beta$(–O) indicates that more quinoid conformations generate. Namely, with the introduction of Fe$_3$O$_4$@RGO nanoparticles, there is a conformation transition of the PEDOT molecules from the coiled benzoid to the extended quinoid structure, which facilitates the carrier transformation and is beneficial for electrical performance. Notably, with the acid treatment, the iron oxide characteristic peaks gradually weaken, indicating a decrease in the Fe$_3$O$_4$ component. This change in Fe$_3$O$_4$@RGO/PEDOT:PSS would favor the formation of porous nanostructures. When compared with the acid-treated PEDOT:PSS, some PEDOT characteristic peaks become indistinct in the composite film because of the decrease in the PEDOT proportion induced by the addition of the Fe$_3$O$_4$ filler. All these phenomena indicate the presence of residual Fe$_3$O$_4$ after acid treatment, and the nanostructures introduced certainly have an influence on the energy performance, which will be further discussed in the subsequent sections.

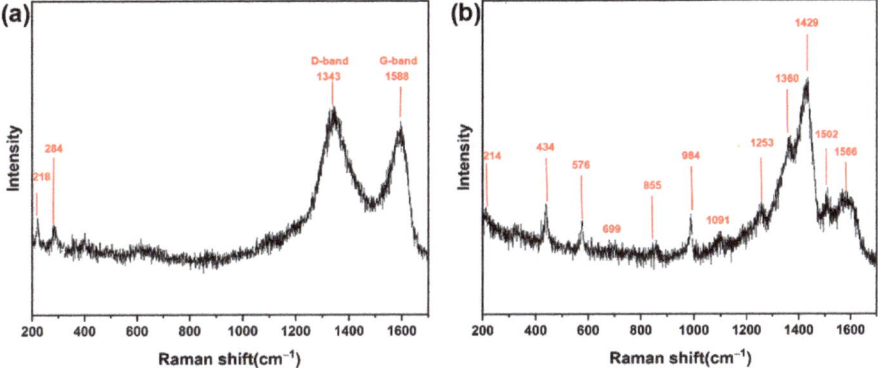

Figure 3. Raman spectra of the obtained hybrid films: (**a**) Fe$_3$O$_4$@GO/PEDOT:PSS films before acid treatment and (**b**) Fe$_3$O$_4$@RGO/PEDOT:PSS films after acid treatment.

In order to further investigate the elemental composition and chemical state of the obtained composite films, X-ray electron spectroscopy (XPS) analysis was conducted, and the results are shown in Figure 4 and Figure S2. As depicted in the XPS survey spectrum (see Figure S2), after acid treatment the porous Fe$_3$O$_4$@RGO/PEDOT:PSS contains elements of Fe, C, O, and S. The high resolution of the Fe2p spectra was decomposed, and the result is shown in Figure 4a. As presented, Fe^{2+} is predominantly correlated with the peaks at 711.2 (Fe2p$_{3/2}$) and 724.5 eV (Fe2p$_{1/2}$), while Fe^{3+} is mainly associated with the peaks at 713.8 (Fe2p$_{3/2}$) and 726.7 eV (Fe2p$_{1/2}$) [37,38]. Additionally, the satellite peaks at 719.1 and 731.8 eV belong to Fe^{2+} and Fe^{3+}, respectively [37]. All these phenomena illustrate the presence of residual Fe_3O_4 even after a relatively lengthy treatment with $HClO_4$ and HI, which serve an important role in electrochemical performance.

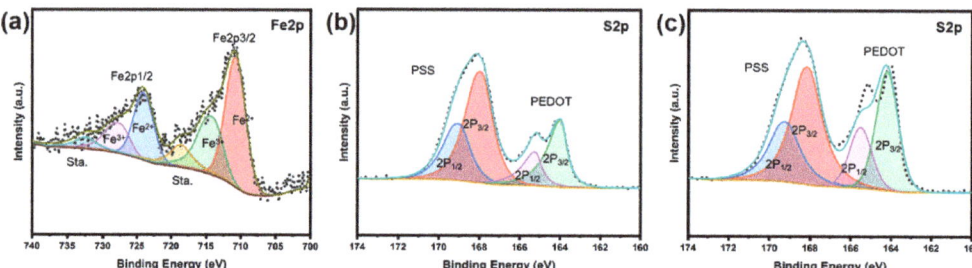

Figure 4. The Fe2p and S2p high-resolution XPS spectra of composite films including fits for the components: (**a**) Fe2p of Fe_3O_4@RGO/PEDOT:PSS after acid treatment; (**b**) S2p of Fe_3O_4@GO/PEDOT:PSS before acid treatment; (**c**) S2p of Fe_3O_4@RGO/PEDOT:PSS after acid treatment.

To deepen understanding of the influence of acid treatment on structural variations in the composite films, the high-resolution S2p spectra for Fe_3O_4@GO/PEDOT:PSS (as prepared) and flexible Fe_3O_4@RGO/PEDOT:PSS (after acid treatment) were analyzed, and the results are displayed in Figure 4b,c. As can be seen, the as-prepared Fe_3O_4@GO/PEDOT:PSS exhibits two distinct types of S element, because the S2p binding energy of the thiophene unit in PEDOT is very different from that of the sulfonate group in PSS. Namely, there are two types of S; one is related to PSS (high-binding-energy region, 171.8~166.5 eV), and the other is associated with PEDOT (low-binding-energy region, 166.5~162.5 eV). For the S element in PSS, S2p peaks can be divided into $S2p_{3/2}$ and $S2p_{1/2}$ peaks, showing peak positions centered at 168.1 eV and 169.2 eV, respectively. Meanwhile, for the S element in PEDOT, S2p peaks can also be classified into $S2p_{3/2}$ and $S2p_{1/2}$ peaks, centered at 164.1 eV and 165.5 eV, respectively. Comparing Figure 4b with Figure 4c, it can be found that the relative content of PSS to PEDOT decreases distinctly. This ratio decrease can be attributed to conformational variations caused by PSS removal during acid treatment. PSS extraction can be generally quantified by figuring the integral area ratio of the characteristic peaks. The calculation results are shown in Figure S3 (see Supporting Information). A PSS/PEDOT surface element ratio of 2.06 can be achieved for the as-obtained Fe_3O_4@GO/PEDOT:PSS composite, whereas the post-treatment using perchloric acid and hydroiodic acid reduces the PSS/PEDOT ratio to 1.48 (for the Fe_3O_4@RGO/PEDOT:PSS sample), confirming the PSS removal effect of the acids' treatment. Because PSS itself is not conductive in nature, its reduction in content can effectively improve the conductive characteristics of the hybrid films, which would not only benefit the enhancement of the electrochemical performance, but also potentially improve the TE properties of the composite electrodes.

3.3. Electrochemical Properties of Fe_3O_4@RGO/PEDOT:PSS Free-Standing Films

CV and GCD were conducted to estimate the electrochemical properties of the electrodes, and the results are presented in Figure 5a,b. It is suggested that the GO skeleton is partially reduced to RGO via HI treatment. Combined with the wrapping effect of the highly conductive PEDOT:PSS, a continuous conductive network is formed, which provides good paths for the transport of ions and rapid redox reactions. This flexible film is expected to be employed as a high-performance film electrode for SCs, working simultaneously as an electrically conducting current collector and active electrode material. As can be observed, the Fe_3O_4@RGO/PEDOT:PSS self-supporting film exhibits a quasi-rectangular CV curve at a scan rate of 50 mV s^{-1} (see Figure 5a). For the Fe_3O_4@RGO electrode, nearly no redox peak can be seen, owing to its relatively low electric conductivity, whereas for the Fe_3O_4@RGO/PEDOT:PSS film, the corresponding redox couples become more clear, which is related to the surface redox reactions between Fe^{2+} and Fe^{3+} [21]. Meanwhile, the area enclosed by the CV curve of the Fe_3O_4@RGO/PEDOT:PSS film triples that of Fe_3O_4@RGO, representing a larger capacitance. Nevertheless, the pristine PEDOT:PSS itself possesses a smaller CV curve area due to its poor electric conductivity. Promisingly, the nonconductive PSS can be partly removed via acid treatment, which is beneficial for improving conductivity. Meanwhile, Fe_3O_4 nanoparticles would be partially etched away via acid treatment, and some macroporous microstructures formed where the reactions take place, benefiting electrolyte transport and the electrochemical redox reactions at the electrolyte/electrode interfaces. The influence of the PEDOT:PSS wrapping effect on the electrochemical performance was further evaluated via GCD measurement, shown in Figure 5b. The specific capacitance of Fe_3O_4@RGO is 71.68 F g^{-1}, which is comparable with a previous report [10], and can be substantially improved with the introduction of PEDOT:PSS and the subsequent acid treatment. As is shown in Figure 5b, the specific capacitance of Fe_3O_4@RGO/PEDOT:PSS can reach a high value of 244.7 F g^{-1} at 1 A g^{-1}. Notably, the shape of the Fe_3O_4@RGO/PEDOT:PSS curve deviates from the ideal triangular, implying the pseudocapacitive performance is contributed from the Fe_3O_4 component. These results are highly consistent with the CV analysis. Therefore, we infer that the introduction of PEDOT:PSS and the subsequent acid treatment contribute to the pseudocapacitance storage of Fe_3O_4, resulting in better SC performance.

The effect of the PEDOT:PSS content on the electrochemical performance was also researched, and the results are given in Figure S4. After PEDOT:PSS addition and the following acid treatment, the PEDOT:PSS molecules wrap intimately around the Fe_3O_4@RGO framework via strong interactions, generating a continuous and conductive network. Combined with the high electric conductivity of PEDOT:PSS, the porous architecture of Fe_3O_4@RGO/PEDOT:PSS facilitates ion transport and fast redox reactions, leading to an enhanced capacitance. Nevertheless, when the addition amount exceeds 28.6 wt% (which becomes 34.06 wt% after acid treatment), the presence of excessive PEDOT:PSS may disrupt the interfacial contact between Fe_3O_4@RGO and the electrolyte, causing a significant reduction in redox activity sites. As a result, the capacitive performance of the flexible film is deteriorated. Specifically, when 71.4 wt% (51.08 wt% was left after acid treatment) Fe_3O_4@RGO was used, the specific capacity possessed a maximum value. As is exhibited in Figure 5c, further CV curves in a wide range of scan rates (from 10 to 100 mV s^{-1}) indicate a good reversibility of the redox reactions. GCD tests of this sample with different current densities were also performed (Figure 5d). They show high specific capacitance of 244.7, 205.0, 181.0, and 146.0 F g^{-1} at 1, 2.5, 5, and 10 A g^{-1}, respectively, which is in accordance with the CV curves. From the relative contents of each component (Figure S5), the quantitative contribution of the Fe_3O_4@RGO and PEDOT:PSS can be determined. In our previous report, an acid-treated PEDOT:PSS film achieved a specific capacity of around 43.2 F g^{-1} at 1 A g^{-1} [33]. According to this value, the specific capacitance of Fe_3O_4@RGO based on its own weight is around 379.0 F g^{-1}. These results are comparable or even better when compared with other Fe_3O_4-based electrodes [18–22,39–41]. The excellent electrochemical performance of the porous Fe_3O_4@RGO/PEDOT:PSS could be ascribed mainly to the acid treatment. First, the nonconductive PSS was effectively removed via $HClO_4$ and HI immersion. Secondly, for the GO component, some functional groups like oxygen-containing groups would favor rapid ion transfer between the film surface and interior [42]. However, the existence of these functional groups is not beneficial for electron transport, resulting in a low electric conductivity. Via HI treatment, the balance between the oxygen-containing functional groups and conductivity could be properly adjusted for enhancement of the electrochemical properties. Last but not least, the interconnected conductive framework was successfully fabricated via acid treatment (Figure 2e–g), ensuring the fast ion and electron transport.

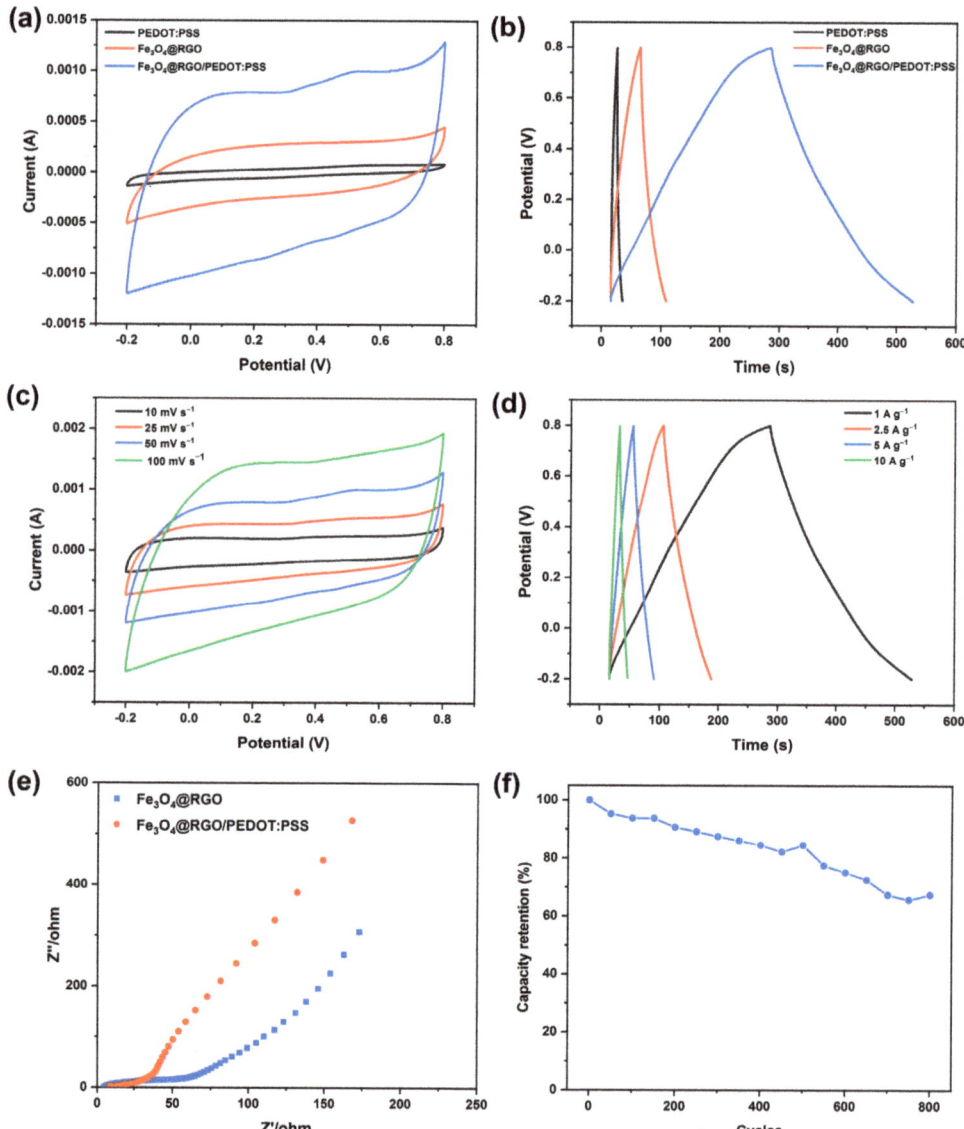

Figure 5. Electrochemical properties of PEDOT:PSS, Fe_3O_4@RGO, and Fe_3O_4@RGO/PEDOT:PSS electrodes: (**a**) CV curves of samples at a scan rate of 50 mV s^{-1}; (**b**) GCD curves of samples at a current density of 1 A g^{-1}; (**c**) CV curves of Fe_3O_4@RGO/PEDOT:PSS at different scan rates; (**d**) CD curves of Fe_3O_4@RGO/PEDOT:PSS at different current densities; (**e**) Nyquist plots of Fe_3O_4@RGO and Fe_3O_4@RGO/PEDOT:PSS composite samples; and (**f**) cycle stabilities of Fe_3O_4@RGO/PEDOT:PSS during the long-term charging/discharging process at a current density of 20 A g^{-1}.

In addition, EIS measurement was conducted to investigate the ion diffusion and electron transfer resistance of the prepared electrodes. As can be seen from Figure 5e, the Nyquist curve mainly consists of two parts: the high-frequency Nyquist curve is composed of a semicircle, while the low-frequency curve is made up of a nearly straight line. Generally, the semicircle diameter at high frequency is related to the charge transfer resistance

(R_{ct}). To be specific, the Fe_3O_4@RGO possesses a charge transfer resistance of around 50 Ω. It is worthy to note that, after being wrapped with PEDOT:PSS, the electrode shows a marked decline in R_{ct} value. Furthermore, the slope of the Fe_3O_4@RGO/PEDOT:PSS electrode at lower frequencies is very similar to that of the Fe_3O_4@RGO, indicating that good diffusive behavior of the electrolyte ions is maintained after PEDOT:PSS addition. All results of these EIS analyses are consistent with those collected from the CV and GCD studies, illustrating that the PEDOT:PSS wrapping indeed facilitates efficient ion transport and consequently enhances its capacitance performance. Moreover, a further cyclic stability test was carried out, and the result reveals that even at a very high current density (20 A g^{-1}), the Fe_3O_4@RGO/PEDOT:PSS integrated electrode film still retains nearly 70% of its original capacitance when it is charged and discharged for 800 cycles (see Figure 5f).

3.4. Thermoelectric Performance of Fe_3O_4@RGO/PEDOT:PSS Free-Standing Films

The excellent performance of the hybrid Fe_3O_4@RGO/PEDOT:PSS electrodes can be attributed to their unique architecture, which may endow this free-standing film with wider application areas. Herein, we also investigate the thermoelectric properties of this Fe_3O_4@RGO/PEDOT:PSS integrated film. For comparison, a Fe_3O_4@RGO sample with a similar thickness (Figure S6) was also prepared. Figure 6 shows the variation in the TE parameters of the composite films as a function of absolute temperature. As can be seen, the σ of the Fe_3O_4@RGO sample reveals a value of 150 S cm^{-1} and keeps almost constant in the tested temperature interval (Figure 6a). With the incorporation of PEDOT:PSS and the following acid treatment, the obtained Fe_3O_4@RGO/PEDOT:PSS integrated film shows a much higher σ value (507.56 S cm^{-1}). With increasing temperature, the σ value just shows a slight decrease, and this variation trend is very similar to those reported in the literature [43,44]. Namely, the electric conductivity is not sensitive to the change in temperature. As shown in Figure 6b, all the composite films exhibit positive Seebeck coefficients, illustrating that the predominant charge carriers are holes in the above obtained samples. The Fe_3O_4@RGO film shows a Seebeck coefficient of 18.96 μV K^{-1} at room temperature. When the test temperature increases, the S demonstrates a slight downward trend and decreases to 15.07 μV K^{-1} at 380 K. In contrast, the free-standing Fe_3O_4@RGO/PEDOT:PSS film possesses an S value of only 13.29 μV K^{-1} at room temperature. Upon increasing the testing temperature from 300 K to 380 K, the Seebeck coefficient presents a visible increasing trend, and achieves an S value of 15.07 μV K^{-1} (the same result with the Fe_3O_4@RGO sample) at 380 K. Taking into account the big variation in electric conductivity, the Fe_3O_4@RGO/PEDOT:PSS film exhibits a much higher power factor when compared with the Fe_3O_4@RGO sample. To be specific (shown in Figure 6c), a maximum PF value of 11.06 μW·m^{-1}·K^{-2} is achieved at 380 K, which is much higher than that of the Fe_3O_4@RGO, by nearly four times.

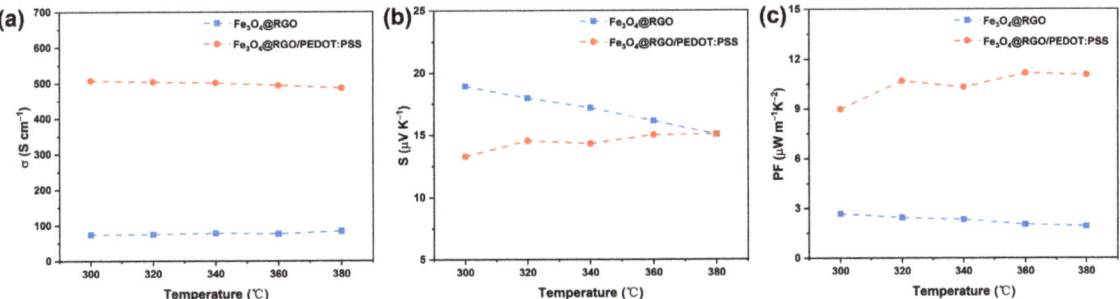

Figure 6. The electrical conductivity (**a**), Seebeck coefficient (**b**), and power factor (**c**) of Fe_3O_4@RGO and Fe_3O_4@RGO/PEDOT:PSS composite films from 300 to 380 K.

4. Conclusions

In summary, we have demonstrated the creation of highly conductive Fe_3O_4@RGO/PEDOT:PSS porous films via a facile but efficient method. Benefiting from the supporting effect of the GO framework and the highly conductive networks arising from PEDOT:PSS wrapping and subsequent acid treatment, the ternary Fe_3O_4@RGO/PEDOT:PSS composites exhibit excellent electrochemical properties: a high specific capacitance of 244.7 F g^{-1} can be achieved at a current density of 1 A g^{-1}; meanwhile, a high level of cycling stability of ~70% is maintained after 800 cycles. The superior electrochemical performance of Fe_3O_4@RGO/PEDOT:PSS hybrid films greatly benefits from their unique structures. Notably, the free-standing flexible hybrid films also show relatively high thermoelectric properties, and a high power factor of 11.06 $\mu W \cdot m^{-1} \cdot K^{-2}$ is reached at 380 K. The novel method proposed in this paper paves an effective way to fabricate and design highly conductive and porous PEDOT:PSS-based composites for electrochemical energy storage and thermoelectric applications.

Supplementary Materials: The following supporting information can be downloaded at: https://www.mdpi.com/article/10.3390/polym15163453/s1, Figure S1: Raman spectra of the PEDOT:PSS film after acid treatment via $HClO_4$; Figure S2: XPS survey spectrum of the porous Fe_3O_4@RGO/PEDOT:PSS; Figure S3: PSS/PEDOT ratio calculated for Fe_3O_4@GO/PEDOT:PSS samples before acid treatment and Fe_3O_4@RGO/PEDOT:PSS after acid treatment.; Figure S4: CV curves of Fe_3O_4@RGO/PEDOT:PSS with different PEDOT:PSS content at a scan rate of 50 mV s^{-1}: a, the content of PEDOT:PSS is 16.7 wt%; b, the content of PEDOT:PSS is 28.6 wt%; c, the content of PEDOT:PSS is 50.0 wt%; Figure S5: TG curves for Fe_3O_4@RGO/PEDOT:PSS flexible film; Figure S6: The thickness of Fe_3O_4@RGO and Fe_3O_4@RGO/PEDOT:PSS films.

Author Contributions: Conceptualization, F.L. (Fuwei Liu) and L.G.; methodology, investigation, and validation, L.G., Q.W., Z.C., J.D. and F.L. (Fuqun Li); writing—original draft preparation, L.G.; writing—review and editing, F.L. (Fuwei Liu); discussion of experiments, H.L., R.L., M.W., J.L. and L.W.; supervision, F.L. (Fuwei Liu). All authors have read and agreed to the published version of the manuscript.

Funding: This research was funded by the National Natural Science Foundation of China (Grant No. U2004174) and the Nanhu Scholars Program for Young Scholars of XYNU.

Institutional Review Board Statement: Not applicable.

Data Availability Statement: All data are contained within the article, and Supplementary Materials are available upon request from the authors.

Acknowledgments: The project was supported by the National Natural Science Foundation of China (Grant No. U2004174) and the Nanhu Scholars Program for Young Scholars of XYNU. We thank the Analysis Testing Center of Xinyang Normal University for the help with characterization analyses. We also appreciate the support and help from Shichao Wang.

Conflicts of Interest: The authors declare no conflict of interest.

References

1. Chen, X.; Xiao, F.; Lei, Y.; Lu, H.; Zhang, J.; Yan, M.; Xu, J. A novel approach for synthesis of expanded graphite and its enhanced lithium storage properties. *J. Energy Chem.* **2021**, *59*, 292–298. [CrossRef]
2. Zhou, Y.; Qi, H.; Yang, J.; Bo, Z.; Huang, F.; Islam, M.S.; Lu, X.; Dai, L.; Amal, R.; Wang, C.H.; et al. Two-birds-one-stone: Multifunctional supercapacitors beyond traditional energy storage. *Energy Environ. Sci.* **2021**, *14*, 1854–1896. [CrossRef]
3. Simon, P.; Gogotsi, Y.; Dunn, B. Where do batteries end and supercapacitors begin? *Science* **2014**, *343*, 1210–1211. [CrossRef] [PubMed]
4. Zhao, M.; Lu, Y.; Yang, Y.; Zhang, M.; Yue, Z.; Zhang, N.; Peng, T.; Liu, X.; Luo, Y. A vanadium-based oxide-nitride heterostructure as a multifunctional sulfur host for advanced Li-S batteries. *Nanoscale* **2021**, *13*, 13085–13094. [CrossRef]
5. Xie, W.; Wang, Q.; Wang, W.; Xu, Z.; Li, N.; Li, M.; Jia, L.; Zhu, W.; Cao, Z.; Xu, J. Carbon framework microbelt supporting SnO(x) as a high performance electrode for lithium ion batteries. *Nanotechnology* **2019**, *30*, 325405. [CrossRef]
6. Peng, T.; Guo, Y.; Zhang, Y.; Wang, Y.; Zhang, D.; Yang, Y.; Lu, Y.; Liu, X.; Chu, P.K.; Luo, Y. Uniform cobalt nanoparticles-decorated biscuit-like VN nanosheets by in situ segregation for Li-ion batteries and oxygen evolution reaction. *Appl. Surf. Sci.* **2021**, *536*, 147982. [CrossRef]

7. Jiang, Y.; Ou, J.; Luo, Z.; Chen, Y.; Wu, Z.; Wu, H.; Fu, X.; Luo, S.; Huang, Y. High Capacitive Antimonene/CNT/PANI Free-Standing Electrodes for Flexible Supercapacitor Engaged with Self-Healing Function. *Small* **2022**, *18*, 2201377. [CrossRef]
8. Zhu, Q.; Zhao, D.; Cheng, M.; Zhou, J.; Owusu, K.A.; Mai, L.; Yu, Y. A New View of Supercapacitors: Integrated Supercapacitors. *Adv. Energy Mater.* **2019**, *9*, 1901081. [CrossRef]
9. Bertana, V.; Scordo, G.; Camilli, E.; Ge, L.; Zaccagnini, P.; Lamberti, A.; Marasso, S.L.; Scaltrito, L. 3D Printed Supercapacitor Exploiting PEDOT-Based Resin and Polymer Gel Electrolyte. *Polymers* **2023**, *15*, 2657. [CrossRef]
10. Chang, J.; Adhikari, S.; Lee, T.H.; Li, B.; Yao, F.; Pham, D.T.; Le, V.T.; Lee, Y.H. Leaf vein-inspired nanochanneled graphene film for highly efficient micro-supercapacitors. *Adv. Energy Mater.* **2015**, *5*, 1500003. [CrossRef]
11. Huang, Y.; Zhu, M.; Meng, W.; Fu, Y.; Wang, Z.; Huang, Y.; Pei, Z.; Zhi, C. Robust reduced graphene oxide paper fabricated with a household non-stick frying pan: A large-area freestanding flexible substrate for supercapacitors. *RSC Adv.* **2015**, *5*, 33981–33989. [CrossRef]
12. Li, X.; Wu, H.; Guan, C.; Elshahawy, A.M.; Dong, Y.; Pennycook, S.J.; Wang, J. (Ni,Co)Se$_2$/NiCo-LDH Core/Shell Structural Electrode with the Cactus-Like (Ni,Co)Se$_2$ Core for Asymmetric Supercapacitors. *Small* **2018**, *15*, 1803895. [CrossRef] [PubMed]
13. Wang, H.; Qiu, F.; Lu, C.; Zhu, J.; Ke, C.; Han, S.; Zhuang, X. A Terpyridine-Fe$^{(2+)}$-Based Coordination Polymer Film for On-Chip Micro-Supercapacitor with AC Line-Filtering Performance. *Polymers* **2021**, *13*, 1002. [CrossRef] [PubMed]
14. Wu, Z.-S.; Wang, D.-W.; Ren, W.; Zhao, J.; Zhou, G.; Li, F.; Cheng, H.-M. Anchoring Hydrous RuO$_2$ on Graphene Sheets for High-Performance Electrochemical Capacitors. *Adv. Funct. Mater.* **2010**, *20*, 3595–3602. [CrossRef]
15. Yu, G.; Hu, L.; Liu, N.; Wang, H.; Vosgueritchian, M.; Yang, Y.; Cui, Y.; Bao, Z. Enhancing the supercapacitor performance of graphene/MnO$_2$ nanostructured electrodes by conductive wrapping. *Nano Lett.* **2011**, *11*, 4438–4442. [CrossRef]
16. Zhang, R.; Wang, X.; Cai, S.; Tao, K.; Xu, Y. A Solid-State Wire-Shaped Supercapacitor Based on Nylon/Ag/Polypyrrole and Nylon/Ag/MnO$_2$ Electrodes. *Polymers* **2023**, *15*, 1627. [CrossRef]
17. Fang, L.; Lan, M.; Liu, B.; Cao, Y. Synthesis and Electrochemical Performance of Flower-like Zn$_{0.76}$Co$_{0.24}$S. *J. Xinyang Norm. Univ. (Nat. Sci. Ed.)* **2022**, *35*, 615–620.
18. Liu, D.; Wang, X.; Wang, X.; Tian, W.; Liu, J.; Zhi, C.; He, D.; Bando, Y.; Golberg, D. Ultrathin nanoporous Fe$_3$O$_4$–carbon nanosheets with enhanced supercapacitor performance. *J. Mater. Chem. A* **2013**, *1*, 1952–1955. [CrossRef]
19. Khoh, W.-H.; Hong, J.-D. Layer-by-layer self-assembly of ultrathin multilayer films composed of magnetite/reduced graphene oxide bilayers for supercapacitor application. *Colloids Surf. A* **2013**, *436*, 104–112. [CrossRef]
20. Li, J.; Chen, Y.; Wu, Q.; Xu, H. Synthesis and electrochemical properties of Fe$_3$O$_4$/MnO$_2$/RGOs sandwich-like nano-superstructures. *J. Alloy. Compd.* **2017**, *693*, 373–380. [CrossRef]
21. Guan, D.; Gao, Z.; Yang, W.; Wang, J.; Yuan, Y.; Wang, B.; Zhang, M.; Liu, L. Hydrothermal synthesis of carbon nanotube/cubic Fe$_3$O$_4$ nanocomposite for enhanced performance supercapacitor electrode material. *Mater. Sci. Eng. B* **2013**, *178*, 736–743. [CrossRef]
22. Ramanathan, S.; SasiKumar, M.; Radhika, N.; Obadiah, A.; Durairaj, A.; Helen Swetha, G.; Santhoshkumar, P.; Sharmila Lydia, I.; Vasanthkumar, S. Musa paradisiaca reduced graphene oxide (BRGO) /MWCNT-Fe$_3$O$_4$ nanocomposite for supercapacitor and photocatalytic applications. *Mater. Today Proc.* **2021**, *47*, 843–852. [CrossRef]
23. Fang, L.; Qiu, Y.; Zhai, T.; Wang, F.; Zhou, H. Synthesis of Ni(OH)$_2$-VS$_2$ Nanocomposite and Their Application in Supercapacitors. *J. Xinyang Norm. Univ. (Nat. Sci. Ed.)* **2017**, *30*, 109–113.
24. Mondal, S.; Thakur, S.; Maiti, S.; Bhattacharjee, S.; Chattopadhyay, K.K. Self-Charging Piezo-Supercapacitor: One-Step Mechanical Energy Conversion and Storage. *ACS Appl. Mater. Interfaces* **2023**, *15*, 8446–8461. [CrossRef] [PubMed]
25. Ouyang, J.; Chu, C.W.; Chen, F.C.; Xu, Q.; Yang, Y. High-Conductivity Poly(3,4-ethylenedioxythiophene):Poly(styrene sulfonate) Film and Its Application in Polymer Optoelectronic Devices. *Adv. Funct. Mater.* **2005**, *15*, 203–208. [CrossRef]
26. Ouyang, J.; Xu, Q.; Chu, C.-W.; Yang, Y.; Li, G.; Shinar, J. On the mechanism of conductivity enhancement in poly(3,4-ethylenedioxythiophene):poly(styrene sulfonate) film through solvent treatment. *Polymer* **2004**, *45*, 8443–8450. [CrossRef]
27. Badre, C.; Marquant, L.; Alsayed, A.M.; Hough, L.A. Highly Conductive Poly(3,4-ethylenedioxythiophene):Poly (styrenesulfonate) Films Using 1-Ethyl-3-methylimidazolium Tetracyanoborate Ionic Liquid. *Adv. Funct. Mater.* **2012**, *22*, 2723–2727. [CrossRef]
28. Fan, Z.; Du, D.; Yu, Z.; Li, P.; Xia, Y.; Ouyang, J. Significant Enhancement in the Thermoelectric Properties of PEDOT:PSS Films through a Treatment with Organic Solutions of Inorganic Salts. *ACS Appl. Mater. Int.* **2016**, *8*, 23204–23211. [CrossRef]
29. Kim, G.H.; Shao, L.; Zhang, K.; Pipe, K.P. Engineered doping of organic semiconductors for enhanced thermoelectric efficiency. *Nat. Mater.* **2013**, *12*, 719–723. [CrossRef]
30. Lee, S.H.; Park, H.; Kim, S.; Son, W.; Cheong, I.W.; Kim, J.H. Transparent and flexible organic semiconductor nanofilms with enhanced thermoelectric efficiency. *J. Mater. Chem. A* **2014**, *2*, 7288–7294. [CrossRef]
31. Takano, T.; Masunaga, H.; Fujiwara, A.; Okuzaki, H.; Sasaki, T. PEDOT Nanocrystal in Highly Conductive PEDOT:PSS Polymer Films. *Macromolecules* **2012**, *45*, 3859–3865. [CrossRef]
32. Crispin, X.; Jakobsson, F.L.E.; Crispin, A.; Grim, P.C.M.; Andersson, P.; Volodin, A.; van Haesendonck, C.; Van der Auweraer, M.; Salaneck, W.R.; Berggren, M. The Origin of the High Conductivity of Poly(3,4-ethylenedioxythiophene)-Poly(styrenesulfonate) (PEDOT-PSS) Plastic Electrodes. *Chem. Mater.* **2006**, *18*, 4354–4360. [CrossRef]
33. Liu, F.; Xie, L.; Wang, L.; Chen, W.; Wei, W.; Chen, X.; Luo, S.; Dong, L.; Dai, Q.; Huang, Y.; et al. Hierarchical Porous RGO/PEDOT/PANI Hybrid for Planar/Linear Supercapacitor with Outstanding Flexibility and Stability. *Nano Micro Lett.* **2020**, *12*, 17. [CrossRef] [PubMed]

34. Bießmann, L.; Saxena, N.; Hohn, N.; Hossain, M.A.; Veinot, J.G.C.; Müller-Buschbaum, P. Highly Conducting, Transparent PEDOT:PSS Polymer Electrodes from Post-Treatment with Weak and Strong Acids. *Adv. Electron. Mater.* **2019**, *5*, 1800654. [CrossRef]
35. Song, H.; Liu, C.; Xu, J.; Jiang, Q.; Shi, H. Fabrication of a layered nanostructure PEDOT:PSS/SWCNTs composite and its thermoelectric performance. *RSC Adv.* **2013**, *3*, 22065–22071. [CrossRef]
36. Ely, F.; Matsumoto, A.; Zoetebier, B.; Peressinotto, V.S.; Hirata, M.K.; de Sousa, D.A.; Maciel, R. Handheld and automated ultrasonic spray deposition of conductive PEDOT:PSS films and their application in AC EL devices. *Org. Electron.* **2014**, *15*, 1062–1070. [CrossRef]
37. Wang, L.; Liu, F.; Pal, A.; Ning, Y.; Wang, Z.; Zhao, B.; Bradley, R.; Wu, W. Ultra-small Fe_3O_4 nanoparticles encapsulated in hollow porous carbon nanocapsules for high performance supercapacitors. *Carbon* **2021**, *179*, 327–336. [CrossRef]
38. Li, X.; Xu, Y.; Wu, H.; Qian, X.; Chen, L.; Dan, Y.; Yu, Q. Porous Fe_3O_4/C nanoaggregates by the carbon polyhedrons as templates derived from metal organic framework as battery-type materials for supercapacitors. *Electrochim. Acta* **2020**, *337*, 135818. [CrossRef]
39. Du, X.; Wang, C.; Chen, M.; Jiao, Y.; Wang, J. Electrochemical Performances of Nanoparticle Fe_3O_4/Activated Carbon Supercapacitor Using KOH Electrolyte Solution. *J. Phys. Chem. C* **2009**, *113*, 2643–2646. [CrossRef]
40. Wang, X.; Jiang, D.; Jing, C.; Liu, X.; Li, K.; Yu, M.; Qi, S.; Zhang, Y. Biotemplate Synthesis of Fe_3O_4/Polyaniline for Supercapacitor. *J. Energy Storage* **2020**, *30*, 101554. [CrossRef]
41. Li, J.; Lu, W.; Yan, Y.; Chou, T.-W. High performance solid-state flexible supercapacitor based on Fe_3O_4/carbon nanotube/polyaniline ternary films. *J. Mater. Chem. A* **2017**, *5*, 11271–11277. [CrossRef]
42. Qiu, C.; Jiang, L.; Gao, Y.; Sheng, L. Effects of oxygen-containing functional groups on carbon materials in supercapacitors: A review. *Mater. Des.* **2023**, *230*, 111952. [CrossRef]
43. Du, Y.; Shi, Y.; Meng, Q.; Shen, S.Z. Preparation and thermoelectric properties of flexible SWCNT/PEDOT:PSS composite film. *Synth. Met.* **2020**, *261*, 116318. [CrossRef]
44. Wei, S.; Huang, X.; Deng, L.; Yan, Z.-C.; Chen, G. Facile preparations of layer-like and honeycomb-like films of poly(3,4-ethylenedioxythiophene)/carbon nanotube composites for thermoelectric application. *Compos. Sci. Technol.* **2021**, *208*, 108759. [CrossRef]

Disclaimer/Publisher's Note: The statements, opinions and data contained in all publications are solely those of the individual author(s) and contributor(s) and not of MDPI and/or the editor(s). MDPI and/or the editor(s) disclaim responsibility for any injury to people or property resulting from any ideas, methods, instructions or products referred to in the content.

Article

Flexible Wet-Spun PEDOT:PSS Microfibers Integrating Thermal-Sensing and Joule Heating Functions for Smart Textiles

Yan Li [1], Hongwei Hu [2,*], Teddy Salim [3], Guanggui Cheng [2,*], Yeng Ming Lam [3] and Jianning Ding [2,4]

1. School of Mechanical Engineering, Jiangsu University of Science and Technology, Zhenjiang 212003, China; yanli@just.edu.cn
2. Institute of Intelligent Flexible Mechatronics, School of Mechanical Engineering, Jiangsu University, Zhenjiang 212013, China; dingjn@ujs.edu.cn
3. School of Materials Science and Engineering, Nanyang Technological University, 50 Nanyang Avenue, Singapore 639798, Singapore; tsalim@ntu.edu.sg (T.S.); ymlam@ntu.edu.sg (Y.M.L.)
4. School of Mechanical Engineering, Yangzhou University, Yangzhou 225009, China
* Correspondence: hwhu@ujs.edu.cn (H.H.); ggcheng@ujs.edu.cn (G.C.)

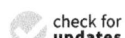

Citation: Li, Y.; Hu, H.; Salim, T.; Cheng, G.; Lam, Y.M.; Ding, J. Flexible Wet-Spun PEDOT:PSS Microfibers Integrating Thermal-Sensing and Joule Heating Functions for Smart Textiles. *Polymers* 2023, *15*, 3432. https://doi.org/10.3390/polym15163432

Academic Editors: Jiangtao Xu and Sihang Zhang

Received: 21 July 2023
Revised: 13 August 2023
Accepted: 15 August 2023
Published: 17 August 2023

Copyright: © 2023 by the authors. Licensee MDPI, Basel, Switzerland. This article is an open access article distributed under the terms and conditions of the Creative Commons Attribution (CC BY) license (https://creativecommons.org/licenses/by/4.0/).

Abstract: Multifunctional fiber materials play a key role in the field of smart textiles. Temperature sensing and active thermal management are two important functions of smart fabrics, but few studies have combined both functions in a single fiber material. In this work, we demonstrate a temperature-sensing and in situ heating functionalized conductive polymer microfiber by exploiting its high electrical conductivity and thermoelectric properties. The conductive polymer microfibers were prepared by wet-spinning the PEDOT:PSS aqueous dispersion with ionic liquid additives, which was used to enhance the electrical and mechanical properties of the final microfibers. The thermoelectric properties of these microfibers were further studied. Due to their excellent flexibility and mechanical properties, these fibers can be easily integrated into commercial fabrics for the manufacture of smart textiles through knitting. We further demonstrated a smart glove with integrated temperature-sensing and in situ heating functions, and further explored thermoelectric fiber-based temperature-sensing array fabric. These works combine the thermoelectric properties and heating function of conductive polymer fibers, providing new insights that enable further development of high-performance, multifunctional wearable smart textiles.

Keywords: thermoelectric fiber; conductive polymer; smart textile; temperature sensor; PEDOT:PSS

1. Introduction

Smart textiles can not only sense environmental changes, but also automatically respond to the surrounding environment or stimuli, such as thermal, chemical, or mechanical changes [1]. They have a wide application prospect in sensing human physiological signals at close range, supporting and assisting daily human activities [2–4]. The development of multifunctional fiber materials is the key to ensuring the functionality and practicality of smart fabrics. Currently, electronic fibers can already be endowed with functions such as high electrical conductivity, strain response, pressure sensing, luminescence, photoelectric sensing, etc. [5–8]. Most of those studies focus on optimizing only a specific function, while research on intricate multifunctional integrated fibers, which could greatly reduce the complexity of device fabrication and system construction, remains scarce.

Conductive polymers (CPs) have received much attention due to their excellent flexibility, electrical properties, and electrochemical activity [9,10]. Poly(3,4-ethylenedioxythiophene)/polystyrene sulfonate (PEDOT/PSS) is one of the most widely used CPs as an intrinsically flexible electronic material, showing great promise in elastic conductors, sensing, displays, and other applications [11,12]. PEDOT:PSS can be attached to fabrics by printing, dyeing, and in situ polymerization to fabricate multifunctional electronic fabrics [13]. In recent years, researchers have prepared pure PEDOT:PSS fibers by wet-spinning, and obtained PEDOT:PSS fibers with high electrical conductivity and excellent mechanical properties by adjusting the spinning dope, coagulation bath, and post-treatment [14–18]. These fibers are further used to prepare highly sensitive sensors, supercapacitors, and organic electrochemical transistors [14,17,18]. On the other hand, these fibers have thermoelectric properties and can be used to harvest wearable energy directly from the heat emitted by the body [19–22]. These studies inspired us to develop a multifunctional fiber based on PEDOT:PSS. Herein, we explored the temperature-sensing and Joule heating functionalized PEDOT:PSS fiber fabric by exploiting their high electrical conductivity and thermoelectric properties. Firstly, PEDOT:PSS fibers were prepared by wet-spinning PEDOT:PSS dispersion with ionic liquid additives, which was used to improve the electrical conductivity and mechanical properties. After studying the thermoelectric properties of these fibers, we further demonstrated a temperature-sensing fabric array based on PEDOT:PSS fibers and its in situ heating capabilities. A temperature sensor array composed of a few simple interlaced fibers was further fabricated to demonstrate their versatile applications. Finally, we demonstrate that thin polymer sheaths can greatly improve the working stability of PEDOT:PSS fibers in different environments. This work integrates the thermoelectric properties and heating functions of conductive polymer fibers, providing a new idea for further development of high-performance, multifunctional wearable smart fabrics.

2. Materials and Methods

2.1. Materials

Ionic liquid 1-butyl-3-methylimidazolium tosylate (BMImOTs, 99%) was purchased from Lanzhou Greenchem ILs (Lanzhou, China); PEDOT:PSS aqueous solution (solid content 1.0–1.3 wt%, PEDOT to PSS ratio 1:2.5, conductivity 700~800 S/cm) was purchased from Shanghai Ouyi Organic Optoelectronic Materials (Shanghai, China).

2.2. Wet-Spinning of PEDOT:PSS Fibers

The PEDOT:PSS (PP) microfibers were fabricated through a wet-spinning process as illustrated in Figure 1a. To enhance the electrical conductivity and mechanical properties, a specifically selected ionic liquid (BMImOTs) was first added to the PP spinning dope to reduce the electrostatic interaction between PEDOT and PSS. The spinning dope was prepared by adding BMImOTs (0–2.6 µg) into the PEDOT:PSS solution (10 mL), so that the BMImOTs accounted for 0 to 5 wt% of the solid content. Then, the solution was rotated and evaporated at 50 °C to obtain PEDOT:PSS solution with concentration of 2.0 to 2.6 wt%. All dispersions were bath-sonicated for 30 min to remove the bubbles prior to fiber spinning. The spinning dope was then loaded into a 3 mL syringe and injected into the sulfuric acid coagulation bath through a needle with an inner diameter of 400 µm. In the sulfuric acid coagulation bath, water and a portion of the PSS in the solution were removed, forming PP microfiber. After entering the coagulation bath, the fibers were immediately collected on the relay roller to prevent fiber entanglement. The fibers were then soaked in the coagulation bath for 3 h before being pulled through a washing bath containing ethanol/water (volume ratio of 3:1) to remove residual sulfuric acid. The fibers were then drawn through two heating plates and finally collected on the roller to obtain PEDOT:PSS fibers. The wet-spinning process used is applicable for continuous production, and in principle, the length of the filament is only limited by the amount of spinning solution. Figure 1b shows ~50 m long microfibers, which have excellent flexibility and can be stitched onto fabrics for use in wearable smart fabrics. Figure 1c shows a microfiber lifting a weight of 10 g, indicating that the microfiber has good mechanical properties.

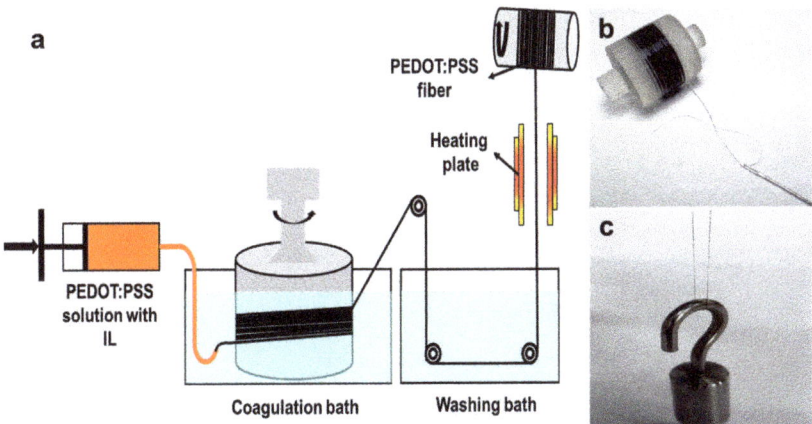

Figure 1. Fabrication process of the PP microfiber using a wet-spinning method (**a**), photo of a roll of PP microfiber (**b**), and a microfiber lifting 10 g weight (**c**).

2.3. Characterization

The mechanical properties of the fibers were measured using a thermomechanical analysis system (TMA, Hitachi 7100E, Hitachi, Tokyo, Japan). A single fiber was used for the tensile strain-stress measurement. The length of the tested fiber sample was approximately 2 cm, and the cross-section area was determined by scanning electron microscope (SEM) image. The electrical conductivity of the PEDOT:PSS fiber was measured using a source meter (Keithley, 2450 SourceMeter, Tektronics, Beaverton, OR, USA). Samples were prepared by placing the fiber on a glass slide with the two ends of the fiber connected to copper wires using silver paste. A low voltage (1–3 V) was applied to the two ends and the resistance was recorded. The conductivity was then calculated by taking into account the sample length and cross area. The Seebeck coefficient was characterized using a home-made device as shown in the supplementary information. Two Peltier modules were used for regulating the temperature difference of the two ends of the fiber, while the thermovoltage was recorded using Keithley 2450. For the above mechanical, electrical, and thermoelectric measurements, at least 10 samples from the same batch of fibers were tested and the average values were obtained. The morphology of the fibers was studied using SEM images (JSM-IT800, JEOL, Tokyo, Japan). Multiple fibers were spread on conductive carbon tape without further treatment. The fibers were cut using scissors for taking the cross-sectional SEM images. The X-ray diffraction (XRD) and wide-angle X-ray scattering (WAXS) were acquired on D8 Advance (Bruker AXS GmbH, Karlsruhe, Germany) and Nano-inXider (Xenocs SAS, Grenoble, France), respectively.

2.4. Preparation of Smart Textiles and Stability Testing

Smart fabrics with thermal-sensing and Joule heating functions were manufactured by weaving PEDOT:PSS microfiber onto commercial gloves. Firstly, ten microfibers were twisted to form a stable fiber yarn. Then, multiple-fiber yarns were sewn onto the glove with a sewing needle. The sensing unit on the index finger consisted of three strands of yarn that detect temperature changes when grasping objects. The two ends of the yarn were placed on the inside and outside of the glove, fixed with silver paste and copper wire, and connected to the thermoelectric potential measurement system. The sensing unit on the back of the glove consisted of six strands of yarn to detect the ambient temperature and provide heating. One end of the bundle was positioned close to the skin, while the other end was exposed to the environment to sense the temperature difference between the environment and the skin of the hand.

To make the temperature sensor array, six bundles of yarn were sewn onto the fabric in parallel three lines, while using silver paste and copper wire to connect both ends of each bundle to a multichannel voltage signal acquisition board.

In order to improve the working stability of the fibers, a thin layer of PDMS was wrapped on the surface of a yarn bundle consisting of 20 PP fibers. The twisted yarn bundle was slowly pulled through the freshly formulated PDMS precursor (Sylgard 184, Dow Corning, Midland, MI, USA). After standing vertically for 15 min, the fiber bundles were treated in an oven at 80 °C for half an hour. The moisture stability of the yarn was tested by dripping water in the middle of the yarn while the conductivity and Seebeck coefficient were measured. The PDMS-wrapped yarn was placed in water and stirred for one hour at a rate of 500 rpm for one wash cycle. After each wash, the conductivity of the fibers and the Seebeck coefficient were measured.

3. Results and Discussion

The morphology and structural characteristics of microfibers have significant influence on their mechanical, electrical, and thermoelectric properties. Figure 2a shows the SEM image of the microfibers, from which it can be seen that the microfibers are in fact mainly flat ribbons with a width of about 40 to 60 μm. Figure 2b shows the cross-section of a single fiber, with a pronounced feature of a flattened ribbon. The microfiber is about 60 μm wide and 10 μm thick, which translates into a cross-sectional area of about 600 μm². The cross-sectional size of the fiber is mainly determined by the dope concentration and the diameter of the injection needle. After rounds of optimization, the needle with a diameter of 400 μm was found to provide fibers with the best mechanical and electrical properties, and was subsequently used throughout the study. The rapid water loss after the injection into the coagulation bath prompts a structural collapse, leading to the formation of flat fibers. Compared with the hollow structure formed by PP fibers in previous studies [23], the presence of ionic liquid in the precursor solution in our work could facilitate a stronger interpolymer interaction and generate a network structure, thus forming dense flat fibers, which is beneficial to improving the mechanical properties of fibers.

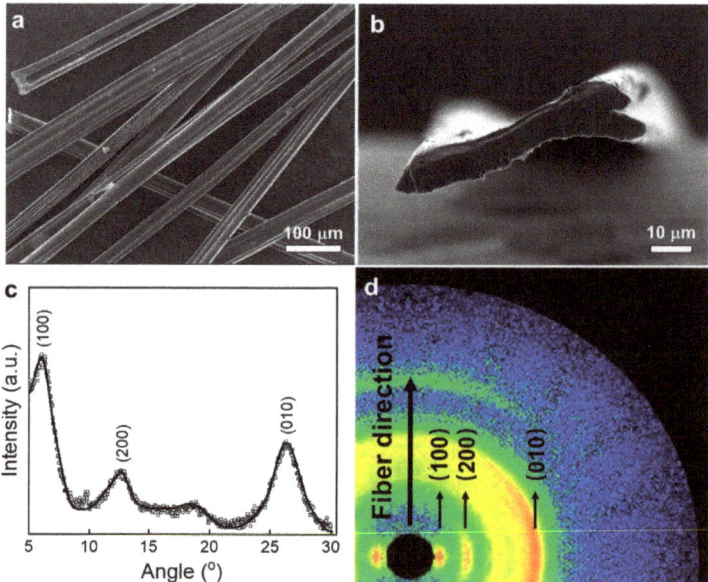

Figure 2. Characterization of the PP fibers: SEM images of the multiple fibers (**a**) and a cross-section of the fiber (**b**), XRD pattern of the fibers (**c**), and WAXS image of the fiber (**d**).

The XRD pattern of the fiber (Figure 2c) clearly shows the crystalline regions as evident in the (100) and (200) diffraction peaks originated from the alternate stacking of PEDOT and PSS, and π-π stacking of the crystallized PEDOT chains (010), which proves that PEDOT and PSS in the fiber have good crystallinity [24,25]. In the WAXS image shown in Figure 2d, the corresponding scattering patterns can also be seen, the plot of which is shown in the supplementary information. Furthermore, these diffraction peaks are mainly located along the radial direction of the fiber, indicating that the fiber has a high degree of anisotropy, which is beneficial to improving the mechanical properties of the fiber in the axial direction [14].

The mechanical properties of the fibers were further tested, and their tensile stress curves are shown in Figure 3a. The maximum strain and strength of the fibers are shown in Figure 3b. It can be seen that the addition of ionic liquid has a significant effect on the mechanical properties of the prepared fibers. A maximum strain of 19% with a breaking strength approaching 206 MPa could be obtained with 2% ionic liquid addition. The specimens nearly doubled their maximum strain and breaking strength compared to those without additives. This result indicates that the addition of ionic liquid can improve the microstructure of the fibers and hence its mechanical properties, which is similar to the effect on PP thin films investigated by Wang et al. [26]. The addition of ionic liquid can reduce the electrostatic interaction between PEDOT and PSS groups, thereby increasing the cross-linking degree of PEDOT and further forming a film with high mechanical properties. However, adding too much ionic liquid could have a detrimental effect, as can be seen from the premature failure of the PP fiber, much sooner than the pristine fiber.

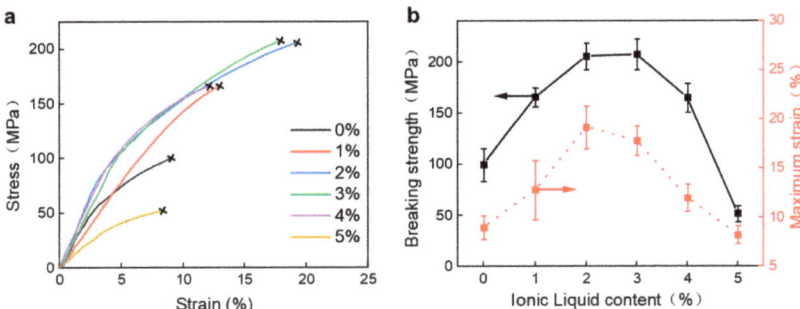

Figure 3. Mechanical properties of the PP fiber: stress–strain curves of the PP fibers with varied ionic liquid content (**a**), breaking strength and maximum strain for the PP fibers (**b**).

The addition of ionic liquids also significantly improved the electrical properties of the fibers. Figure 4a shows the electrical conductivity of fibers formed from different ionic liquids. It can be seen that by adding 2% ionic liquid, the conductivity of the fiber is significantly increased from 1170 S/cm to 2258 S/cm. The trends in conductivity are generally similar to those in mechanical properties. It is plausible that the structural changes responsible for the improvement in mechanical properties also have direct consequences on the electrical properties. We further tested the thermoelectric properties of the fibers. Figure 4b shows the Seebeck coefficient and power coefficient of the fibers. The Seebeck coefficient of fibers with ionic liquid addition showed a downward trend with the increase in ionic liquid addition, mainly because the doping level of PEDOT was increased with the addition of ionic liquid, thereby reducing its Seebeck coefficient [27]. The power factor of the fiber, calculated from the conductivity and the Seebeck coefficient, is shown in Figure 4b, from which it can be seen that the power factor is mainly dominated by the conductivity of the fiber, that is, at the optimum ionic liquid concentration of 2%, the power coefficient also reaches a maximum value of 90 $\mu W/m \cdot K^2$. Therefore, due to their excellent mechanical and electrical properties, we selected fibers with 2% ionic liquids to be integrated into the prototype device in the follow-up studies. Figure 4c shows the thermoelectric potential of the fibers at different temperature differences. It can be seen that the thermoelectric potential exhibits a strong linear relationship with the temperature difference, which is highly favorable for temperature-sensing applications which will be discussed later. We further tested the thermoelectric output properties of the fibers, as shown in Figure 4d. At a temperature difference of 20 °C, the open-circuit voltage and short-circuit current of the fiber reached 0.42 mV and 32 µA, respectively, and its maximum power reached 3.5 nW.

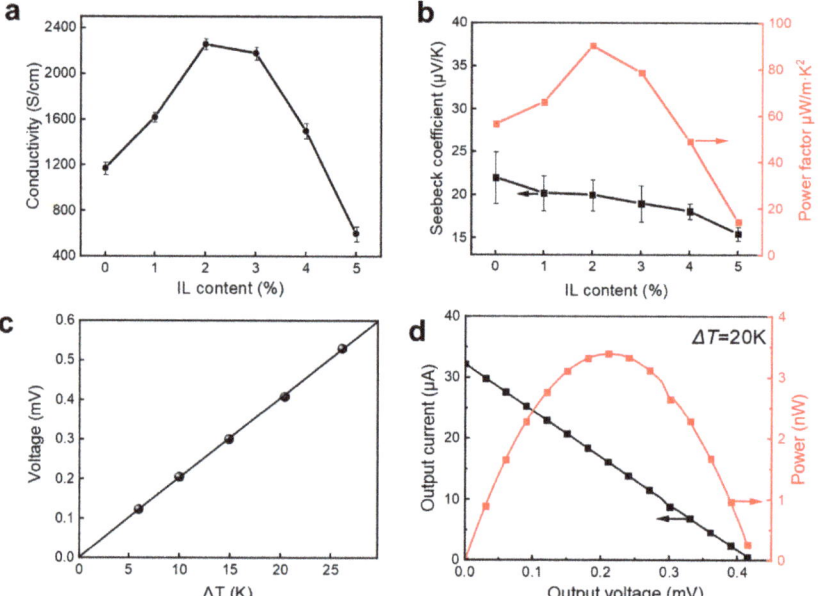

Figure 4. Electrical and thermoelectric properties of PP fibers: electrical conductivity (**a**), Seebeck co-efficient and power factor (**b**), thermoelectric voltage at varied DT (**c**), and output power at ΔT = 20 K (**d**).

To demonstrate the application of PP fibers in wearable smart sensing, we weave the PP microfiber bundles into gloves, as shown in Figure 5a. Due to the good flexibility, the microfibers fit well with the existing glove fibers. Three PP microfiber bundles are woven on the inside of the glove fingers, and six bundles on the back of the hand. The two ends of each fiber and the contact electrodes are, respectively, distributed on the inner and outer sides of the glove to realize the temperature measurement function by using the temperature difference between both ends. The fibers on the finger can measure the temperature difference between the object and the hand when grasping the object, while the fibers on the back of the hand can not only measure the temperature difference between the environment and the hand itself, but it can also provide a heating function when needed. Figure 5b shows the responses of the glove to the cold and hot water cups. It can be seen that the thermoelectric potential generated by the fiber sensor has a good corresponding relationship with the temperature, and the response speed is almost synchronous. Figure 5c shows the thermoelectric potential of the glove at an ambient temperature of 18 °C. Since the temperature of the human hand is relatively constant, the ambient temperature can be easily inferred from the reading of the thermoelectric potential. On the other hand, due to the relatively low resistance and good environmental stability of PP fibers, PP fibers can also achieve Joule heating by applying a bias voltage. Figure 5d shows the temperature change that occurs when a voltage of 1 V to 5 V was applied to the PP fibers on a smart glove, where the ambient temperature was about 0 °C. The higher the voltage, the more pronounced the Joule heating effect. When voltages of 3 V, 4 V, and 5 V were applied, the surface temperature could be increased to 11 °C, 21 °C, and 38 °C, respectively. This suggests that these low-voltage electrical heating fabrics can be combined with commercial portable battery packs to provide heat to the human body in low-temperature environments. Combining the thermoelectric sensing function of the fiber with the heating function, our PP-fiber-embedded prototype gloves can activate the heating function upon sensing a low ambient temperature, thereby providing a timely thermal protection for the human hand.

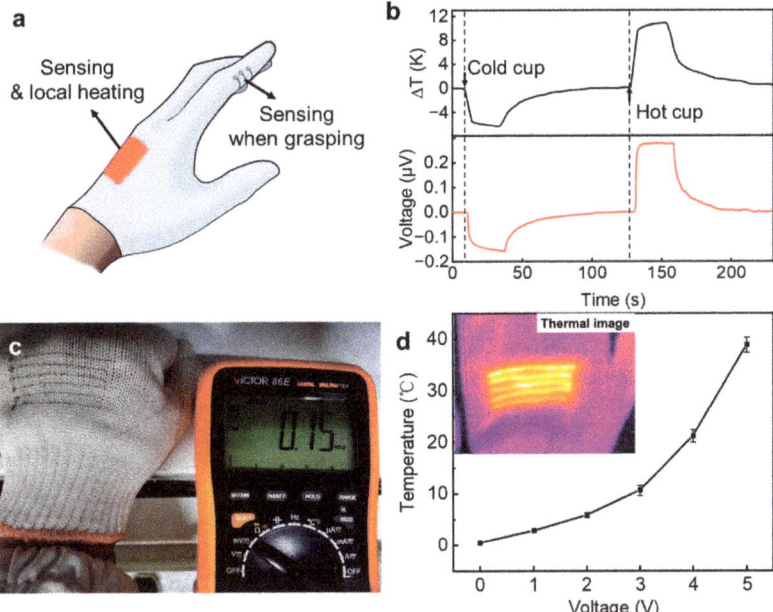

Figure 5. Demonstration of smart glove based on PP fibers: multiple sensing and heating functions integrated on fabric glove (**a**), sensing of cold cup and hot cup when grasping (**b**), thermoelectric voltage generated when temperature drops to 18 °C (**c**), temperature raising by applying various voltages to the PP fibers (**d**), inset shows the thermal image of smart glove with heating area.

Fiber-shaped thermoelectric materials offer a new possibility for building smart fabrics with temperature-location sensing. Both ends of the fiber can simultaneously sense the temperature changes, and then sense where the temperature changes occur. Taking advantage of this feature of PP fibers, we constructed a 3 × 3 sensing array for position and temperature sensing, as shown in Figure 6a. The position points are connected by fibers, and by analyzing and comparing the voltage signals collected by the fibers, the positions of the sensed temperature and the temperature difference value can be obtained. Figure 6b shows the thermoelectric potential of the fibers when the finger touches nine different locations (labeled A to I). It can be seen that touching points A and D will generate opposite thermoelectric potentials at the two ends of fiber 1; while touching point D, a thermoelectric potential will also be generated on fiber 4. Therefore, the location and temperature of the object can be restored by comparing the thermoelectric potential generated at different points. Figure 6c shows a hot metal rod and two cold metal weights placed on the thermoelectric fiber fabric, and Figure 6d shows the temperature at nine points on the fabric deduced from the generated thermoelectric potential, which is consistent with the measured results. This result provides a new idea for the further construction of intelligent sensing arrays of thermoelectric fibers.

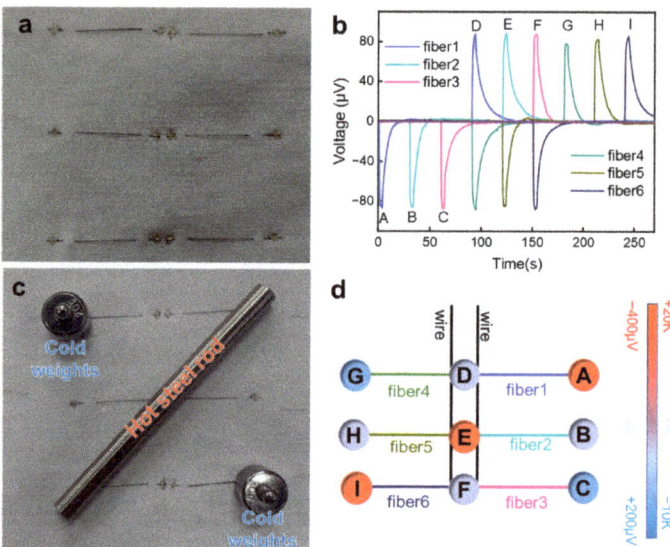

Figure 6. Thermoelectric fiber array for location and temperature sensing: distribution of six fibers and the nine sensing points (**a**), thermoelectric voltages generated from the nine points (labeled from A to I) when touched with finger (**b**), sensing of multiple objects with varied temperatures (**c**), and their location and temperature obtained from the sensing array (**d**).

The electrical properties of conductive polymers are often influenced by ambient humidity, which affects their working stability. In this work, since the thermoelectric and heating functions of conductive polymer fibers can be carried out by means of heat exchange, they do not need to be exposed directly to the environment. We wrap a thin layer of hydrophobic PDMS outside the PP fiber yarn to reduce the influence of moisture on the fiber properties. As shown in Figure 7a, PDMS forms an extremely thin protective layer (about a few microns) around the fiber yarn, which does not seriously affect the heat conduction between the PP fiber and the external environment. As shown in Figure 7b and c, the conductivity and Seebeck coefficient of uncoated PP fibers decreased significantly when wet. In contrast, the conductivity and thermoelectric coefficient of the wrapped PP fiber, even if immersed in water, are hardly affected. We then washed the wrapped fibers several times in water. As shown in Figure 7d, after five cycles of washing, the conductivity and thermoelectric coefficient of the fiber remain above 95% of the original value. The margin loss may be due to the penetration of trace amounts of water molecules into the fiber from the crack or junction. Nevertheless, we believe that PDMS wrapping can significantly improve the working stability of PP fibers, providing a material basis for the realization of multifunctional smart fabrics.

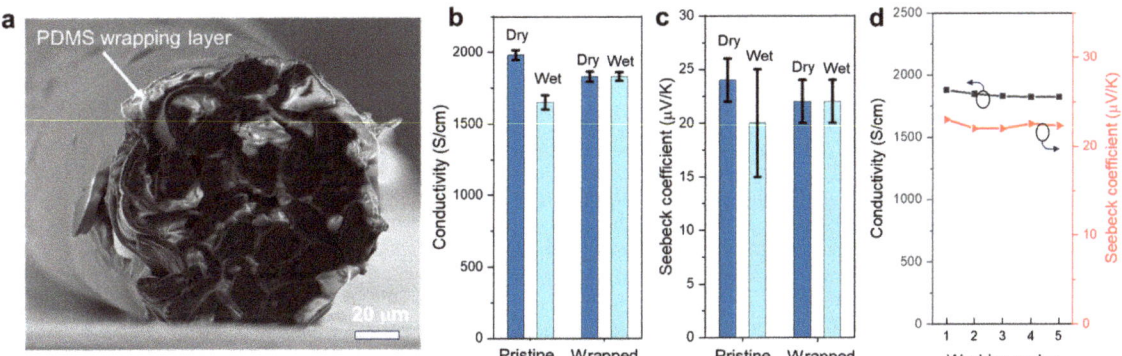

Figure 7. Improving the stability of PP fibers by PDMS wrapping: cross-sectional SEM image showing the PDMS wrapping on the PP fiber yarns (**a**), conductivity (**b**) and Seebeck coefficient (**c**) of the pristine PP fibers and wrapped PP fibers in dry and wet conditions, and the performance change of the PP fibers after multiple washing cycles (**d**).

4. Conclusions

In summary, a temperature-sensing and in situ heating functionalized microfiber fabric has been demonstrated by exploiting the high electrical conductivity and thermoelectric properties of the conductive polymer microfibers. By adding ionic liquid into the wet-spinning dope, both the conductivity and mechanical properties of the fabricated microfibers were improved. We further demonstrated a multifunctional fabric based on conductive polymer microfibers equipped with both temperature-sensing and in situ heating capabilities. In addition, a sensor array consisting of a few simple interlaced microfibers was manufactured to collect information on the temperature and position of the object. This work integrates the thermoelectric properties and heating functions of PEDOT:PSS, providing a new idea for further development of high-performance, multifunctional wearable smart fabrics.

Supplementary Materials: The following supporting information can be downloaded at: https://www.mdpi.com/article/10.3390/polym15163432/s1, Figure S1: The homemade device used for measuring the Seebeck coefficient of PEDOT:PSS fibers: photo of the setup (a) and its schematic diagram (b). Figure S2: Plot of the integration of the scattering intensity in the horizontal direction in the WAXS image.

Author Contributions: Conceptualization, Y.L. and H.H.; methodology, Y.L.; validation, H.H., G.C. and J.D.; formal analysis, T.S.; investigation, Y.M.L.; data curation, T.S.; writing—original draft preparation, Y.L.; writing—review and editing, H.H. and Y.M.L.; supervision, G.C.; project administration, J.D.; funding acquisition, H.H. All authors have read and agreed to the published version of the manuscript.

Funding: This research was funded by the China Postdoctoral Science Foundation, grant no. 2022M711372; Postdoctoral Research Program of Jiangsu Province, grant no. 2021K544C; Key R&D Program of Zhenjiang city, grant no. GY2022025; General Program of Natural Science Foundation for Higher Education in Jiangsu Province, grant no. 21KJB510004, and Major Program of National Natural Science Foundation of China (NSFC) for Basic Theory and Key Technology of Tri-Co Robots, grant no. 92248301.

Institutional Review Board Statement: Not applicable.

Data Availability Statement: The data that support the findings of this study are available from the authors upon reasonable request.

Conflicts of Interest: The authors declare no conflict of interest.

References

1. Shi, Q.; Sun, J.; Hou, C.; Li, Y.; Zhang, Q.; Wang, H. Advanced Functional Fiber and Smart Textile. *Adv. Fiber Mater.* **2019**, *1*, 3–31. [CrossRef]
2. Shi, J.; Liu, S.; Zhang, L.; Yang, B.; Shu, L.; Yang, Y.; Ren, M.; Wang, Y.; Chen, J.; Chen, W.; et al. Smart Textile-Integrated Microelectronic Systems for Wearable Applications. *Adv. Mater.* **2020**, *32*, 1901958. [CrossRef] [PubMed]
3. Li, Z.; Zhu, M.; Shen, J.; Qiu, Q.; Yu, J.; Ding, B. All-Fiber Structured Electronic Skin with High Elasticity and Breathability. *Adv. Funct. Mater.* **2020**, *30*, 1908411. [CrossRef]
4. Hu, J.; Meng, H.; Li, G.; Ibekwe, S.I. A Review of Stimuli-Responsive Polymers for Smart Textile Applications. *Smart Mater. Struct.* **2012**, *21*, 053001. [CrossRef]
5. Cherenack, K.; Van Pieterson, L. Smart Textiles: Challenges and Opportunities. *J. Appl. Phys.* **2012**, *112*, 091301. [CrossRef]
6. Wang, L.; Fu, X.; He, J.; Shi, X.; Chen, T.; Chen, P.; Wang, B.; Peng, H. Application Challenges in Fiber and Textile Electronics. *Adv. Mater.* **2020**, *32*, 1901971. [CrossRef]
7. Yang, C.; Cheng, S.; Yao, X.; Nian, G.; Liu, Q.; Suo, Z. Ionotronic Luminescent Fibers, Fabrics, and Other Configurations. *Adv. Mater.* **2020**, *32*, 2005545. [CrossRef] [PubMed]
8. Bayindir, M.; Abouraddy, A.F.; Arnold, J.; Joannopoulos, J.D.; Fink, Y. Thermal-Sensing Fiber Devices by Multimaterial Codrawing. *Adv. Mater.* **2006**, *18*, 845–849. [CrossRef]
9. Chen, J.; Zhu, Y.; Huang, J.; Zhang, J.; Pan, D.; Zhou, J.; Ryu, J.E.; Umar, A.; Guo, Z. Advances in Responsively Conductive Polymer Composites and Sensing Applications. *Polym. Rev.* **2021**, *61*, 157–193. [CrossRef]
10. Collins, G.E.; Buckley, L.J. Conductive Polymer-Coated Fabrics for Chemical Sensing. *Synth. Met.* **1996**, *78*, 93–101. [CrossRef]
11. Kayser, L.V.; Lipomi, D.J. Stretchable Conductive Polymers and Composites Based on PEDOT and PEDOT:PSS. *Adv. Mater.* **2019**, *31*, 1806133. [CrossRef] [PubMed]
12. Fan, X.; Nie, W.; Tsai, H.; Wang, N.; Huang, H.; Cheng, Y.; Wen, R.; Ma, L.; Yan, F.; Xia, Y. PEDOT:PSS for Flexible and Stretchable Electronics: Modifications, Strategies, and Applications. *Adv. Sci.* **2019**, *6*, 1900813. [CrossRef]
13. Tseghai, G.B.; Mengistie, D.A.; Malengier, B.; Fante, K.A.; Van Langenhove, L. PEDOT:PSS-Based Conductive Textiles and Their Applications. *Sensors* **2020**, *20*, 1881. [CrossRef]
14. Zhang, J.; Seyedin, S.; Qin, S.; Lynch, P.A.; Wang, Z.; Yang, W.; Wang, X.; Razal, J.M. Fast and Scalable Wet-Spinning of Highly Conductive PEDOT:PSS Fibers Enables Versatile Applications. *J. Mater. Chem. A* **2019**, *7*, 6401–6410. [CrossRef]
15. Jalili, R.; Razal, J.M.; Innis, P.C.; Wallace, G.G. One-Step Wet-Spinning Process of Poly(3,4-Ethylenedioxythiophene): Poly(Styrenesulfonate) Fibers and the Origin of Higher Electrical Conductivity. *Adv. Funct. Mater.* **2011**, *21*, 3363–3370. [CrossRef]
16. Zhou, J.; Li, E.Q.; Li, R.; Xu, X.; Ventura, I.A.; Moussawi, A.; Anjum, D.H.; Hedhili, M.N.; Smilgies, D.M.; Lubineau, G.; et al. Semi-Metallic, Strong and Stretchable Wet-Spun Conjugated Polymer Microfibers. *J. Mater. Chem. C* **2015**, *3*, 2528–2538. [CrossRef]
17. Yuan, D.; Li, B.; Cheng, J.; Guan, Q.; Wang, Z.; Ni, W.; Li, C.; Liu, H.; Wang, B. Twisted Yarns for Fiber-Shaped Supercapacitors Based on Wetspun PEDOT:PSS Fibers from Aqueous Coagulation. *J. Mater. Chem. A* **2016**, *4*, 11616–11624. [CrossRef]
18. Sarabia-Riquelme, R.; Andrews, R.; Anthony, J.E.; Weisenbergera, M.C. Highly conductive wet-spun PEDOT:PSS fibers for applications in electronic textiles. *J. Mater. Chem. C* **2020**, *8*, 11618–11630. [CrossRef]
19. Wen, N.; Fan, Z.; Yang, S.; Zhao, Y.; Cong, T.; Xu, S.; Zhang, H.; Wang, J.; Huang, H.; Li, C.; et al. Highly Conductive, Ultra-Flexible and Continuously Processable PEDOT:PSS Fibers with High Thermoelectric Properties for Wearable Energy Harvesting. *Nano Energy* **2020**, *78*, 105361. [CrossRef]
20. Wen, N.; Fan, Z.; Yang, S.; Zhao, Y.; Li, C.; Cong, T.; Huang, H.; Zhang, H.; Guan, X.; Pan, L. High-Performance Stretchable Thermoelectric Fibers for Wearable Electronics. *Chem. Eng. J.* **2021**, *426*, 130816. [CrossRef]
21. Jin, S.; Sun, T.; Fan, Y.; Wang, L.; Zhu, M.; Yang, J.; Jiang, W. Synthesis of Freestanding PEDOT:PSS/PVA@Ag NPs Nanofiber Film for High-Performance Flexible Thermoelectric Generator. *Polymer* **2019**, *167*, 102–108. [CrossRef]
22. Kim, Y.; Lund, A.; Noh, H.; Hofmann, A.I.; Craighero, M.; Darabi, S.; Zokaei, S.; Park, J.I.; Yoon, M.H.; Müller, C. Robust PEDOT:PSS Wet-Spun Fibers for Thermoelectric Textiles. *Macromol. Mater. Eng.* **2020**, *305*, 1900749. [CrossRef]
23. Ruan, L.; Zhao, Y.; Chen, Z.; Zeng, W.; Wang, S.; Liang, D.; Zhao, J. A Self-Powered Flexible Thermoelectric Sensor and Its Application on the Basis of the Hollow PEDOT: PSS Fiber. *Polymers* **2020**, *12*, 553. [CrossRef] [PubMed]
24. Xu, S.; Hong, M.; Shi, X.L.; Wang, Y.; Ge, L.; Bai, Y.; Wang, L.; Dargusch, M.; Zou, J.; Chen, Z.G. High-Performance PEDOT:PSS Flexible Thermoelectric Materials and Their Devices by Triple Post-Treatments. *Chem. Mater.* **2019**, *31*, 5238–5244. [CrossRef]
25. Kim, S.M.; Kim, C.H.; Kim, Y.; Kim, N.; Lee, W.J.; Lee, E.H.; Kim, D.; Park, S.; Lee, K.; Rivnay, J.; et al. Influence of PEDOT:PSS Crystallinity and Composition on Electrochemical Transistor Performance and Long-Term Stability. *Nat. Commun.* **2018**, *9*, 3858. [CrossRef] [PubMed]
26. Wang, Y.; Zhu, C.; Pfattner, R.; Yan, H.; Jin, L.; Chen, S.; Molina-Lopez, F.; Lissel, F.; Liu, J.; Rabiah, N.I.; et al. A Highly Stretchable, Transparent, and Conductive Polymer. *Sci. Adv.* **2017**, *3*, e1602076. [CrossRef]
27. Bubnova, O.; Berggren, M.; Crispin, X. Tuning the Thermoelectric Properties of Conducting Polymers in an Electrochemical Transistor. *J. Am. Chem. Soc.* **2012**, *134*, 16456–16459. [CrossRef]

Disclaimer/Publisher's Note: The statements, opinions and data contained in all publications are solely those of the individual author(s) and contributor(s) and not of MDPI and/or the editor(s). MDPI and/or the editor(s) disclaim responsibility for any injury to people or property resulting from any ideas, methods, instructions or products referred to in the content.

Article

Flexible BaTiO$_3$-PDMS Capacitive Pressure Sensor of High Sensitivity with Gradient Micro-Structure by Laser Engraving and Molding

Jiayi Li [1,2], Shangbi Chen [3], Jingyu Zhou [1,2], Lei Tang [1,2], Chenkai Jiang [1,2], Dawei Zhang [1,2] and Bin Sheng [1,2,*]

[1] School of Optical Electrical and Computer Engineering, University of Shanghai for Science and Technology, Shanghai 200093, China; 2035051414@st.usst.edu.cn (J.L.); 2133305789@st.usst.edu.cn (J.Z.); leitangstark@foxmail.com (L.T.); 212180323@st.usst.edu.cn (C.J.); dwzhang@usst.edu.cn (D.Z.)
[2] Shanghai Key Laboratory of Modern Optical Systems, Engineering Research Center of Optical Instruments and Systems, Shanghai 200093, China
[3] Inertial Technology Division, Shanghai Aerospace Control Technology Institute, Shanghai 201109, China; chensb@mail.ustc.edu.cn
* Correspondence: bsheng@usst.edu.cn

Citation: Li, J.; Chen, S.; Zhou, J.; Tang, L.; Jiang, C.; Zhang, D.; Sheng, B. Flexible BaTiO$_3$-PDMS Capacitive Pressure Sensor of High Sensitivity with Gradient Micro-Structure by Laser Engraving and Molding. Polymers 2023, 15, 3292. https://doi.org/10.3390/polym15153292

Academic Editors: Jiangtao Xu and Sihang Zhang

Received: 17 July 2023
Revised: 30 July 2023
Accepted: 2 August 2023
Published: 3 August 2023

Copyright: © 2023 by the authors. Licensee MDPI, Basel, Switzerland. This article is an open access article distributed under the terms and conditions of the Creative Commons Attribution (CC BY) license (https://creativecommons.org/licenses/by/4.0/).

Abstract: The significant potential of flexible sensors in various fields such as human health, soft robotics, human–machine interaction, and electronic skin has garnered considerable attention. Capacitive pressure sensor is popular given their mechanical flexibility, high sensitivity, and signal stability. Enhancing the performance of capacitive sensors can be achieved through the utilization of gradient structures and high dielectric constant media. This study introduced a novel dielectric layer, employing the BaTiO$_3$-PDMS material with a gradient micro-cones architecture (GMCA). The capacitive sensor was constructed by incorporating a dielectric layer GMCA, which was fabricated using laser engraved acrylic (PMMA) molds and flexible copper-foil/polyimide-tape electrodes. To examine its functionality, the prepared sensor was subjected to a pressure range of 0–50 KPa. Consequently, this sensor exhibited a remarkable sensitivity of up to 1.69 KPa^{-1} within the pressure range of 0–50 KPa, while maintaining high pressure-resolution across the entire pressure spectrum. Additionally, the pressure sensor demonstrated a rapid response time of 50 ms, low hysteresis of 0.81%, recovery time of 160 ms, and excellent cycling stability over 1000 cycles. The findings indicated that the GMCA pressure sensor, which utilized a gradient structure and BaTiO$_3$-PDMS material, exhibited notable sensitivity and a broad linear pressure range. These results underscore the adaptability and viability of this technology, thereby facilitating enhanced flexibility in pressure sensors and fostering advancements in laser manufacturing and flexible devices for a wider array of potential applications.

Keywords: flexible capacitance pressure sensor; gradient micro-structure; polymer; barium titanate; laser engraving

1. Introduction

In recent times, the pressure sensor garnered significant attention due to its wide-ranging applications in diverse fields such as soft robotics [1,2], human–machine interaction [3], electronic skin [4–6], tactile and touch sensing applications [7,8], contactless sensing [9], and information communication [10]. Presently, pressure sensors can be primarily categorized into resistance [11–13], capacitance [11,14,15], piezoelectric [11,16], and triboelectric types [11,17] based on distinct transduction mechanisms. Capacitive sensors are favored among these types due to their simple structure, ease of fabrication, low-energy consumption, and ability to precisely modify device design through analysis of the governing equation. In terms of performance, they demonstrate notable attributes such as high sensitivity and rapid response times. Additionally, these sensors have been proven to replicate the sensing behavior of human skin, encompassing strain sensitivity, pressure detection, and proximity sensing [18]. Furthermore, capacitive pressure sensors exhibit low power consumption and can be engineered to be unaffected by temperature variations [18–22]. Consequently, they are deemed attractive. To cater to various applications, capacitive sensors are required to exhibit both high sensitivity and a wide linear range. The sensitivity of traditional capacitive sensors utilizing solid dielectric layers is hindered by their restricted deformation capacity [23–25]. In order to enhance sensitivity, extensive research has been conducted on diverse micro-structures such as micro-spheres [26], micro-pillars [27], porous structures [28], micro-pyramids [29], nanoparticles [30–32], and micro-array structures [33,34]. However, it has been observed that these micro-structures primarily operate within the low-pressure range, thereby diminishing their sensitivity in the high-pressure range and consequently limiting the linear range [21,35–39]. Previous studies indicated that the utilization of gradient structures can effectively enhance linearity and substantially augment sensitivity [21]. It is important to acknowledge that the sensitivity of sensor components featuring microstructure effectively trades off against hysteresis. According to scholarly research, the implementation of a random distribution of pixels featuring gradient structures has been found to effectively diminish interface adhesion [40]. Consequently, this approach enables the sensor to sustain high sensitivity levels while minimizing hysteresis.

A flexible material possessing a high permittivity is crucial for the development of flexible capacitive sensors and charge storage devices. Currently, ferroelectric polymer materials, including poly-(vinylidene fluoride)(PVDF) and poly-(vinylidene fluoride-trifluoroethylene)[P(VDF-TrFE)], have been successfully employed to achieve high permittivity. Nevertheless, these polymers exhibit temperature instability and can lead to device corrosion due to the formation of hydrogen fluoride [41]. Several studies investigated the advancement of composite materials possessing superior dielectric properties in order to address the aforementioned constraints of organic polymers. The incorporation of ceramic fillers exhibiting high permittivity represents a prevalent approach for enhancing the dielectric permittivity of polymers [42,43]. For instance, flexible materials like polydimethylsiloxane (PDMS) were combined with $Pb(Zr, Ti)O_3$ and MXene($Ti_3C_2T_x$) to fabricate dielectric layers [25,43]. The dielectric properties of these materials are influenced by the attributes of the constituents, the morphology and size of the additives, and the concentration of the additive [41]. $BaTiO_3$ is a material possessing high permittivity and readily accessible [44]; thus, we opted for $BaTiO_3$ powder blended with PDMS as the material for fabricating a flexible dielectric layer.

The dielectric layer of flexible pressure sensors is typically fabricated through the primary mold method, which traditionally involves lithography or the utilization of natural molds such as lotus leaves and petals [45–50]. However, the photo-lithography manufacturing process is intricate, expensive, and time-consuming, while the micro-structured surface created with natural molds is uneven and lacks controllable aspect ratios. In order to overcome these constraints, Valliammai Palaniappan et al. conducted a study to ascertain the viability of employing a laser-assisted engraving technique for the fabrication of PDMS dielectric layers. This method proved to be relatively uncomplicated, user-friendly, and time-efficient, requiring minimal preparation time [46,51–53]. Furthermore, during the laser-assisted production of the primary mold, the aspect ratio of the model can be readily manipulated by adjusting the power and scanning speed of the laser beam [46,54].

This study introduces a novel approach to capacitive pressure sensing by utilizing a $BaTiO_3$-PDMS dielectric layer with a gradient micro-cones architecture (GMCA). The GMCA is achieved by laser carving a micro-cones hole array with varying heights on a black acrylic plate, which serves as a template for the deposition of the $BaTiO_3$-PDMS dielectric layer. The pressure sensor is formed by sandwiching the dielectric layer between flexible electrodes composited with copper foil and polyimide tape. Consequently, the sensor exhibits a notable sensitivity of up to 1.69 KPa^{-1} within the pressure range of 0–50 KPa, ensuring the preservation of high pressure-resolution throughout the entire pressure spectrum. Additionally, the sensor demonstrates a rapid response time of 50 ms, minimal hysteresis of 0.81%, swift recovery time of 160 ms, and exceptional cycling stability over 1000 cycles. Consequently, this sensor possesses the capability to serve as a dependable motion recorder for comprehensive detection of physiological signals, including pulse, sound vibration, and joint flexion, among others.

2. Materials and Methods

2.1. Chemicals and Materials

A main mold was manufactured using a black poly (methyl methacrylate) (PMMA) with a thickness of approximately 5mm, sourced from Tian Gong company in Zhenjiang, Jiangsu, China. The barium titanate powder, with a particle size smaller than 3 µm, was obtained from Macklin company in Shanghai, China. The dielectric layers were manufactured using PDMS (SYLGARD 184 Silicone) from the DOW Chemical company in the Midland, MI, USA. The bonding process utilized ecoflex 00-30 from SMOOTH-ON Company in the Macungie, PA, USA, which included liquid A and liquid B. The flexible electrodes were manufactured using copper foil sourced from Zhengying Company in Anhui, China, and polyimide tape from Ubisoft Corporation, Hangzhou, Zhe Jiang, China.

2.2. Sensor Fabrication

The sensor was comprised of two flexible electrode plates and a dielectric with a gradient micro-structure in the center (Figure 1a). The manufacturing process of a rectangular pressure sensor measuring 30 × 25 mm is depicted in Figure 1. Using AutoCAD™, a 3 × 3 array consisting of circles with a diameter of 1 mm was designed and then imported into the laser machine (K3020, Julong Laser Co., Ltd., Liaocheng, Shandong, China). The formation of the micro-cone exhibited a strong correlation with the laser power. The cross section intensity of the laser adhered to a Gaussian distribution, which played a significant role in the genesis of micro-cones [55,56]. As the laser ablation process advanced on the board, the gradual decoking resulted in a progressive decrease in the critical point's radius, where the laser-induced ablation occurred, ultimately leading to the emergence of the micro-cone. Initially, a scanning speed of 100 mm/s was employed, along with a specific amount of light power, to carve a hole array consisting of 9 circular holes on the acrylic plate (Figure 1b). Next, the exit position of the laser beam and the light power should be adjusted to proceed with the creation of another hole array. This process should be repeated until a main mode consisting of 20 hole arrays, arranged in a 5 × 4 configuration, is achieved. The resulting main module, denoted as M, is depicted in Figure 1c. To prepare the PDMS, the elastomer base and hardener should be mixed in a ratio of 10:1, followed by thorough stirring with a glass rod for approximately 20 min. Subsequently, the obtained PDMS should be mixed with $BaTiO_3$ powder in a mass ratio of 10:1 to obtain $BaTiO_3$-PDMS, ensuring even distribution of the components. The mixed $BaTiO_3$-PDMS composite was subjected to vacuum treatment using a vacuum machine (DZF-6053, Yong Guangming, Beijing, China) for approximately 30 min to eliminate any surface bubbles. Subsequently, the $BaTiO_3$-PDMS composite was poured onto the acrylic main mold M (Figure 1d) and subjected to vacuum treatment for an additional 30 min. The vacuum machine was then heated to 60 °C for approximately one hour to facilitate curing (Figure 1e). Following the curing process, the $BaTiO_3$-PDMS dielectric layer GMCA was carefully removed from the main molds M (Figure 1f). It is noteworthy that the utilization of lasers can be enhanced more effectively by employing black acrylic panels as opposed to conventional transparent acrylic panels, resulting in a higher aspect ratio of the tapered hole at equivalent optical power levels.

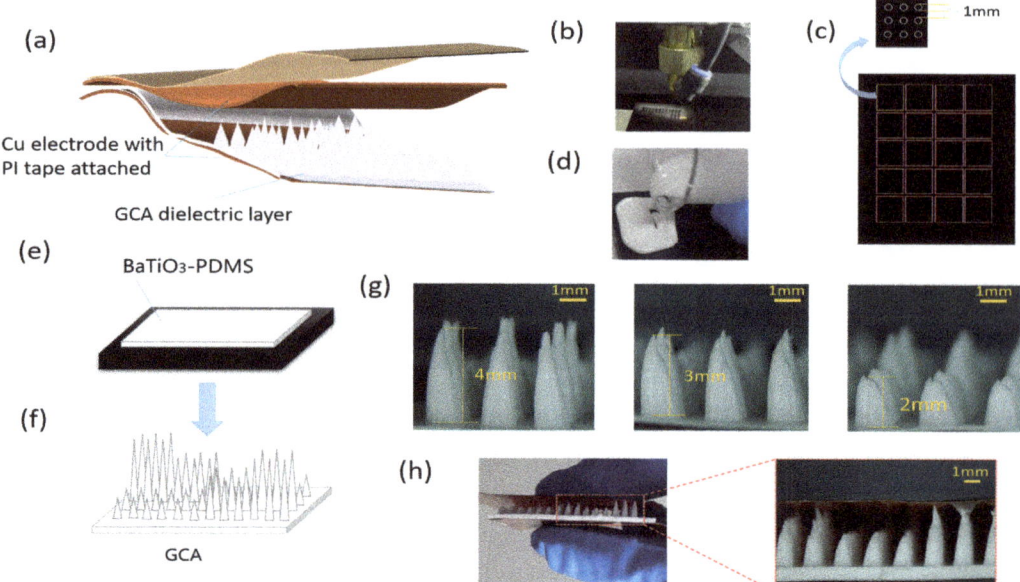

Figure 1. The manufacturing process of the pressure sensors. (**a**) Schematic diagram of the overall structure of the sensor (only some micro-structure pixels are shown in the figure). (**b**) Laser engraved hole array on the black acrylic board. (**c**) The main module. (**d**) The BaTiO$_3$-PDMS was poured onto the acrylic main mold, and then it was vacuumed. (**e**) Heat at 60 °C for about one hour and wait for the dielectric layer to form. (**f**) The BaTiO$_3$-PDMS dielectric layer GMCA was peeled-off from the main molds. (**g**) Digital microscope images and their heights of A, B, C three types of pixels. (**h**) The side view of the pressure sensor (PS4).

The electrode was composed of a copper foil that possessed adhesive on one side and was manufactured to have a thickness of 5 μm. Positioned above the dielectric layer, the lower surface of the micro-cone exhibited a seamless and uninterrupted plane, while the upper portion of the cone featured spaced-out spikes. The copper foil electrode can be affixed directly onto the smooth layer, while the spike tips can be gently positioned downwards onto the copper foil coated with a thin layer of ecoflex. This ecoflex substance was created by combining part A and part B in equal proportions of 1:1, followed by a vacuum process. Simultaneously, for the purpose of enhancing the electrode's stability, it was possible to affix a layer of polyimide tape (thickness of 0.1 mm) onto the external surface of the electrode, thereby reducing the susceptibility of the copper foil to deformation. Following the curing of ecoflex, a capacitive sensor comprising two copper foil electrodes and a dielectric layer in between can be achieved.

To establish the gradient structure, the set layer power was modified, resulting in varying hole arrays corresponding to the layer power. In this experimental study, the nomenclature A, B, and C was assigned to the hole arrays based on their respective heights, with A representing the highest and C denoting the lowest. Considering the fabrication error of ~0.3 mm for laser engraving processing, the determination of the heights of three distinct micro cone structures (A, B, and C) was conducted to facilitate the establishment of gradient structures. Subsequently, the distance between the sensor electrode plates was ascertained. The composition of a GMCA, comprising multiple hole arrays, can be succinctly represented as AaBbCc, where the lowercase letters serve as subscripts indicating the number of hole arrays within the GMCA structure. Three arrays of holes with varying heights were randomly distributed on the main mode. The stochastic arrangement of pixels exhibiting varying heights facilitates the compatibility of the pressure sensor with diverse pixel configurations [41]. In order to demonstrate the significance of gradient structure and investigate the values of parameters a, b, and c, we also fabricated dielectric layers without gradient structure, which exhibited a highly uniform distribution of micro-cones architecture (MCA). The production method described above resulted in the creation of five dielectric layers, along with their respective parameters, as presented in Table 1. The images captured using a high-resolution digital microscope can be observed in Figure 1g, while the side view of the pressure sensor (PS4) that was fabricated is depicted in Figure 1h.

Table 1. Dielectric layer and its main mode engraving parameters.

Sensor Number	Dielectric Layer	Laser Parameters		Main Module	Micro-Structures Height (μm)
		Speed (mm/s)	Power (%)		
PS1	MCA(A_{20})	100	30	M_1	4500
PS2	MCA(B_{20})	100	40	M_2	3500
PS3	MCA(C_{20})	100	50	M_3	2500
PS4	GMCA($A_3B_6C_{11}$)	100	30, 40, 50	M_4	4500
PS5	GMCA($A_2B_5C_{13}$)	100	30, 40, 50	M_5	4500

2.3. Experiment Setup

The experimental setup is depicted in Figure 2. The sensor was positioned on the pressure tester platform (Zhiqu company, Guangzhou, Guangdong, China) and subjected to a pressure range of 0–50 Kpa. The two electrodes of the capacitance sensor were connected to the digital bridge (LCR) instrument of Company Tonghui (Changzhou, Jiangsu, China) in order to measure the sensor's capacitance response to varying pressure. A computer was linked to the LCR meter through a USB connection, facilitating post-processing and data analysis. All experiments were conducted under ambient room temperature conditions.

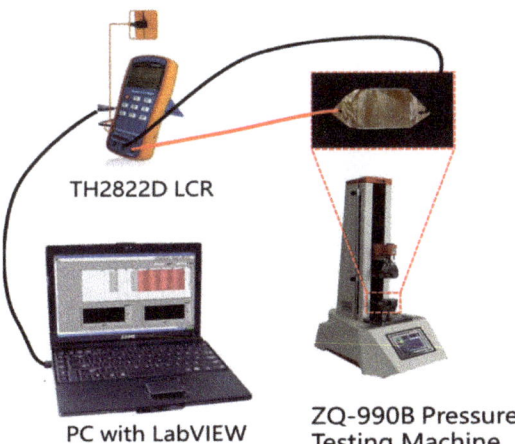

Figure 2. Experiment setup. The sensor was placed on the pressure testing platform and was connected to the LCR instrument, which was connected to the computer. When the pressure testing machine starts, LCR can measure the capacitance of the capacitor and upload it to the computer to record the data.

3. Results and Discussion

3.1. Capacitive Pressure Sensor Response

The working principle of capacitive pressure sensors can be explained by Equation (1) [21]:

$$C = \frac{\varepsilon_0 \varepsilon_r A}{d} \qquad (1)$$

where C, ε_r, ε_0, A, and d are the capacitance, effective dielectric constant, permittivity of free space, device contact area, and the thickness of dielectric material, respectively.

When a normal force is exerted on an electrode, the dielectric layer undergoes compression, leading to an increase in capacitance. Consequently, for a fixed overlapping area, the extent of deformation caused by a specific force on the dielectric layer directly influences the distance between the two electrodes, thereby resulting in a significant alteration in capacitance. The determination of sensitivity is elucidated in Equation (2) [21].

$$S = \frac{\partial(\Delta C/C_0)}{\partial P} \qquad (2)$$

where C and C_0 are the resultant capacitance and the initial capacitance without loading the pressure (P), respectively. Figure 3a displays the curve representing the alteration in relative capacitance of sensors PS1–PS5 as pressure increases. When comparing the three curves of PS1–PS3, it becomes evident that an increase in the height of the micro cone structure corresponds to an increase in sensor sensitivity. The spacing between the electrode plates plays a crucial role in enabling a high aspect ratio for the dielectric layer's micro-cone, thereby ensuring a heightened level of sensor sensitivity. In comparison to sensors lacking gradient structure dielectric layers (PS1, PS2, PS3), sensors PS4 and PS5 exhibit a more significant relative change in capacitance, suggesting an enhanced sensitivity to some extent through the utilization of gradient structure. The capacitive responses of sensors PS1–PS5 were examined across three distinct pressure ranges: 0–2 KPa, 2–15 KPa, and 15–50 KPa, based on the observed trend in the curve. Under low pressure conditions, the sensitivities of PS4 and PS5 were 1.69 KPa^{-1} and 1.24 KPa^{-1}, respectively, with linear correlation coefficients of 0.99 for both. However, the linearity of PS1-PS3, which lack a gradient structure, was less than 0.99. In comparison, PS1 exhibited a higher sensitivity of 1.31 KPa^{-1} due to its excellent aspect ratio of the dielectric layer MCA (A20). As the pressure increased to the range of 2–15 KPa, the sensitivities of PS4 and PS5 became 0.42 KPa^{-1} and 0.41 KPa^{-1}, respectively, both surpassing those of PS1, PS2, and PS3 without a gradient structure. Even under high pressure conditions (15–50 KPa), PS4 and PS5 still maintained a certain level of sensitivity. In comparison to the sensors with sensitivity of 2.21×10^{-6} KPa^{-1} in previous studies [51], our pressure sensors (PS4 and PS5) with the dielectric layers of gradient structures exhibited considerable higher sensitivity up to 1.69 KPa^{-1}.

The dielectric response of the dielectric layer is predominantly influenced by the alterations in contact area under different pressures [25]. As the dielectric layer of the micro cone structure was compressed, the rate of change in contact area gradually diminished, resulting in a decrease in sensitivity. The PS1–PS3, lacking a gradient structure, exhibited remarkable sensitivity at low pressure; however, its sensitivity significantly declined as pressure increased, thereby hindering the attainment of a satisfactory dielectric response across all pressure ranges. Hence, the selection of a gradient structure facilitated the sequential contact of the electrode with the micro cone structure, ensuring a sustained high dielectric response across diverse pressure ranges. The dielectric behavior of GMCA dielectric in distinct pressure regions was influenced by the various types of micro-cone pixels, contingent upon their height gradient.

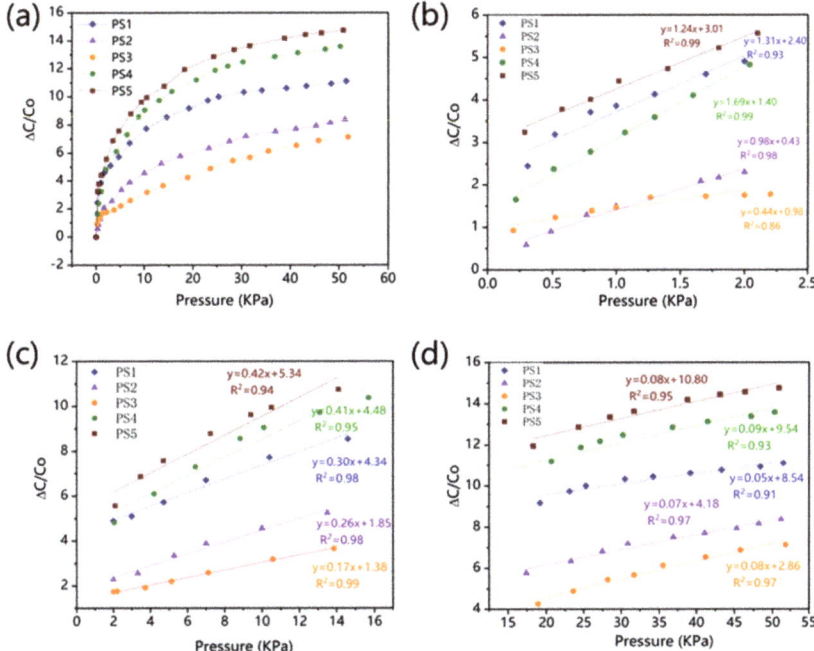

Figure 3. (**a**) Relative capacitance change of pressure sensors PS1–PS5. (**b**) Linear fitting of relative capacitance changes in the pressure range of 0–2 KPa. (**c**) Linear fitting of relative capacitance changes in the pressure range of 2–15 KPa. (**d**) Linear fitting of relative capacitance changes in the pressure range of 15–50 KPa.

Under low pressures, micro-cone pixels with comparatively lower height (e.g., B and C) became connected to the air components due to their separation from the upper electrode. As a result, the limited capabilities of pixels B and C significantly constrained the dielectric properties of the GMCA layer, as they were determined by the tallest pixel (pixel A). Following the compression of pixels A, pixels B and C can subsequently come into contact with the upper electrode, thereby transforming their series connection with the air components into a parallel connection. Consequently, in the medium and high-pressure regions (Figure 4d), pixels B and C will exert a dominant influence on the dielectric behavior of the GMCA dielectric.

In principle, changes in the height and number of gradient micro-cone pixels will affect the dielectric behavior of GMCA dielectrics. When no external pressure was applied, a base capacitance of 6pF was measured. The dynamic response of the fabricated pressure sensor (PS4, PS5) for the applied pressure ranges are shown in Figure 4a,b. For PS4, the capacitance of the pressure sensor was increased from the base capacitance of 6 pF to 34.9 pF when the pressure was increased from 0 Pa to 2 KPa, respectively. In addition, it was observed that the capacitance was increased from 34.9 pF to 68.3 pF and 68.3 pF to 87.5 pF when the applied pressure was increased from 2 KPa to 15 KPa, and 15KPa to 50 KPa, respectively. The pressure ranges of 0 KPa to 2 KPa, 2 KPa to 15 KPa, and 15 KPa to 50 KPa resulted in overall relative capacitance changes of 482%, 1038%, and 1358%, respectively. However, for PS4, these values were 538%, 1110%, and 1476%, respectively. Figure 4c demonstrates that PS5, with more high micro cone pixels, exhibited a more significant relative change in capacitance and better overall linearity compared to PS4. Additionally, PS4 demonstrated higher sensitivity under low pressure compared to PS5. Hence, the utilization of PS4 under low pressure conditions can enhance sensitivity, whereas employing PS5 across a broader pressure range can amplify the obtained outcomes. Consequently, PS4 was chosen for subsequent experimentation and implementation in this study.

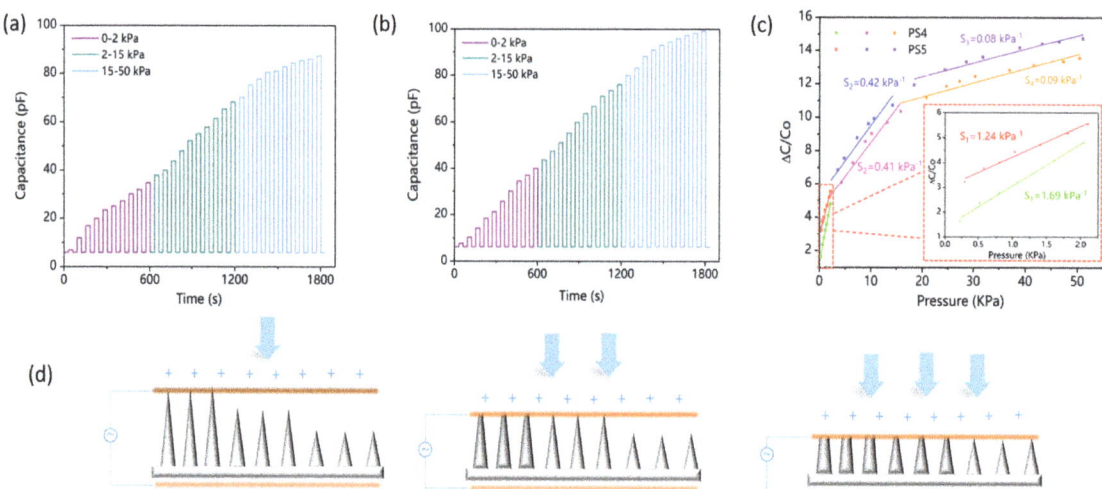

Figure 4. (**a**) Dynamic capacitive response of PS4. (**b**) Dynamic capacitive response of PS5. (**c**) Capacitive sensitivity of PS4 and PS5. (**d**) Working mechanism of the GMCA dielectric.

3.2. The Impact of $BaTiO_3$ on Sensor Performance and Content Determination

Based on the findings of Equations (1) and (2), enhancing the effective dielectric constant of the dielectric layer can significantly enhance the sensor's sensitivity. The investigation demonstrated that the inclusion of $BaTiO_3$ particles in the composite yielded a substantial dielectric constant. Consequently, the $BaTiO_3$-PDMS material was chosen for the fabrication of the capacitors' dielectric layer in this experiment. Prior research outcomes indicated that the dielectric constant of $BaTiO_3$ is contingent upon the crystal's grain size. The dielectric constant of BaTiO3 was determined to be 1750 for particle diameters ranging from 20 μm to 50 μm, and 5000 for a diameter of 1.1 μm. As the grain size of the $BaTiO_3$ crystal decreased, its permittivity value also decreased significantly. When the particle diameter was less than 100 nm, the permittivity value became extremely low [44,57]. Consequently, for this experiment, $BaTiO_3$ powder with particle sizes in the micrometer range was chosen as the composite material.

To ascertain the most favorable blending ratio of PDMS and $BaTiO_3$ powder, we fabricated $BaTiO_3$-PDMS thin films with varying mass ratios and, subsequently, assessed their dielectric constants, as depicted in Figure 5a. As the mass proportion of $BaTiO_3$ in the mixture augments, the dielectric constant of the material also increased, thereby suggesting that the proportion of $BaTiO_3$ can be maximized to enhance the dielectric characteristics of the dielectric layer. However, the addition of $BaTiO_3$ powder in increasing quantities results in a decrease in the fluidity of the mixed solution during preparation, leading to a hardened material that poses challenges in shaping the microstructure of the dielectric layer. This is illustrated in Figure 5c, where the higher pixel parts in the micro-structure were not formed when a 5:1 mixing ratio was employed. Considering both practicality and the level of difficulty in preparation, we opted for a 10:1 mass ratio of PDMS and $BaTiO_3$ to fabricate the dielectric layer GMCA. Figure 5b illustrates the relative capacitance change curves of GMCA pressure sensors fabricated from pure PDMS devoid of $BaTiO_3$ addition, as well as composite materials with PDMS to $BaTiO_3$ mass ratios of 20:1 and 10:1, correspondingly. The sensitivity of these sensors within the pressure range is documented in Table 2. Evidently, the utilization of a composite material comprising $BaTiO_3$ and PDMS exhibited a notable enhancement in sensor sensitivity when compared to pure PDMS materials. This method, which is both simple and easy to operate, deserves attention as a means to enhance the sensor's sensitivity. Figure 5d illustrates the XRD plot of PDMS to BaTiO$_3$ at a mass ratio of 10:1. Additionally, Figure 5e presents the scanning electron microscope image of GMCA with a mass ratio of PDMS to BaTiO$_3$ of 10:1. The particle size structure of BaTiO$_3$ is depicted in Figure 5f, while the electron microscopy-based elemental analysis results are displayed in Figure 5g,h, showcasing the distribution of Ba, Ti, and C elements.

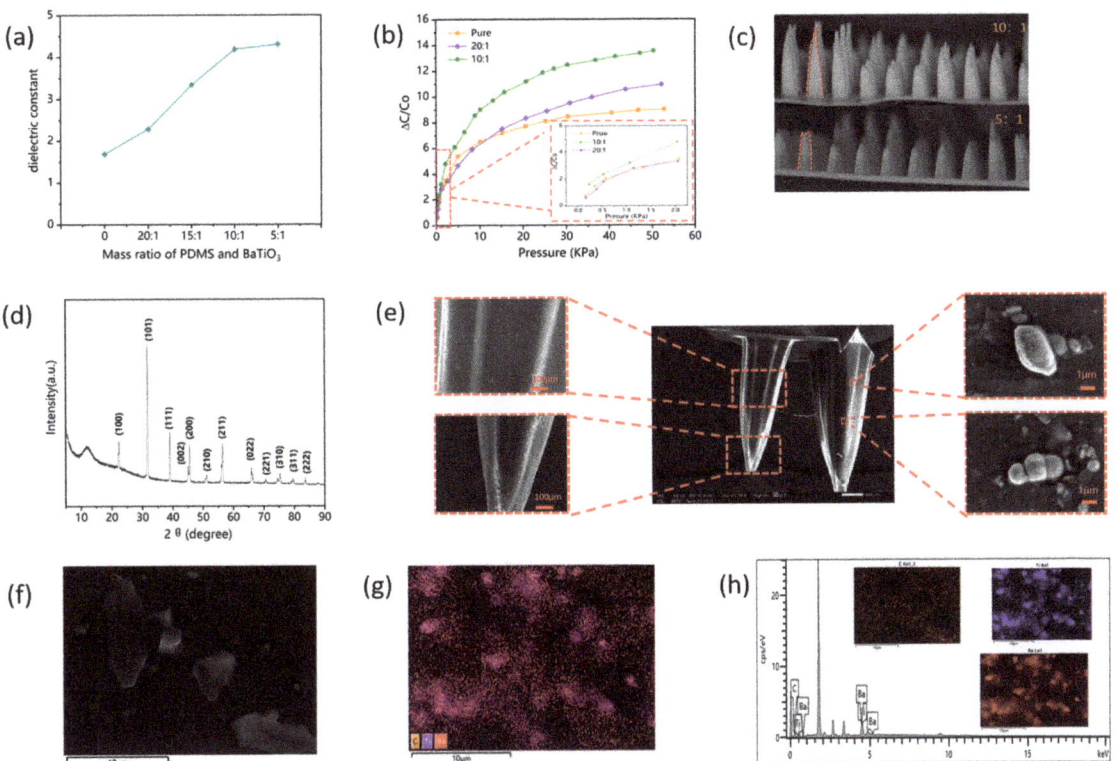

Figure 5. (a) Dielectric constant of different PDMS and BaTiO$_3$ mixed mass ratios. (b) Relative capacitance variation of different PDMS and BaTiO$_3$ mixed mass ratios. (c) Microscopic images of GMCA dielectric layers prepared at a mass ratio of 10:1 and 5:1 for PDMS to BaTiO$_3$. (d) The XRD plot of PDMS to BaTiO$_3$ at a mass ratio of 10:1. (e) The electron microscope scanning image of GMCA with a mass ratio of PDMS to BaTiO$_3$ of 10:1. (f) The plot of particle size structure of BaTiO$_3$. (h) The elemental analysis results of Ba, Ti, C. (g) The map sum spectrum.

Table 2. Sensitivity of sensors with different mixing ratios of dielectric layers.

Mass Ratio of PDMS and BaTiO$_3$	0–2 Kpa	2–15 Kpa	15–50 Kpa
pure	1.30	0.17	0.04
20:1	1.37	0.27	0.08
10:1	1.69	0.41	0.09

3.3. Hysteresis Response

The hysteresis response of the PS4 was investigated for the pressure range of 0 to 15 Kpa. Step-wise pressures increasing from 0 to 15 Kpa and then decreasing from 15 Kpa to 0 Pa, in steps of 1 Kpa, were applied with 3 cycles per step (Figure 6a). The capacitive response of the pressure sensor was measured and the maximum hysteresis (MH) was mathematically calculated using Equation (3) [51].

$$MH(\%) = \frac{[x_2 - x_1] \times 100}{x_p - x_b} \tag{3}$$

where x_1 and x_2 are the capacitances measured for an applied pressure during stepwise increase and decrease, respectively. x_p is the peak capacitance at 15 KPa and x_b is the base capacitance (at 0 KPa). A *MH* of 2.91% was calculated at an applied pressure of 5 KPa where x_1 and x_2 were measured to be 45.9 pF and 47.7 pF, respectively (Figure 7b). As shown in Figure 6b, a minimum hysteresis of 0.18% was calculated at an applied pressure of 14 KPa (where x_1 = 65.9 pF and x_2 = 66.4 pF). This indicated that pressure sensor has relatively better recovery and elasticity characteristics.

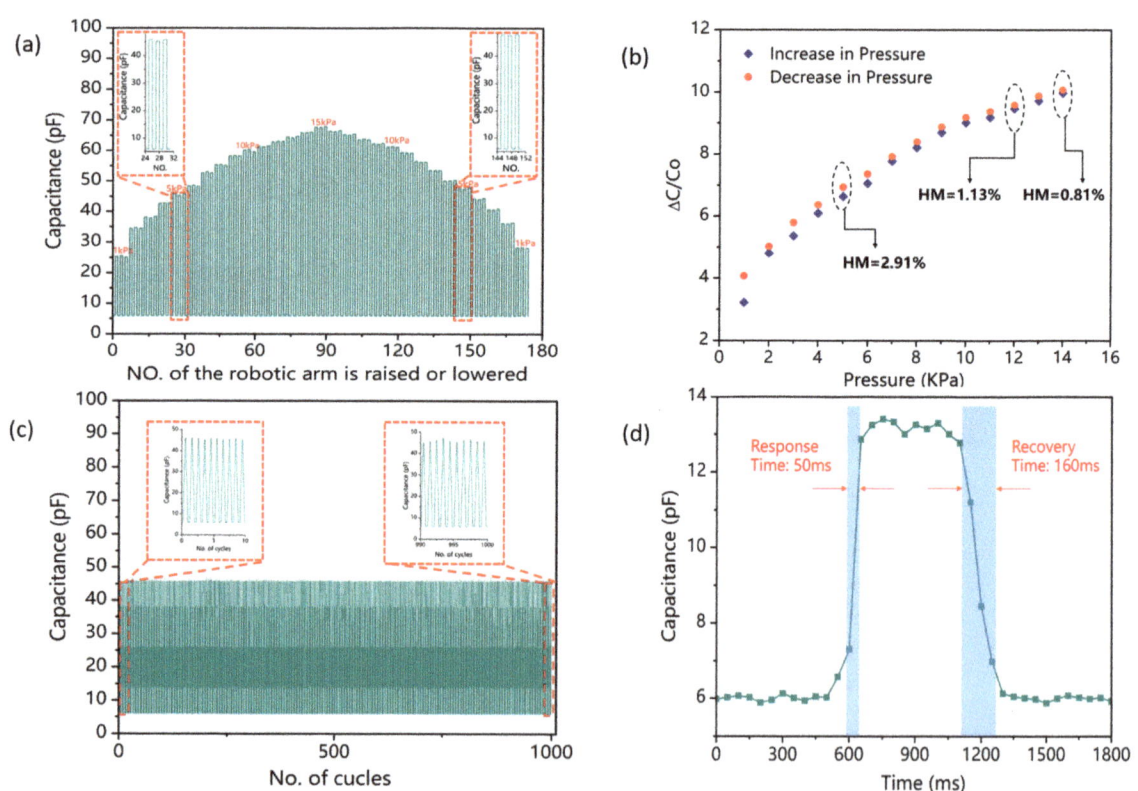

Figure 6. (**a**) Stepwise pressure response of GMCA pressure sensor. (**b**) Hysteresis of GMCA pressure sensor. (**c**) Repeatability of GMCA pressure sensor for 1000 cycles at 5 KPa. (**d**) Response and recovery time of GMCA pressure sensor at 100 Pa of applied pressure.

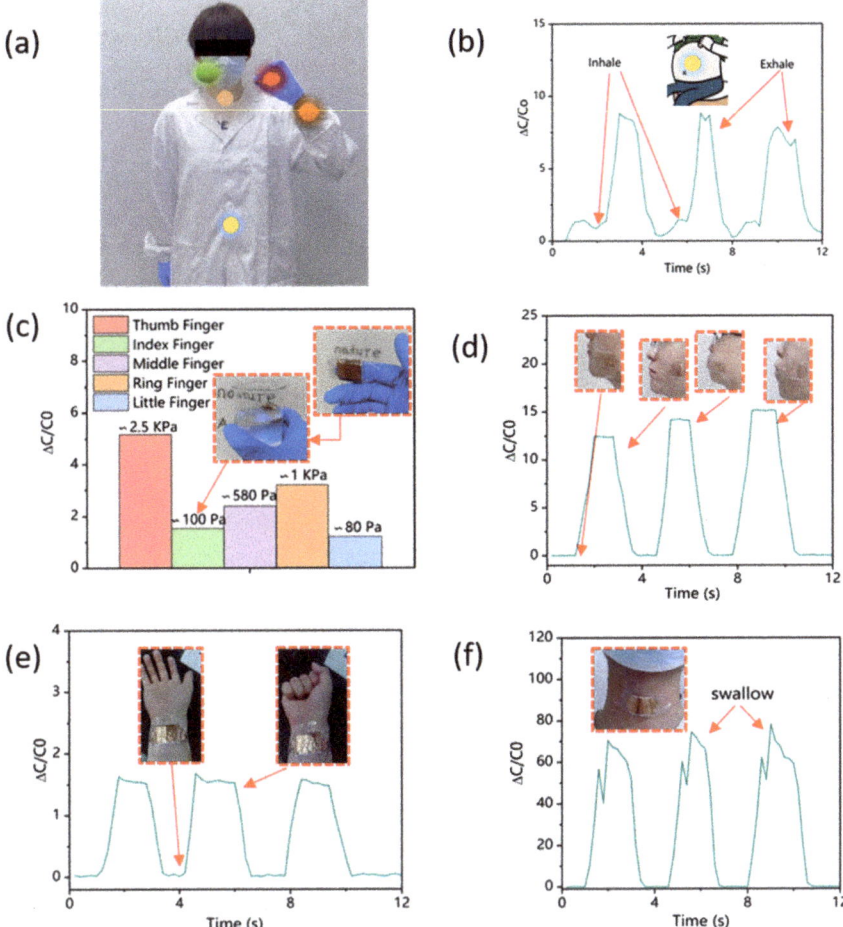

Figure 7. (**a**) Monitoring human motion with the GMCA pressure sensor at different sites. (**b**) The capacitance changes in abdominal pressure during breathing. (**c**) The relative capacitance changes of sensor for thumb, index, middle, ring, and little finger while holding the beaker. (**d**) GMCA sensor recognizes facial expressions. (**e**) Capacitive response of pressure sensor for monitoring hand closing and opening gesture. (**f**) GMCA pressure sensor detects the swallowing movements.

3.4. Repeatability

The repeatability test was performed on the pressure sensor for a 1000 loading and unloading cycles of 5 KPa pressure. Figure 6c shows the repeatability of the sensor response; the inset shows the capacitive response of the sensor for 10 cycles between 0–10 and 990–1000 at applied pressure of 5 KPa. It was observed that the capacitance response of the pressure sensor over the 1000 cycles increased from a base value of ~6.0 pF to ~45.9 pF, demonstrating a maximum change of 665 ± 20%. From the results, it can be concluded that the pressure sensor demonstrated high repeatability and durability. It can be inferred that our sensor possessed a robust operational lifespan and was capable of withstanding the conditions associated with multiple repetitive tests in a general usage environment.

3.5. Response and Recovery Time

The response and the recovery time of the GMCA pressure sensor was measured by subjecting it to 100 Pa applied pressure. The capacitance of the pressure sensor was increased from 6 pF (base value) to 13.25 pF resulting in a relative capacitance change of 121%. The response time of the pressure sensor (time taken for the capacitance to reach from 10% (T10% = 595 ms) to 90% (T90% = 645 ms) of the total capacitance change (7.25 pF) was calculated to be 50 ms. Similarly, the 100 Pa applied pressure was removed on the pressure sensor and a recovery time of 160 ms was obtained (Figure 6d).

3.6. Comparison of the Performance Indicators of Various Microstructure Capacitive Pressure Sensors

Table 3 presents a comprehensive overview of the performance indicators exhibited by diverse microstructure capacitive pressure sensors thus far. The dimensions, aspect ratio, and magnitude of the microstructure play a pivotal role in influencing several performance indicators, such as sensitivity, hysteresis effect, and repeatability. This research endeavor successfully attained heightened sensitivity and reduced response time across a broad pressure spectrum by incorporating high dielectric constant $BaTiO_3$ particles and implementing gradient heights. Furthermore, the gradient micro-structure-based sensor developed in this study demonstrated applicability in pressure sensing up to 50 KPa.

Table 3. Summary of the performance indicators of various microstructure capacitive pressure sensors.

Micro-Structure	Fabrication Method	Mechanism	Pressure Range	Sensitivity	Response Time	References
Tilted micropillar	Photolithography	Capacitive	0–40 KPa	0.42 KPa^{-1}	70 ms (8 KPa)	[58]
Semi sphere	Electroless plating	Capacitive	0–10 KPa	0.13 KPa^{-1}	–	[59]
Micro-square	Photolithography	Capacitive	0.5 Pa–3 KPa	0.185 KPa^{-1}	–	[60]
Micro-porous	Chemical method	Capacitive	<0.02 KPa	1.18 KPa^{-1}	150 ms (0.6 KPa)	[61]
Micro-pyramid	Laser patterning	Capacitive	0–0.1 KPa 0.1–1 KPa 1–10 KPa	0.221% Pa^{-1} 0.033% Pa^{-1} 0.011% Pa^{-1}	50 ms (0.02 KPa)	[51]
Gradient micro-cone	Laser engraving and molding	Capacitive	0–2 KPa 2–15 KPa 15–50 KPa	1.69 KPa^{-1} 0.41 KPa^{-1} 0.09 KPa^{-1}	50 ms (0.1 KPa)	This work

3.7. Application Demonstration

The potential application of the GMCA pressure sensor in wearable robotics was examined as an application demonstrator. The results, depicted in Figure 7a–f, indicate that GMCA pressure sensors have the ability to detect various human movement behaviors, including facial expressions, breathing, grasping heavy objects, swallowing, and more. Specifically, Figure 7b displays the alterations in abdominal pressure during breathing. To measure these changes, the GMCA sensor was affixed to the abdomen of a volunteer who then performed breathing exercises. The compression of the sensor during exhalation and inhalation allowed for the measurement of corresponding changes in capacitance. Hold a beaker containing water using a finger connected to a GMCA pressure sensor and measure the alteration in its capacitance. Figure 7c illustrates the variation in capacitance response among different fingers, revealing that the thumb, when gripping a heavy object, exhibited the greatest change, suggesting that it experienced the highest pressure. The capacitance of the GMCA pressure sensor fluctuated rapidly and then stabilized as facial expressions altered (cheeks protrude), remaining relatively constant. Once the expression returned to its original state, the capacitance promptly reverted to its initial value (Figure 7d). The GMCA pressure sensor was affixed to the wrist of participants in order to monitor the capacitance response when performing hand opening and closing gestures. As depicted in Figure 7e, the observed change in capacitance during these gestures corresponded to a pressure of approximately 0.1 KPa. Furthermore, the GMCA pressure sensor was capable of detecting swallowing movements, with the capacitance changes observed during three distinct swallowing movements being largely consistent (Figure 7f). The experimental findings presented herein showcase the extensive potential of GMCA pressure sensors in the domains of biosensing and wearable sensing devices for human–machine interfaces.

4. Conclusions

This study presented a novel capacitive pressure sensor based on laser-assisted engraving technology, referred to as GMCA sensor. The dielectric layer of the sensor was fabricated using $BaTiO_3$-PDMS material with a gradient structure, aiming to enhance the sensitivity and linearity of the sensor. Furthermore, the impact of incorporating BaTiO3-mixed PDMS material on the performance of the sensor was investigated, and a mass ratio of 10:1 was identified as the optimal configuration. The performance of the GMCA pressure sensor was evaluated by subjecting it to a range of pressures from 0 to 10 KPa. The pressure sensor demonstrated a sensitivity of 1.69 KPa^{-1}, 0.41 KPa^{-1}, and 0.09 KPa^{-1} within the pressure ranges of 0 Pa to 2 KPa, 2 KPa to 15 KPa, and 15 KPa to 50 KPa, respectively. The utilization of flexible PDMS facilitated the pressure sensor to showcase notable attributes such as a rapid response time of 50 ms, minimal hysteresis, a recovery time of 160ms, as well as exceptional repeatability and durability across a broad pressure spectrum. The findings indicate that the GMCA pressure sensor, which utilized a gradient structure and $BaTiO_3$-PDMS material, exhibited notable sensitivity and a broad linear pressure range. These outcomes underscore the adaptability and viability of this technology, thereby enabling enhanced flexibility in pressure sensors and fostering advancements in laser manufacturing and flexible devices for a wider range of potential applications. Furthermore, it is anticipated that our GMCA-based capacitive sensor design holds potential for utilization as a wearable sensor for health monitoring and noninvasive detection on the skin, offering promising strategies for future applications. Subsequent investigations will encompass the examination of environmental variables, such as humidity and temperature, and their influence on the performance of the sensor.

Author Contributions: Conceptualization, J.L. and B.S.; methodology, J.L.; software, J.Z.; validation, J.L., B.S. and S.C.; formal analysis, J.L.; investigation, J.L.; resources, B.S.; data curation, J.L.; writing—original draft preparation, J.L. and B.S.; writing—review and editing, J.L. and B.S.; visualization, L.T. and C.J.; supervision, D.Z. and B.S.; project administration, B.S.; funding acquisition, B.S. All authors have read and agreed to the published version of the manuscript.

Funding: This research was funded by the Natural Science Foundation of Shanghai (19ZR1436100) and the National Natural Science Foundation of China (11105149).

Institutional Review Board Statement: Not applicable.

Data Availability Statement: Data will be made available on request.

Acknowledgments: The authors express gratitude to the editors and the reviewers for their constructive and helpful review comments.

Conflicts of Interest: The authors declare no conflict of interest.

References

1. Araromi, O.A.; Graule, M.A.; Dorsey, K.L. Ultra-Sensitive and Resilient Compliant Strain Gauges for Soft Machines. *Nature* **2020**, *587*, 219–224. [CrossRef]
2. Zhan, Z.; Lin, R.; Tran, V.-T.; An, J.; Wei, Y.; Du, H.; Tran, T.; Lu, W. Self-Powered Nanofluidic Pressure Sensor with A Linear Transfer Mechanism. *ACS Appl. Mater. Interfaces* **2017**, *9*, 37921–37928. [CrossRef] [PubMed]
3. Pyo, S.; Lee, J.; Bae, K.; Sim, S.; Kim, J. Recent Progress in Flexible Tactile Sensors for Human-Interactive Systems: From Sensors to Advanced Applications. *Adv. Mater.* **2021**, *33*, 2005902. [CrossRef] [PubMed]
4. Wadm, J.; Qi, K.; Qin, Y.; Chin, A.; Serrano-García, W.; Baskar, C.; Wang, H.; He, J.; Cui, S. Significance of Nanomaterials in Wearables: A Review on Wearable Actuators and Sensors. *Adv. Mater.* **2019**, *31*, e1805921. [CrossRef]

5. Pang, Y.; Zhang, K.; Yang, Z.; Jiang, S.; Ju, Z.; Li, Y.; Wang, X.; Wang, D.; Jian, M.; Zhang, Y.; et al. Epidermis Microstructure Inspired Graphene Pressure Sensor with Random Distributed Spinosum for High Sensitivity and Large Linearity. *ACS Nano* **2018**, *12*, 2346–2354. [CrossRef]
6. Wang, X.; Liu, Z.; Zhang, T. Flexible Sensing Electronics for Wearable/Attachable Health Monitoring. *Small* **2017**, *13*, 1602790. [CrossRef]
7. Yao, S.; Ren, P.; Song, R.; Liu, Y.; Huang, Q.; Dong, J.; O'Connor, B.T.; Zhu, Y. Nanomaterial-Enabled Flexible and Stretchable Sensing Systems: Processing, Integration, and Applications. *Adv. Mater.* **2020**, *32*, 1902343. [CrossRef]
8. Soni, M.; Dahiya, R. Soft eSkin: Distributed Touch Sensing with Harmonized Energy and Computing. *Philos. Trans. Royal Soc. A Math. Phys. Eng. Sci.* **2020**, *378*, 2164. [CrossRef]
9. Wang, N.; Tong, J.; Wang, J.; Wang, Q.; Chen, S.; Sheng, B. Polyimide-Sputtered and Polymerized Films with Ultrahigh Moisture Sensitivity for Respiratory Monitoring and Contactless Sensing. *ACS Appl. Mater. Interfaces* **2022**, *14*, 11842–11853. [CrossRef]
10. Ji, B.; Zhou, Q.; Hu, B.; Zhong, J.; Zhou, J.; Zhou, B. Bio-Inspired Hybrid Dielectric for Capacitive and Triboelectric Tactile Sensors with High Sensitivity and Ultrawide Linearity Range. *Adv. Mater.* **2021**, *33*, 2100859. [CrossRef]
11. Ruth, S.R.A.; Feig, V.R.; Tran, H.; Bao, Z. Microengineering Pressure Sensor Active Layers for Improved Performance. *Adv. Funct. Mater.* **2020**, *30*, 2003491. [CrossRef]
12. Tian, H.; Shu, Y.; Wang, X.-F.; Ali Mohammad, M.; Bie, Z.; Xie, Q.-Y.; Li, C.; Mi, W.-T.; Yang, Y.; Ren, T.-L. A Graphene-Based Resistive Pressure Sensor with Recordhigh Sensitivity in a Wide Pressure Range. *Sci. Rep.* **2015**, *5*, 8603. [CrossRef] [PubMed]
13. Xiao, F.; Jin, S.; Zhang, W.; Zhang, Y.; Zhou, H.; Huang, Y. Wearable Pressure Sensor Using Porous Natural Polymer Hydrogel Elastomers with High Sensitivity Over a Wide Sensing Range. *Polymers* **2023**, *15*, 2736. [CrossRef] [PubMed]
14. Ali, S.; Maddipatla, D.; Narakathu, B.-B.; Chlaihawi, A.-A.; Emamian, S.; Janabi, F. Flexible Capacitive Pressure Sensor Based on PDMS Substrate and Ga–In Liquid Metal. *IEEE Sens. J.* **2019**, *19*, 97–104. [CrossRef]
15. Mitrakos, V.; Macintyre, L.; Denison, F.; Hands, P.; Desmulliez, M. Design. Manufacture and Testing of Capacitive Pressure Sensors for Low-Pressure Measurement Ranges. *Micromachines* **2017**, *8*, 41. [CrossRef]
16. Ke, K.; Sang, Z.; Manas-Zloczower, I. Hybrid Systems of Three-Dimensional Carbon Nanostructures with Low Dimensional Fillers for Piezoresistive Sensors. *Polym. Compos.* **2020**, *41*, 468–477. [CrossRef]
17. Xiao, J.; Xiong, Y.; Chen, J.; Zhao, S.; Chen, S.; Xu, B.; Sheng, B. Ultrasensitive and Highly Stretchable Fibers with Dual Conductive Microstructural Sheaths for Human Motion and Micro Vibration Sensing. *Nanoscale* **2022**, *14*, 1962–1970. [CrossRef]
18. Kang, S.; Lee, J.; Lee, S.; Kim, S.; Kim, J.-K.; Algadi, H.; Al-Sayari, S.; Kim, D.-E.; Kim, D.; Lee, T. Highly Sensitive Pressure Sensor Based on Bioinspired Porous Structure for Real-Time Tactile Sensing. *Adv. Electron. Mater.* **2016**, *2*, 1600356. [CrossRef]
19. Niu, H.; Yue, W.; Li, Y.; Yin, F.; Gao, S.; Zhang, C.; Kan, H.; Yao, Y.; Jiang, C.; Wang, C. Ultrafast-Response/Recovery Capacitive Humidity Sensor Based on Arc-Shaped Hollow Structure with Nanocone Arrays for Human Physiological Signals Monitoring. *Sens. Actuators B-Chem.* **2021**, *334*, 129637. [CrossRef]
20. Mao, Y.; Robust, B.J. Wearable Pressure Sensor Assembled from Agnw-Coated PDMS Micropillar Sheets with High Sensitivity and Wide Detection Range. *ACS Appl. Nano Mater.* **2019**, *2*, 3196–3205. [CrossRef]
21. Ha, K.-H.; Huh, H.; Li, Z.; Lu, N. Soft Capacitive Pressure Sensors: Trends, Challenges, and Perspectives. *ACS Nano* **2022**, *16*, 3442–3448. [CrossRef] [PubMed]
22. Wang, J.; Suzuki, R.; Shao, M.; Gillot, F.; Shiratori, S. Capacitive Pressure Sensor with Wide-Range, Bendable, and High Sensitivity Based on the Bionic Komochi Konbu Structure and Cu/Ni Nanofiber Network. *ACS Appl. Mater. Interfaces* **2019**, *11*, 11928–11935. [CrossRef]
23. Bai, N.; Wang, L.; Wang, Q.; Deng, J.; Wang, Y.; Lu, P.; Huang, J.; Li, G.; Zhang, Y.; Yang, J.; et al. Graded Intrafillable Architecture-Based Iontronic Pressure Sensor with Ultra-Broad-Range High Sensitivity. *Nat. Commun.* **2020**, *11*, 209. [CrossRef] [PubMed]
24. Ji, B.; Zhou, Q.; Wu, J.; Gao, Y.; Wen, W.; Zhou, B. Synergistic Optimization Toward the Sensitivity and Linearity of Flexible Pressure Sensor Via Double Conductive Layer and Porous Microdome Array. *ACS Appl. Mater. Interfaces* **2020**, *12*, 31021–31035. [CrossRef] [PubMed]
25. Ji, B.; Zhou, Q.; Lei, M.; Ding, S.; Song, Q.; Gao, Y.; Li, S.; Xu, Y.; Zhou, Y. Gradient Architecture-Enabled Capacitive Tactile Sensor with High Sensitivity and Ultrabroad Linearity Range. *Small* **2021**, *17*, 2103312. [CrossRef] [PubMed]
26. Pan, L.; Chortos, A.; Yu, G.; Wang, Y.; Isaacson, S.; Allen, R.; Shi, Y.; Dauskardt, R.; Bao, Z. An Ultra-Sensitive Resistive Pressure Sensor Based on Hollow-Sphere Microstructure Induced Elasticity in Conducting Polymer Film. *Nat. Commun.* **2013**, *5*, 3002. [CrossRef]
27. Luo, Y.; Shao, J.; Chen, S.; Chen, X.; Tian, H.; Li, X.; Wang, L.; Wang, D.; Lu, B. Flexible Capacitive Pressure Sensor Enhanced by Tilted Micropillar Arrays. *ACS Appl. Mater. Interfaces* **2019**, *11*, 17796–17803. [CrossRef]
28. Masihi, S.; Panahi, M.; Maddipatla, D.; Bose, A.-K.; Palaniappan, V.; Narakathu, B.-B.; Hanson, A.-J. A Novel Printed Fabric Based Porous Capacitive Pressure Sensor for Flexible Electronic Applications. In Proceedings of the IEEE Sensors, Montreal, QC, Canada, 27–30 October 2019; pp. 1–4. [CrossRef]
29. Luo, S.; Yang, J.; Song, X.; Zhou, X.; Yu, L.; Sun, T.; Yu, C.; Huang, D.; Du, C.; Wei, D. Tunable-Sensitivity Flexible Pressure Sensor Based on Graphene Transparent Electrode. *Solid-State Electron.* **2018**, *145*, 29–33. [CrossRef]
30. Kim, H.; Kim, G.; Kim, T.; Lee, S.; Kang, D.; Hwang, M.-S.; Chae, Y.; Kang, S.; Lee, H.; Park, H.-G.; et al. Transparent, Flexible, Conformal Capacitive Pressure Sensors with Nanoparticles. *Small* **2018**, *14*, 1703432. [CrossRef]

31. Li, F.; Kong, Z.; Wu, J.-H.; Ji, X.-Y.; Liang, J. Advances in Flexible Piezoresistive Pressure Sensor. *Acta Phys. Sin.* **2021**, *70*, 100703. [CrossRef]
32. Hao, D.; Yang, R.; Yi, N. Highly sensitive piezoresistive pressure sensors based on laser-induced graphene with molybdenum disulfide nanoparticles. *Sci. China Technol.* **2021**, *64*, 2408–2414. [CrossRef]
33. Zhou, Q.; Ji, B.; Wei, Y.; Hu, B.; Gao, Y.; Xu, Q.; Zhou, J.; Zhou, B. A Bio-Inspired Cilia Array as The Dielectric Layer for Flexible Capacitive Pressure Sensors with High Sensitivity and a Broad Detection Range. *J. Mater. Chem. A* **2019**, *7*, 27334–27346. [CrossRef]
34. Mannsfeld, S.; Tee, B.; Stoltenberg, R.; Chen, C.; Barman, S.; Muir, B.; Sokolov, A.; Reese, C.; Bao, Z. Highly Sensitive Flexible Pressure Sensors with Microstructured Rubber Dielectric Layers. *Nat. Mater.* **2010**, *9*, 859–864. [CrossRef]
35. Wang, D.; Sheng, B.; Peng, L.; Huang, Y.; Ni, Z. Flexible and Optical Fiber Sensors Composited by Graphene and PDMS for Motion Detection. *Polymers* **2019**, *11*, 1433. [CrossRef]
36. Wang, Q.; Tong, J.; Wang, N.; Chen, S.; Sheng, B. Humidity Sensor of Tunnel-Cracked Nickel@Polyurethane Sponge for Respiratory and Perspiration Sensing. *Sens. Actuators B Chem.* **2021**, *330*, 129322. [CrossRef]
37. Jiang, N.; Chang, X.; Hu, D.; Chen, L.; Wang, Y.; Chen, J.; Zhu, Y. Flexible, Transparent, and Antibacterial Ionogels Toward Highly Sensitive Strain and Temperature Sensors. *Chem. Eng. J.* **2021**, *424*, 130418. [CrossRef]
38. Qin, J.; Yin, L.-J.; Hao, Y.-N.; Zhong, S.-L.; Zhang, D.-L.; Bi, L.; Zhang, Y.-X.; Zhao, Y.; Dang, Z.-M. Flexible and Stretchable Capacitive Sensors with Different Microstructures. *Adv. Mater.* **2021**, *33*, e2008267. [CrossRef]
39. Choi, J.; Kwon, D.; Kim, K.; Park, J.; Orbe, D.-D.; Gu, J.; Ahn, J.; Cho, I.; Jeong, Y.; Oh, Y.; et al. Synergetic Effect of Porous Elastomer and Percolation of Carbon Nanotube Filler toward High Performance Capacitive Pressure Sensors. *ACS Appl. Mater. Interfaces* **2020**, *12*, 1698–1706. [CrossRef]
40. Cheng, W.; Wang, J.; Ma, Z.; Yan, K.; Wang, Y.; Wang, H.; Li, S.; Li, Y.; Pan, L.; Shi, Y. Flexible Pressure Sensor with High Sensitivity and Low Hysteresis Based on a Hierarchically Microstructured Electrode. *IEEE Electron. Device Lett.* **2018**, *39*, 288–291. [CrossRef]
41. Chhetry, A.; Sharma, S.; Yoon, H.; Ko, S.; Park, J.Y. Enhanced Sensitivity of Capacitive Pressure and Strain Sensor Based on CaCu$_3$Ti$_4$O$_{12}$ Wrapped Hybrid Sponge for Wearable Applications. *Adv. Funct. Mater.* **2020**, *30*, 1910020. [CrossRef]
42. Liang, J.; Zhao, Z.; Tang, Y.; Liang, Z.; Sun, L.; Pan, X.; Wang, X.; Qiu, J. A Wearable Strain Sensor Based on Carbon Derived from Linen Fabrics. *New Carbon. Mater.* **2020**, *35*, 522–530. [CrossRef]
43. Sharma, S.; Chhetry, A.; Sharifuzzaman, M.; Yoon, H.; Park, J.-Y. Wearable Capacitive Pressure Sensor Based on MXene Composite Nanofibrous Scaffolds for Reliable Human Physiological Signal Acquisition. *ACS Appl. Mater. Interfaces* **2020**, *12*, 22212–22224. [CrossRef]
44. Arlt, G.; Hennings, D.; With, G.-D. Dielectric Properties of Fine-Grained Barium Titanate Ceramics. *J. Appl. Phys.* **1985**, *58*, 1619–1625. [CrossRef]
45. Li, T.; Luo, H.; Qin, L.; Wang, X.; Xiong, Z.; Ding, H.; Gu, Y.; Liu, Z.; Zhang, T. Flexible Capacitive Tactile Sensor Based on Micropatterned Dielectric Layer. *Small* **2016**, *12*, 5042–5048. [CrossRef]
46. Maddipatla, D.; Narakathu, B.; Turkani, V.-S. A Flexible Copper Based Electrochemical Sensor Using Laser-Assisted Patterning Process. In Proceedings of the IEEE Sensors, New Delhi, India, 28–31 October 2018; pp. 1–4. [CrossRef]
47. Shelly, M.; Lee, S.; Suarato, G.; Meng, Y.; Pautot, S. *Photolithography-Based Substrate Microfabrication for Patterning Semaphorin 3A to Study Neuronal Development, in Semaphorin Signaling*; Humana Press: New York, NY, USA, 2017; pp. 321–343.
48. Tee, B.C.-K.; Chortos, A.; Dunn, R.R.; Schwartz, G.; Eason, E.; Bao, Z. Tunable Flexible Pressure Sensors using Microstructured Elastomer Geometries for Intuitive Electronics. *Adv. Funct. Mater.* **2014**, *24*, 5427–5434. [CrossRef]
49. Luo, S.; Zhou, X.; Tang, X.; Li, J.; Wei, D.; Tai, G.; Chen, Z.; Liao, T.; Fu, J.; Wei, D.; et al. Microconformal Electrode-Dielectric Integration for Flexible Ultrasensitive Robotic Tactile Sensing. *Nano Energy* **2021**, *80*, 105580. [CrossRef]
50. Yang, J.-C.; Kim, J.-O.; Oh, J.; Kwon, S.-Y.; Sim, J.-Y.; Kim, D.-W.; Choi, H.-B.; Park, S. Microstructured Porous Pyramid-Based Ultrahigh Sensitive Pressure Sensor Insensitive to Strain and Temperature. *ACS Appl. Mater. Interfaces* **2019**, *11*, 19472–19480. [CrossRef]
51. Palaniappan, V.; Panahi, M.; Masihi, S.; Maddipatla, D.; Bose, A.-K. Laser-Assisted Fabrication of a Highly Sensitive and Flexible Micro Pyramid-Structured Pressure Sensor for E-Skin Applications. *IEEE Sens. J.* **2020**, *20*, 7605–7613. [CrossRef]
52. Wang, J.; Wang, N.; Xu, D.; Tang, L.; Sheng, B. Flexible Humidity Sensors Composed With Electrodes of Laser Induced Graphene and Sputtered Sensitive Films Derived from Poly(Ether-Ether-Ketone). *Sens. Actuators B Chem.* **2023**, *375*, 132846. [CrossRef]
53. Shi, S.; Liang, J.; Qu, C.; Chen, S.; Sheng, B.; Fabric, R. Treated with Carboxymethylcellulose and Laser Engraved for Strain and Humidity Sensing. *Micromachines* **2022**, *13*, 1309. [CrossRef]
54. Bose, A.-K.; Beaver, C.L.; Narakathu, B.; Rossbach, S.; Bazuin, B.-J.; Atashbar, M.-Z. Development of Flexible Microplasma Discharge Device for Sterilization Applications. In Proceedings of the IEEE Sensors, New Delhi, India, 28–31 October 2018; pp. 1–4. [CrossRef]
55. Mannion, P.-T.; Magee, J.; Coyne, E.; O'Connor, G.-M.; Glynn, T.-J. The Effect of Damage Accumulation Behaviour on Ablation Thresholds and Damage Morphology in Ultrafast Laser Micro-Machining of Common Metals in Air. *Appl. Surf. Sci.* **2004**, *233*, 275–287. [CrossRef]
56. Qu, C.; Lu, M.; Zhang, Z.; Chen, S.; Liu, D.; Zhang, D.; Wang, J.; Sheng, B. Flexible Microstructured Capacitive Pressure Sensors Using Laser Engraving and Graphitization from Natural Wood. *Molecules* **2023**, *28*, 5339. [CrossRef] [PubMed]
57. Kinoshita, K.; Yamaji, A. Grain-Size Effects on Dielectric Properties in Barium Titanate Ceramics. *J. Appl. Phys.* **1976**, *47*, 371–373. [CrossRef]

58. Shi, H.; Pinto, T.; Zhang, Y.; Wang, C.; Tan, X. Screen-Printed Soft Capacitive Sensors for Spatial Mapping of Both Positive and Negative Pressures. *Adv. Funct. Mater.* **2019**, *29*, 1809116. [CrossRef]
59. Kim, Y.; Yang, H.; Oh, J.-H. Simple Fabrication of Highly Sensitive Capacitive Pressure Sensors Using a Porous Dielectric Layer with Coneshaped Patterns. *Mater. Des.* **2020**, *197*, 109203. [CrossRef]
60. Zeng, X.; Wang, Z.; Zhang, H.; Yang, W.; Xiang, L.; Zhao, Z.; Peng, L.-M.; Hu, Y. Tunable, Ultrasensitive, and Flexible Pressure Sensors Based on Wrinkled Microstructures for Electronic Skins. *ACS Appl. Mater. Inter.* **2019**, *11*, 21218–21226. [CrossRef]
61. Lee, B.-Y.; Kim, J.; Kim, H.; Kim, C.; Lee, S.-D. Low-Cost Flexible Pressure Sensor Based on Dielectric Elastomer Film with Micro-Pores. *Sens. Actuators A Phys.* **2016**, *240*, 103–109. [CrossRef]

Disclaimer/Publisher's Note: The statements, opinions and data contained in all publications are solely those of the individual author(s) and contributor(s) and not of MDPI and/or the editor(s). MDPI and/or the editor(s) disclaim responsibility for any injury to people or property resulting from any ideas, methods, instructions or products referred to in the content.

Article

A Modified Constitutive Model for Isotropic Hyperelastic Polymeric Materials and Its Parameter Identification

Wei Wang, Yang Liu * and Zongwu Xie

State Key Laboratory of Robotics and Systems, Harbin Institute of Technology, Harbin 150001, China; wangwei1023@stu.hit.edu.cn (W.W.); xiezongwu@hit.edu.cn (Z.X.)
* Correspondence: liuyanghit@hit.edu.cn

Abstract: Given the importance of hyperelastic constitutive models in the design of engineering components, researchers have been developing the improved and new constitutive models in search of a more accurate and even universal performance. Here, a modified hyperelastic constitutive model based on the Yeoh model is proposed to improve its prediction performance for multiaxial deformation of hyperelastic polymeric materials while retaining the advantages of the original Yeoh model. The modified constitutive model has one more correction term than the original model. The specific form of the correction term is a composite function based on a power function represented by the principal stretches, which is derived from the corresponding residual strain energy when the Yeoh model predicts the equibiaxial mode of deformation. In addition, a parameter identification method based on the cyclic genetic-pattern search algorithm is introduced to accurately obtain the parameters of the constitutive model. By applying the modified model to the experimental datasets of various rubber or rubber-like materials (including natural unfilled or filled rubber, silicone rubber, extremely soft hydrogel and human brain cortex tissue), it is confirmed that the modified model not only possesses a significantly improved ability to predict multiaxial deformation, but also has a wider range of material applicability. Meanwhile, the advantages of the modified model over most existing models in the literatures are also demonstrated. For example, when characterizing human brain tissue, which is difficult for most existing models in the literature, the modified model has comparable predictive accuracy with the third-order Ogden model, while maintaining convexity in the corresponding deformation domain. Moreover, the effective prediction ability of the modified model for untested equi-biaxial deformation of different materials has also been confirmed using only the data of uniaxial tension and pure shear from various datasets. The effective prediction for the untested equibiaxial deformation makes it more suitable for the practice situation where the equibiaxial deformation of certain polymeric materials is unavailable. Finally, compared with other parameter identification methods, the introduced parameter identification method significantly improves the predicted accuracy of the constitutive models; meanwhile, the uniform convergence of introduced parameter identification method is also better.

Keywords: hyperelastic; polymeric materials; constitutive model; strain energy density function; parameter identification

Citation: Wang, W.; Liu, Y.; Xie, Z. A Modified Constitutive Model for Isotropic Hyperelastic Polymeric Materials and Its Parameter Identification. *Polymers* **2023**, *15*, 3172. https://doi.org/10.3390/polym15153172

Academic Editors: Fahmi Zaïri, Jiangtao Xu and Sihang Zhang

Received: 14 April 2023
Revised: 24 June 2023
Accepted: 24 July 2023
Published: 26 July 2023

Copyright: © 2023 by the authors. Licensee MDPI, Basel, Switzerland. This article is an open access article distributed under the terms and conditions of the Creative Commons Attribution (CC BY) license (https://creativecommons.org/licenses/by/4.0/).

1. Introduction

Hyperelastic polymeric materials, which possess exceptional hyperelastic properties, have been widely utilized in various fields such as medical devices [1], flexible electrodes [2], and soft robots [3]. The widespread applications of hyperelastic polymeric materials have also spurred a lot of research into characterizing their hyperelastic properties [4]. In particular, carrying out the analysis of the 3D stress–strain of complex elastic components relying on finite element analysis (FEA) has become an indispensable part in the process of product design in recent decades. However, the reliability and accuracy of results strongly depend on the performance of the constitutive model predicting the mechanical behavior of the material. Therefore, a constitutive model that can more accurately predict the mechanical behavior of material is the key to making the analytical results based on finite element analysis more realistic.

Currently, constitutive models characterizing hyperelastic properties are usually built by the strain energy density function. In light of the different ways of constructing the strain energy density function, the hyperelastic constitutive models can be divided into two overall categories: the first category encompass statistical mechanics models based on the idealized network structure of molecular chains, while the other category comprise phenomenological models based on continuum mechanics.

For statistical mechanical models, it is usually constructed based on the assumption of the network structure of molecular chains inside the polymer [5]. Although the parameters of these models usually have practical physical context, the statistical mechanical models with better representational ability often have a more complex form. So, they are not very good for providing an analytical solution, and they are not good for numerical solutions [6], as well as having a large corresponding calculation [7]. In addition, some statistical mechanical models are not well suited for dealing with some important issues such as irreversible deformation and inelastic volume expansion [8].

For phenomenological models, they are usually constructed based on the strain invariants or principal stretches. Although the parameters of these phenomenological models often do not have practical physical context, they have been widely studied and applied because of their relative advantages (including easy-to-obtain parameters, relatively high computational efficiency, no requirement for understanding the microstructure of the materials, and wider material applicability). The Mooney model is the earliest constitutive model used to represent the mechanical behavior of polymeric rubber. It has a linearized form, and it usually has better performance in the case with small deformation [9]. Taking into account the limited capabilities of the Mooney model, Rivlin extended it in the form of polynomial series to obtain a generalized polynomial model [10]. Because the polynomial model has higher-order terms of the strain invariants, it is more suitable for the case with large deformation. However, the model with higher-order terms usually requires more parameters, which easily make the model unstable. In line with the structure of polynomial model, researchers explored different orders and combinations of strain invariants, and derived many other models [11–14]. In particular, when the second strain invariant in the polynomial model is ignored, the reduced polynomial model is obtained [15]. Even though the polynomial model and reduced polynomial model can well characterize the hyperelastic properties of filled and unfilled rubber by retaining higher-order terms [16], they may have difficulties in solving numerical problems [17]. Therefore, the second-order polynomial model or the third-order reduced polynomial model is generally used. Except for some of the above constitutive models on account of the polynomial form of strain invariants, there are also some constitutive models based on the logarithm or exponential form of strain invariants. Notwithstanding, comparative studies show that these models do not have a particular advantage in characterizing hyperelastic properties [18,19]. Besides, the eigenvalues of the strain tensor are related to the principal stretches, so there are also some constitutive models constructed directly based on the principal stretches [20–22]. These models are usually composed of special functions related to the principal stretches, and they show good performance in characterizing the hyperelastic properties.

Although there are already many classical constitutive models to characterize the hyperelastic property, there are still continuously improved and completely new models emerging. The reason for this may be that the deficiencies of existing models still exist [23] and the importance of hyperelastic models in designing engineering components still motivates researchers to develop more general and robust constitutive models [24,25]. As presented in some reviews [7,26], the performances of these existing models are different from each other, and not all models can effectively characterize the multiaxial deformation of hyperelastic material based solely on a single set of model parameters, and most of these models are applied to a specific type of hyperelastic material. Furthermore, many models with relatively few parameters cannot reliably predict the whole range of strain and different modes of deformation. Considering that hyperelastic materials often exhibit multiaxial states of deformation in practical applications, it is unquestionable that the ability of a model to characterize multiaxial deformation should be evaluated first. However, evaluating the ability of a model to characterize multiaxial deformation often requires the simultaneous use of experimental data from three modes of homogenous deformation (including uniaxial tension, equibiaxial tension, and pure shear) to calibrate the parameters of the model. This may be prohibitively expensive and even infeasible in practice [15]. Due to limited hardware conditions or limitations in the tensile strength of the target hyperelastic material, some pure deformation mode data of materials are simply not measurable, especially for the equibiaxial tension. So, there are also some researchers evaluating the performance of existing models from another dimension. As presented in the comparative studies in [15,27], the parameters of these models were calibrated only using data from a single pure mode of deformation (such as uniaxial tension). The corresponding comparison results highlight the effective characterization ability of Yeoh model [28] for untested deformation modes (exceeding the well-known Ogden model [21]). Although some studies believe that it may not be reliable for a model only calibrated through the data of a single mode to describe the multiaxial mode of deformation [6,29], as an auxiliary evaluation method, it is also being pursued by researchers [18,30–32]. Especially in [3], researchers refer to the model's ability to effectively predict the untested mode of deformation as the property of deformation-mode independency. As described above, this property is particularly useful for real-world engineering applications where the available experimental data are limited [32].

Considering the pursuit of a universal and robust hyperelastic constitutive model, this study aims to improve the Yeoh model to compensate for its shortcoming in performance and applicability. The idea of focusing on improving the Yeoh model is not arbitrary, but rather based on its unique advantages in characterizing untested modes of deformation as explained above. The reason why we want to take advantage of this advantage is because we have indeed encountered the situation in practice where the equibiaxial deformation mode of a used polymer material is not measurable. With the widespread application of different soft polymer materials in engineering design, we believe that the probability of this situation occurring will be higher. So, the actual demand also drives us to carry out this work. Furthermore, it is well known that the Yeoh model will underestimate the equibiaxial mode of deformation when characterizing multiaxial state of stress; hence, it should be further improved in order to have more accurate characterization results. Hitherto, there are several sporadic improved Yeoh models. Earlier, Yeoh proposed to improve the fitting accuracy of the Yeoh model to the data from simple shear by adding an additional exponential term related to the first principal invariant [33]. The comparative studies confirmed that the improved model has a poorer performance for predicting the biaxial behavior of natural rubber (using Treloar's data), and the comprehensive performance of the improved model is not as good as the original Yeoh model [34]. This may be because the improved model is still not related to the second principal invariant or the added additional exponential item is inappropriate. Relevant study has confirmed that the model containing the second principal invariant is important for improving the prediction accuracy of the model, especially for the equibiaxial deformation [35]. Based on this, a recently improved Yeoh model adds the square root of the second principal invariant as a correction term to the original Yeoh model [30]. The research results confirm that the modified model has significantly improved predictive performance for equibiaxial tension, but the entire modification process is based on uniaxial data, which may lead to insufficient improvement in the model's actual predictive ability for multiaxial deformations. In addition, it has not yet imposed effective constraints on the parameters of the modified model to ultimately make the modified model convex and stable. In addition to these two modified Yeoh models mentioned above, Hohenberger et al. also replaced all the determined orders in the Yeoh model with undetermined coefficients to expand it into a generalized Yeoh model, so as to describe the low and high strain nonlinearity of highly filled high damping rubber [36]. This generalized Yeoh model has been proven to have improved prediction accuracy for uniaxial tension and compression, but its characterization accuracy for multiaxial deformation has not been determined. To sum up, there is still a certain distance between the current modified Yeoh models and the universal and robust model; so, it is still meaningful to improve the Yeoh model again.

In this study, we use a thoughtful correction term to modify the Yeoh model, so as to improve its ability for characterizing multiaxial deformations. This correction term is derived from the corresponding residual strain energy when the Yeoh model predicts the equibiaxial mode of deformation, and its specific form is a composite function based on a power function represented by the principal stretches. The strain energy density function of the modified model is represented by the first strain invariant and the principal stretches, and it contains only five parameters to be identified. In the case of specific parameters, the model can be degraded to the neo-Hookean model, the Mooney–Rivlin model, the Yeoh model, and the Biderman model. Therefore, the modified model can also be regarded as the parent model of these four classical models. In addition, we also introduce a special parameter identification method based on the cyclic genetic-pattern search algorithm in order to improve the predicted accuracy of constitutive models. For demonstrating the modified model's robustness and generality, the modified model is applied to six different types of hyperelastic materials. During this process, in addition to comparing the performance of the modified constitutive model with some landmark constitutive models in the literatures, the stability and convexity of the modified constitutive model is also verified. Finally, the validity of the introduced parameter identification method is also confirmed by comparing it with four existing parameter identification methods.

2. Materials and Methods

2.1. Experimental Data

Experimental data involving simple modes of deformation like uniaxial tension (UT), equibiaxial tension (ET) and pure shear (PS) form the basis for parameter identification of the constitutive model. And, it is a common practice to contrast the predicted results of model and experimental data for examining the performance of the model in different ranges of deformation and different modes of deformation. Here, considering the widespread use of the experimental dataset about natural unfilled vulcanized rubber obtained by Treloar [37], we take the lead in using it to obtain the correction term of the Yeoh model. Then, we will analyze the performance of the modified model based on this dataset. Given that Treloar's dataset contains a large range of deformation, the dataset is divided into three ranges of deformation for providing a more detailed evaluation of the model's ability to describe different ranges of deformation [26]. That is, the experimental data in each mode of deformation (UT, ET and PS) will be divided into three ranges of deformation based on the principal stretch, namely, small deformation ($1 < \lambda < \lambda_{max}/3$), medium deformation ($1 < \lambda < 2\lambda_{max}/3$) and large deformation ($1 < \lambda < \lambda_{max}/3$). λ_{max}, respectively, represents the maximum principal stretch of UT, ET and PS. Although the cutoff ranges of data in different deformation ranges under the same mode of deformation are different, they all belong to the complete stress–strain curve under the corresponding deformation mode. In addition, these data with different ranges of deformation are smoothed and homogenized to obtain more data points and minimize measurement noise as much as possible, thereby obtaining more accurate results of parameter estimation. The specific data points can be found in Table S1.

Furthermore, in order to fully verify the characterization ability of the modified model for the multiaxial deformation of different materials, we also considered experimental datasets from another five different types of rubber-like materials (including isoprene vulcanized rubber [38], unfilled silicone rubber [17], poly-acrylamide hydrogel [39], carbon-black-filled styrene butadiene rubber [40] and human brain cortex tissue [41]), which have been also used in other literatures [24,25,42]. Except for the dataset of human brain cortex tissue, the other four datasets all contain data from three deformation modes of UT, ET and PS (see Tables S2–S6 in the Supporting Information for details). Overall, we used a total of six experimental datasets, which is extremely rare in other studies. Based on the six datasets, the comprehensive applicability of the modified model proposed in this study to different types of rubber-like materials will be fully demonstrated.

In addition to applying the proposed model to more material datasets, comparing the modified model proposed in this study with some commonly used or newly established models in existing studies is also beneficial for visually demonstrating the capabilities of the proposed model. To this effect, we selected a total of eight constitutive models that exist in the literature for comparison. The selection of these constitutive models for comparison is not arbitrary. Firstly, considering that this study is an improvement on the Yeoh model [28], the Yeoh model and the other three improved Yeoh models (including the modified Yeoh model proposed by Yeoh [33], generalized Yeoh model proposed by Hohenberger [36] and Melly model proposed by Melly [30]) are first added to the comparison list to verify that the improvement in this study is more effective. Secondly, in order to investigate whether the improved model in this study also holds an advantage in the current existing catalog of constitutive models, the powerful and universal third-order Ogden model [21] and the Alexander model with the best performances in the category of phenomenological models [26,43] have been added to the comparison list. Finally, under the persuasion of the reviewer, a model (we call it the Anssari–Benam model) recently proposed by him in [24] and a model (we call it the modified Anssari–Benam model) obtained by improving the former based on our correction term in our study have also been added to our comparison list. We believe that by comparing our proposed modified model with the aforementioned models on different material datasets, the performance and even advantages of our modified model will be presented more intuitively.

2.2. Preliminaries

In view of the assumptions of isotropy and incompressibility for hyperelastic polymeric materials, the constitutive equation of the isotropic and incompressible hyperelastic polymeric materials can be expressed as follows:

$$\sigma = -p\boldsymbol{I} - 2\frac{\partial W}{\partial I_2}\boldsymbol{B}^{-1} + 2\frac{\partial W}{\partial I_1}\boldsymbol{B} \tag{1}$$

where

$$\begin{aligned} I_1 &= tr\boldsymbol{B} = \lambda_1^2 + \lambda_2^2 + \lambda_3^2 \\ I_2 &= \tfrac{1}{2}\left[(tr\boldsymbol{B})^2 - tr\boldsymbol{B}^2\right] = \lambda_1^2\lambda_2^2 + \lambda_2^2\lambda_3^2 + \lambda_3^2\lambda_1^2 \end{aligned} \tag{2}$$

In the above equations, I_1 and I_2 are the first and second invariants of the left Cauchy–Green strain tensor $\boldsymbol{B}$, respectively; λ_1, λ_2 and λ_3 are principal stretches; p is an undetermined pressure which is independent of deformation; the strain energy density function W is only a function of the first and second invariants corresponding to tensor-type arguments.

For isotropic hyperelastic polymeric materials, the principal axis of the Cauchy stress tensor σ, left Cauchy–Green strain tensor $\boldsymbol{B}$ and its inverse $\boldsymbol{B}^{-1}$ are the same, so the component form of the constitutive equation of the hyperelastic polymeric materials in its two principal directions can be expressed as follows:

$$\sigma_\alpha = -p - 2\frac{\partial W}{\partial I_2}\bullet\frac{1}{\lambda_\alpha^2} + 2\frac{\partial W}{\partial I_1}\bullet\lambda_\alpha^2 \tag{3}$$

$$\sigma_\beta = -p - 2\frac{\partial W}{\partial I_2}\bullet\frac{1}{\lambda_\beta^2} + 2\frac{\partial W}{\partial I_1}\bullet\lambda_\beta^2 \tag{4}$$

where λ_α and λ_β are, respectively, the principal stretch of main direction α and β.

The strain energy density function also has the following relationships to the principal stretches:

$$\frac{\partial W}{\partial \lambda_\alpha} = 2\lambda_\alpha \frac{\partial W}{\partial I_1} - \frac{2}{\lambda_\alpha^3}\frac{\partial W}{\partial I_2} \tag{5}$$

$$\frac{\partial W}{\partial \lambda_\beta} = 2\lambda_\beta \frac{\partial W}{\partial I_1} - \frac{2}{\lambda_\beta^3}\frac{\partial W}{\partial I_2} \tag{6}$$

According to Equations (3)–(6), the following expression can be obtained by eliminating p:

$$\sigma_\alpha - \sigma_\beta = \lambda_{\overline{\alpha}}\frac{\partial W}{\partial \lambda_{\overline{\alpha}}} - \lambda_{\overline{\beta}}\frac{\partial W}{\partial \lambda_{\overline{\beta}}} \tag{7}$$

where $\overline{\alpha}$ and $\overline{\beta}$ are not summation indices; so, do not sum them up.

2.3. The Modified Strain Energy Density Function

As stated in the introduction, the third-order polynomial based on the first strain invariant (Yeoh model) not only has a simple form, but can also reproduce the inverse S-shape of the stress–strain curve of the hyperelastic polymeric materials under different modes of deformation [28] and it also has the ability to effectively predict untested deformation modes of natural unfilled vulcanized rubber [15,27]. However, it is always underestimating the equibiaxial mode of deformation (as shown in Figure 1). This underestimation will greatly reduce its ability to predict multiaxial deformation, thus hindering its application in actual engineering [30]. Some studies have shown that the second strain invariant is also important and making the model include the second strain invariant can achieve higher precision for predicting the multiaxial stress state [35]. For example, because the Biderman model has an additional term (including the second strain invariant) compared to the Yeoh model, the Biderman model has better prediction accuracy for the equibiaxial deformation of natural unfilled vulcanized rubber [44]. Nonetheless, the Biderman model may incorrectly predict the deformation mode without the corresponding calibrated data [30]. Hence, the benefits obtained by simply adding the second strain invariant to the Yeoh model are limited.

In order to more effectively improve the characterization ability of the Yeoh model for the equibiaxial mode of deformation, the corresponding residual strain energy is calculated (as shown in Figure 1d). Based on the analysis of residual strain energy, we found that a composite function based on a power function represented by the principal stretches can perfectly predict the residual strain energy (as shown in Figure 1d). Considering the function predicting the residual strain energy as a correction term, a modified strain energy density function is proposed in the following form:

$$W = C_{10}(I_1-3) + C_{20}(I_1-3)^2 + C_{30}(I_1-3)^3 + \frac{\alpha}{\beta}\left[(\lambda_1\lambda_2)^\beta + (\lambda_2\lambda_3)^\beta + (\lambda_1\lambda_3)^\beta - 3\right] \tag{8}$$

The correction term of modified strain energy density function can be understood as a generalization of the first-order form about the second strain invariant. When $\beta = 2$, the correction term degenerates to the first-order form about the second strain invariant. In this point, the modified model becomes the Biderman model. In addition, the modified strain energy function can also be reduced to the neo-Hookean model ($C_{20} = C_{30} = \alpha = 0$ MPa), Yeoh model ($\alpha = 0$ MPa) and Mooney–Rivlin model ($C_{20} = C_{30} = 0$ MPa, $\beta = 2$) under the condition of a specific coefficient. Therefore, the modified model can also be regarded as the parent model of these four classical models. Furthermore, reducing the higher-order terms of third-order polynomials can also make Equation (8) turn into other forms, whose performance will be explained later.

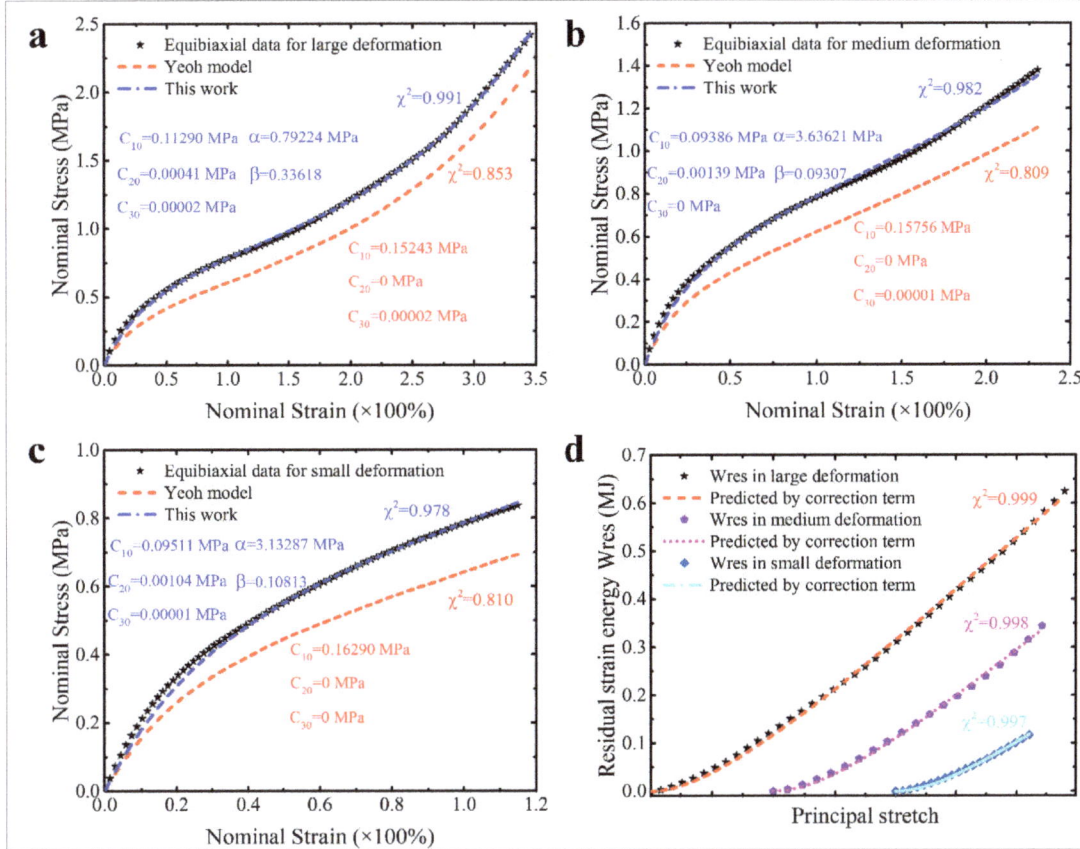

Figure 1. The characterization ability of Yeoh model for equibiaxial deformation mode under different ranges of deformation: (**a**) large deformation; (**b**) medium deformation; (**c**) small deformation; (**d**) residual strain energy under different ranges of deformation. The model parameters here are obtained by using experimental data of natural unfilled vulcanized rubber from uniaxial tension, equibiaxial tension and pure shear at the same time.

2.3.1. Constitutive Restrictions of the Proposed Strain Energy Density Function

Considering the material objectivity and the consistency between the general theory of isotropic elasticity and its classical linear theory, six general postulates for the strain energy density function are proposed by Ogden [45]. Based on these postulates, the modified strain energy density function should satisfy the following relationships:

$$\frac{\partial^2 W(\lambda_1 = 1, \lambda_2 = 1)}{\partial \lambda_i^2} = 8C_{10} + 2\alpha\beta > 0 \tag{9}$$

$$\det[H_{ij}] > 0, \\ H_{ij} = \frac{\partial^2 W(\lambda_1 = 1, \lambda_2 = 1)}{\partial \lambda_i \partial \lambda_j}, i, j = 1, 2 \tag{10}$$

In view of the above formulas, and in order to make the model consistent with its four degraded models, here, the following constraint for the coefficient of strain energy density function is kept:

$$C_{10} > 0; \ 4C_{10} + \alpha\beta > 0 \tag{11}$$

Furthermore, some researchers gradually regard the poly-convexity as a fundamental mathematical constraint to guarantee the existence of a solution for the constitutive model in the boundary-value problem [46]. Here, the stricter condition of convexity, which requires the Hessian matrix of Equation (8) to be positive definite [42], is adopted. According to the components in the Hessian matrix (see Equation (S1) in the supporting information), H is symmetric. Therefore, the conditions for making H positive definite are as follows:

$$\begin{cases} \frac{\partial^2 W}{\partial \lambda_1^2} > 0, \quad \frac{\partial^2 W}{\partial \lambda_2^2} > 0 \\ \frac{\partial^2 W}{\partial \lambda_1^2} \cdot \frac{\partial^2 W}{\partial \lambda_2^2} - \frac{\partial^2 W}{\partial \lambda_1 \partial \lambda_2} \cdot \frac{\partial^2 W}{\partial \lambda_2 \partial \lambda_1} > 0 \end{cases} \quad (12)$$

Hence, as long as the inequality (12) is satisfied, the constitutive model with calibrated parameters will possess the poly-convexity.

2.3.2. The Stability Criterion of the Proposed Strain Energy Density Function

A stable constitutive model obeying the laws of thermodynamics is a prerequisite for subsequent finite element analysis. An unstable strain energy function will adversely affect the nonlinear numerical algorithms in the finite element codes [47]. If the obtained constitutive model is unstable, it is necessary to re-determine the parameters of the model or even replace other models.

According to the Drucker stability criterion used in the commercial finite element software [47], the relation between changes in the principal stress and changes in the principal strain can be described by the following matrix equation:

$$\begin{bmatrix} d\sigma_1 \\ d\sigma_2 \end{bmatrix} = \begin{bmatrix} D_{11} & D_{12} \\ D_{21} & D_{22} \end{bmatrix} \begin{bmatrix} d\varepsilon_1 \\ d\varepsilon_2 \end{bmatrix} \quad (13)$$

If the hyperelastic constitutive model is stable, it is only required that the coefficient matrix (D) of the above matrix equation is positive definite. The components of the coefficient matrix (D) are as shown in Equation (S2) in the supporting information.

2.4. Analytical Stress Formulations for Standard Tests

The mechanical behavior of hyperelastic polymeric materials can be determined by standard tests involving simple modes of deformation such as uniaxial tension (UT), equibiaxial tension (ET) and pure shear (PS). The relationships between nominal stress and principal stretches under these simple modes of deformation can be derived from Equations (7) and (8) as follows:

$$T_{UT} = \frac{1}{\lambda} \cdot \left[C_{10} + 2C_{20} \cdot (I_1 - 3) + 3C_{30} \cdot (I_1 - 3)^2 \right] \cdot \left(2\lambda^2 - \frac{2}{\lambda} \right) + \frac{1}{\lambda} \cdot \left[\alpha \cdot \left(\lambda^{\frac{\beta}{2}} - \lambda^{-\beta} \right) \right] \quad (14)$$

$$T_{ET} = \frac{1}{\lambda} \cdot \left[C_{10} + 2C_{20} \cdot (I_1 - 3) + 3C_{30} \cdot (I_1 - 3)^2 \right] \cdot \left(2\lambda^2 - \frac{2}{\lambda^4} \right) + \frac{1}{\lambda} \cdot \left[\alpha \cdot \left(\lambda^{2\beta} - \lambda^{-\beta} \right) \right] \quad (15)$$

$$T_{PS} = \frac{1}{\lambda} \cdot \left[C_{10} + 2C_{20} \cdot (I_1 - 3) + 3C_{30} \cdot (I_1 - 3)^2 \right] \cdot \left(2\lambda^2 - \frac{2}{\lambda^2} \right) + \frac{1}{\lambda} \cdot \left[\alpha \cdot \left(\lambda^{\beta} - \lambda^{-\beta} \right) \right] \quad (16)$$

2.5. Parameter Identification of Model

Identifying the parameters of the constitutive model is an important step during the characterization of the hyperelasticity. Although some optimization algorithms have been used to identify the parameters of constitutive models, these used methods still have their own problems. For example, the damped least squares algorithm based on nonlinear least squares [48] and the sequential quadratic programming algorithm based on multi-objective optimization strategy [49] are all affected by initial values. The genetic algorithm using multi-objective optimization strategy has deficiencies in terms of local search capability [50]. Although the hybrid optimization algorithm based on the pattern search and damped least squares can improve the local search capability of the single-optimization algorithm, its global search capability is not as strong as the genetic algorithm [51]. In order to obtain more accurate parameters of the model, the sensitivity of the identification method to the initial values and its search capability need to be considered.

On these grounds, this study introduces a parameter identification method of the constitutive model based on the cyclic genetic-pattern search algorithm, and its framework is shown in Figure 2. In order to contain the experimental data with different modes of deformation during the process of parameter identification as much as possible and effectively weigh the importance of different modes of deformation, a weighted multi-objective optimization strategy is adopted. To better measure the error between the predicted results of the model and experimental data, the objective function is defined as shown in Equation (17). For obtaining the global optimal solution during solving the optimization problem as much as possible, the hybrid optimization algorithm based on genetic-pattern search is employed. The method utilizes the genetic algorithm for global coarse search and utilizes the pattern search algorithm for local fine search. The combination of coarse search and fine search makes the result closer to the ideal optimal value. Moreover, for further improving the accuracy of the model and reducing the uncertainty caused by the random search, the cyclic item based on the genetic algorithm is added to the process of identification. Before the cyclic condition terminates, the genetic algorithm independently operates multiple times, and its corresponding results are also stored in sequence. Afterwards, the optimal set of parameters is extracted from the results of multiple runs as the initial values for the subsequent pattern search algorithm, thereby completing local fine search. The parameters obtained by the pattern search algorithm are the final optimized parameters, which will be directly used to draw the prediction curve and relative error curve of the model, so as to intuitively evaluate the performance of the model. In order to effectively evaluate the performance of constitutive model, the goodness of fit described in Equation (18) will be used to measure the capability of the model to characterize different modes of deformation, and the total error described in Equation (17) will be used to measure the overall predicted accuracy of the model. The smaller the total error is, the higher the overall predicted accuracy of the model is. And the closer the goodness of fit is to 1, the better the model characterizes the corresponding modes of deformation.

$$min: Error_{total} = \sum_{p=1}^{3} w_p \bullet \left(\sqrt{\sum_{i=1}^{m_p} \left(T_{pi}^{model} - T_{pi}^{experiment} \right)^2} / \sqrt{\sum_{i=1}^{m_p} \left(T_{pi}^{experiment} \right)^2} \right) \qquad (17)$$

$$\chi_p^2 = 1 - \sqrt{\sum_{i=1}^{m_p} \left(T_{pi}^{model} - T_{pi}^{experiment} \right)^2} / \sqrt{\sum_{i=1}^{m_p} \left(T_{pi}^{experiment} \right)^2} \qquad (18)$$

where p represents the simple mode of deformation such as uniaxial tension ($p = 1$), equibiaxial tension ($p = 2$) and pure shear ($p = 3$), respectively; m_p is the number of the experiment data for the pth mode of deformation; $Error_{total}$ corresponds to the total error between the predicted results of the model and experimental data; T_{pi}^{model} and $T_{pi}^{experiment}$ are the prediction stress of the model and experimental stress under pth mode of deformation, respectively; and w_p is the weight for different modes of deformation, which satisfies $w_1 + w_2 + w_3 = 1$. Specifically, when experimental data from three deformation modes are used for identifying the parameters of the model simultaneously, the corresponding weight $w_p = 1/3$; when experimental data with only two modes of deformation are used for parameter identification simultaneously, the corresponding weight $w_p = 1/2$ (the weight of the unused mode of deformation is 0); when experimental data with only one mode of deformation is used, the corresponding weight $w_p = 1$ (the weight of the other two unused modes of deformation is 0). χ_p^2 is the goodness of fit for the pth mode of deformation.

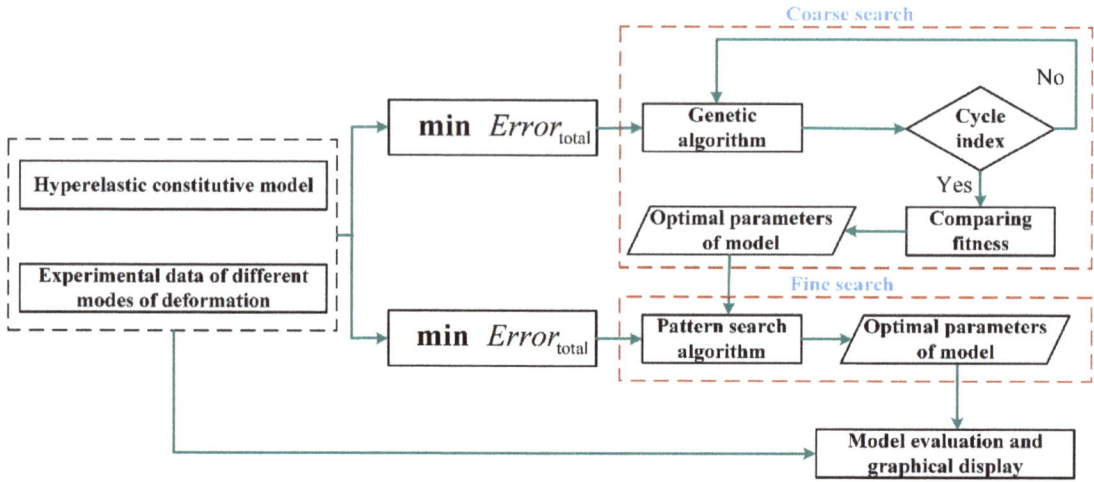

Figure 2. The framework of parameter identification of constitutive model based on cyclic genetic-pattern search algorithm.

3. Results and Discussion

3.1. Parameter Identification and Validation of Model

3.1.1. Predictive Ability of the Modified Model to the Single Deformation Mode

The ability of a constitutive model to accurately predict a single mode of deformation is its foundation to effectively predict multiaxial modes of deformation. In view of this, the single set of deformation data (UT, ET or PS) of natural vulcanized rubber from a large deformation range are first used to identify the parameters of the proposed modified model, respectively. Figure 3 presents the predictive effect of the modified model calibrated based on the single deformation data on the corresponding single deformation mode. It is clear that the independent predictive effect of the modified model is excellent for the two deformation modes of equibiaxial tension and pure shear, and their goodness of fit reaches 0.99 (see Figure 3c,e). Although its independent prediction effect for the deformation mode of uniaxial tension is slightly inferior, its prediction error for uniaxial tension does not exceed 4% (see Figure 3a). Based on this, we can conclude that this modified model has the ability to independently predict three deformation modes of uniaxial tension, equibiaxial tension and pure shear, which lays the foundation for its ability to effectively predict the multiaxial deformation of hyperelastic material. Here, the modified models, respectively calibrated based on the tested data from uniaxial tension, equibiaxial and pure shear, are also used to predict the other two untested deformation modes. As shown in Figure 3b,d,f, the modified model calibrated based on single set of tested data can also reasonably predict the other two untested deformation modes. As we explained in the introduction, although researchers have mixed opinions on the method to calibrate the parameters of the model based solely on data from a single deformation mode, the ability to effectively predict untested deformation modes is extremely helpful for characterizing an unmeasurable deformation mode of certain polymeric materials in practice. Anyway, compared with the results of other studies [30], the results here indicate that the additional correction term we added enhances the model's ability to characterize untested deformation modes. This point will also continue to be discussed in a more reliable manner in the subsequent content.

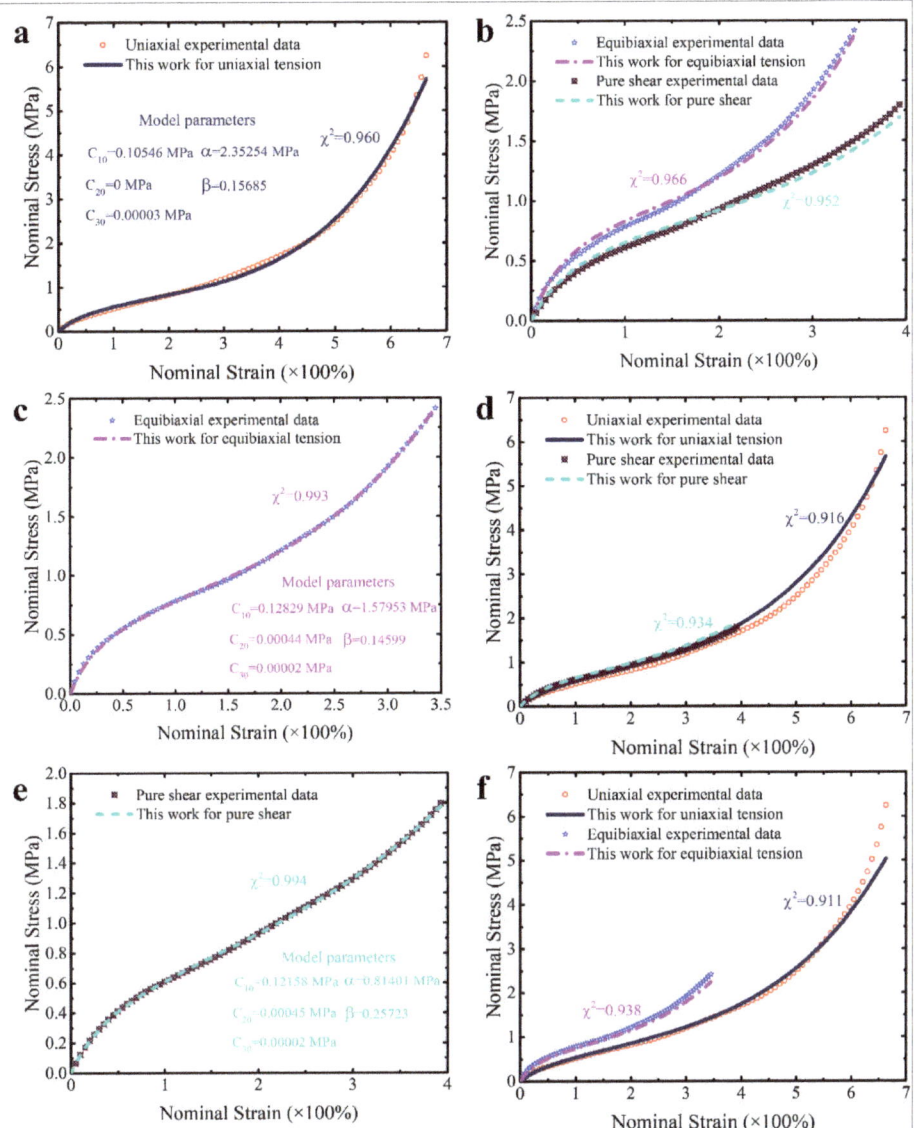

Figure 3. The predictive effect of the modified model on the single deformation mode. (**a**,**b**) The predictive effect of the modified model calibrated based on the single set of experimental data of uniaxial tension. (**c**,**d**) The predictive effect of the modified model calibrated based on the single set of experimental data of equibiaxial tension. (**e**,**f**) The predictive effect of the modified model calibrated based on the single set of experimental data of pure shear.

3.1.2. Predictive Ability of the Modified Model to the Multiaxial Deformation Mode

For verifying the prediction ability of the modified model for the multiaxial deformation mode, the data of multiple deformation modes (UT, ET and PS) from the same material dataset are simultaneously used to identify the parameters of the modified model. The following will present the predictive effects of the modified constitutive model on the multi axial deformation state of different materials in sequence. And these results are also compared with the results of other existing models (as mentioned in Section 2.1).

Natural Vulcanized Rubber

As mentioned earlier, the widespread use of Treloar's experimental dataset on the natural vulcanized rubber has made it a barometer for evaluating the ability of constitutive models to characterize hyperelastic materials. Using this dataset to calibrate the modified model proposed in this study will facilitate an intuitive comparison with some models in the literature (including those models that are not on our comparison list). Table 1 summarizes the quantitative results of different constitutive models for characterizing the hyperelastic property of the natural vulcanized rubber. It can be seen from the corresponding total error that the proposed constitutive model is not only more capable in characterizing the states of multiaxial stress than the Yeoh model and the corresponding improved models (such as the modified Yeoh model, generalized Yeoh model and Melly model), but also performs better than the Anssari–Benam model newly proposed by a reviewer. And its total predicted error for different ranges of deformation is all less than 3%. In addition, the performance of the model proposed in this study also outperforms the performance of the modified Anssari–Benam model. The modified Anssari–Benam model is made by us under the persuasion of a reviewer, which is an improvement on the original Anssari–Benam model based on our correction term. Although the modified Anssari–Benam model showed significant improvement in characterizing medium and small deformations compared to the original model, it showed significant bias in characterizing uniaxial tension under large deformation. Compared with the excellent third-order Ogden model and Alexander model, our model outperforms the third-order Ogden model in characterizing large deformation, but is slightly weaker than the Alexander model. The reason why it is weaker than the Alexander model is because the goodness of fit of the proposed model for UT under large deformation is slightly lower than the Alexander model. Based on the fact that industrial rubber materials typically experience a strain range of 0–100% [32,36,52], we believe that this weakness does not affect the use of the modified model proposed in this study. Moreover, the quantitative data in Table 1 also prove that the modified model's ability to characterize medium to small deformations is comparable to the third-order Ogden model. From a quantitative point of view, the total prediction errors of the two models are within the range of 1–1.5%, and the corresponding difference between their goodness of fit under different modes of deformation is not more than 0.5%. The overall performance of the Alexander model in characterizing medium deformation is slightly weaker than that of our proposed model.

Table 1. Total error and goodness of fit of different models on natural vulcanized rubber [37].

Models	Large Deformation				Medium Deformation				Small Deformation			
	Error	UT	ET	PS	Error	UT	ET	PS	Error	UT	ET	PS
Ogden-N3	0.035	0.927	0.987	0.981	0.011	0.988	0.986	0.994	0.012	0.988	0.979	0.996
Alexander	0.017	0.982	0.984	0.982	0.016	0.981	0.987	0.985	0.013	0.987	0.980	0.996
Yeoh	0.083	0.928	0.853	0.969	0.093	0.940	0.809	0.973	0.117	0.872	0.810	0.967
Melly	0.033	0.944	0.969	0.988	0.021	0.984	0.967	0.987	0.020	0.986	0.966	0.988
Modified Yeoh	0.077	0.938	0.843	0.989	0.084	0.942	0.809	0.996	0.097	0.929	0.782	0.999
Generalized Yeoh	0.084	0.914	0.861	0.973	0.086	0.943	0.807	0.993	0.110	0.914	0.791	0.967
Anssari-Benam	0.079	0.966	0.828	0.968	0.093	0.941	0.806	0.973	0.099	0.930	0.781	0.993
Modified Anssari–Benam	0.164	0.608	0.954	0.945	0.014	0.986	0.981	0.990	0.013	0.987	0.976	0.997
This work	0.024	0.947	0.991	0.991	0.012	0.991	0.982	0.991	0.013	0.987	0.978	0.997

Notes: Error indicates the total error of different models; UT, ET and PS represent the goodness of fit of different models for uniaxial tension, equibiaxial tension and pure shear, respectively. The model parameters are shown in Tables S7–S9 in supporting information.

In order to make these results more visible and clear, Figures 4–6 present the prediction curves and corresponding prediction error curves of the modified constitutive model within different deformation ranges for different deformation modes, respectively. Obviously, these qualitative curves also reflect the good predicted capability of the modified model. Especially from the corresponding relative error curves, it can be seen that the relative error of the proposed model for predicting different deformation modes under different deformation ranges is comparable to the third-order Ogden model and Alexander model (their average relative errors are less than 5%), and all the average relative errors of our proposed model do not exceed 3.5%. Even for predicting the uniaxial tension under large deformation, the average relative error of the modified model proposed in this study is only 3.3%.

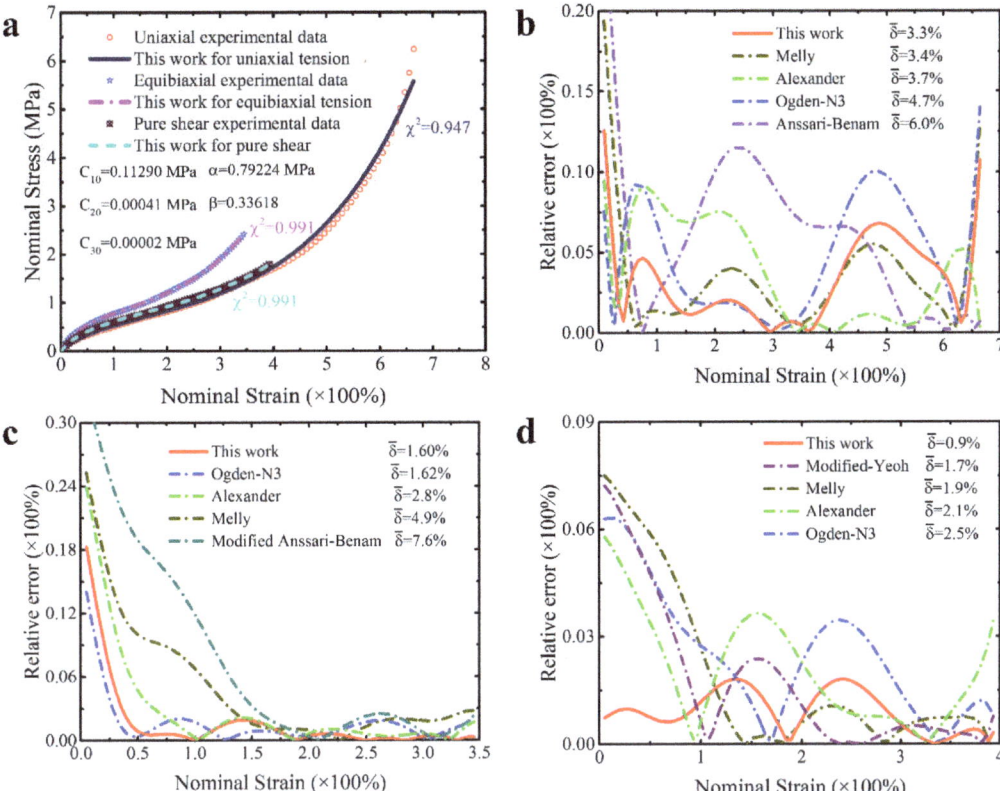

Figure 4. The prediction performance and corresponding prediction error curves of modified constitutive model in large deformation ranges for different deformation modes. (**a**) Prediction curves for different deformation modes; (**b**) relative error of different models in predicting uniaxial deformation mode; (**c**) relative error of different models in predicting equibiaxial deformation mode; (**d**) relative error of different models in predicting the deformation mode of pure shear. Note: Only the top five models are presented in Figure 4b–d. $\bar{\delta}$ means the average relative error.

Further comparing the predicted results of the modified constitutive model with the Yeoh model, modified Yeoh model and generalized Yeoh model, it can be found that the overall predicted accuracy of the proposed model is several times higher than theirs in all ranges of deformation. In particular, the predicted accuracy on the equibiaxial mode of deformation is improved most obviously (as shown in Figure 1a–c). Continuing to compare with the new Melly model (it uses $\sqrt{I_2}$ as the correction term to improve the performance of the Yeoh model), the modified constitutive model proposed in this study is better in both overall prediction accuracy and goodness of fit for different deformation modes (the relative error curves shown in Figures 4–6 can also directly reflect this point). Based on these findings, we have reason to believe that the correction term expressed in the form of a power function based on the principal stretches is beneficial in improving the ability of the model to characterize the state of multiaxial stress.

Focusing on the coefficient of the proposed model in Figures 4–6, it is clear that the coefficient of the third-order term of the first strain invariant is a small value in the case of large deformation, while it changes to zero in the case of medium or small deformation. This shows that the coefficient of the third-order term of the first strain invariant cannot be ignored under the condition of large deformation. It can be speculated that the coefficient should affect the capability of the proposed model to characterize large deformation. In other words, the proposed constitutive model needs five parameters when characterizing the large deformation, while it requires only four parameters when it characterizes a small or medium deformation, which makes the proposed model more simple.

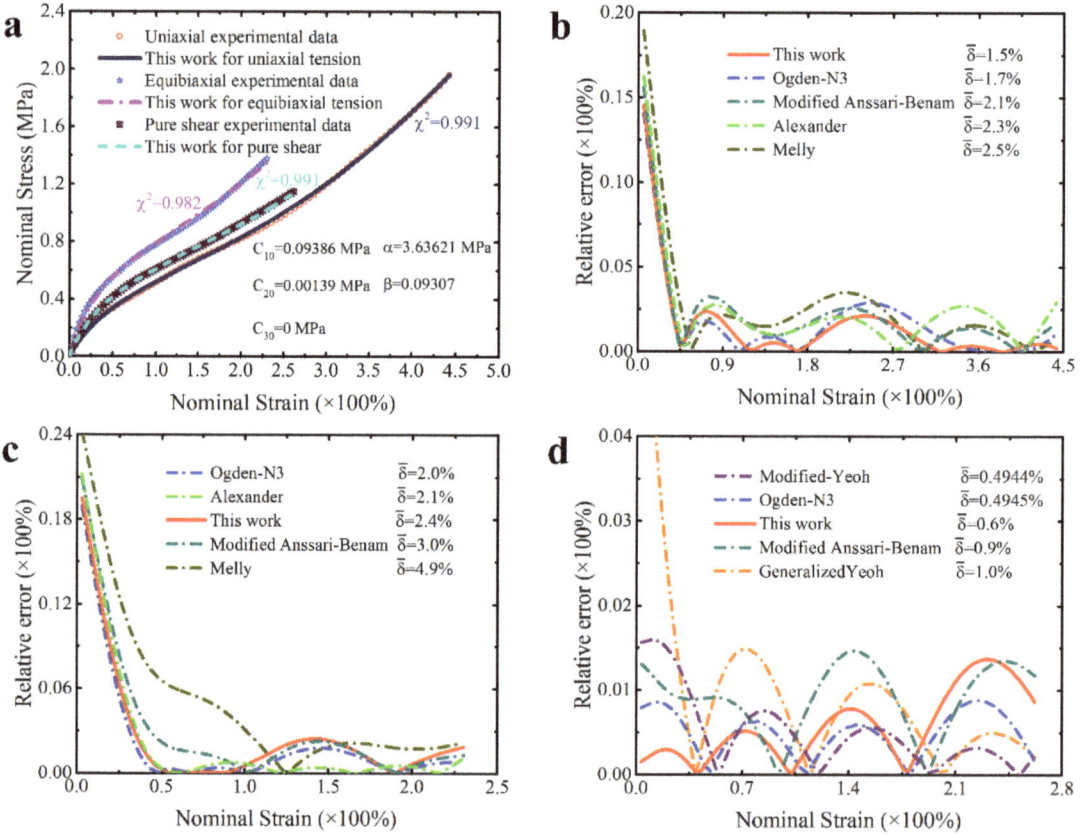

Figure 5. The prediction performance and corresponding prediction error curves of modified constitutive model in medium deformation ranges for different deformation modes. (**a**) Prediction curves for different deformation modes; (**b**) relative error of different models in predicting uniaxial deformation mode; (**c**) relative error of different models in predicting equibiaxial deformation mode; (**d**) relative error of different models in predicting the deformation mode of pure shear. Note: Only the top five models are presented in Figure 5b–d. $\bar{\delta}$ means the average relative error.

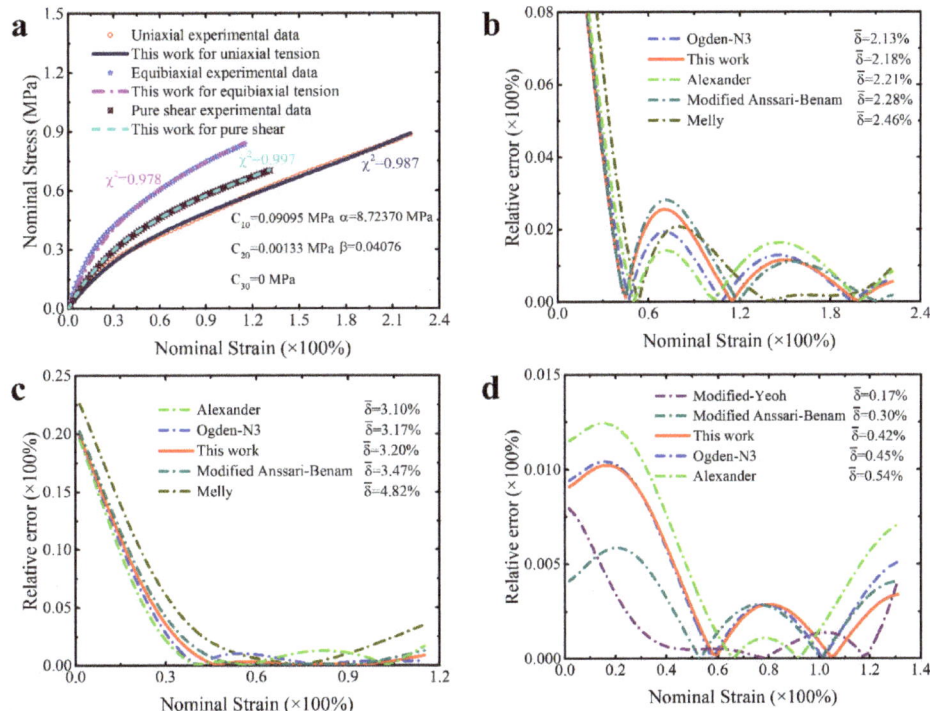

Figure 6. The prediction performance and corresponding prediction error curves of modified constitutive model in small deformation ranges for different deformation modes. (**a**) Prediction curves for different deformation modes; (**b**) relative error of different models in predicting uniaxial deformation mode; (**c**) relative error of different models in predicting equibiaxial deformation mode; (**d**) relative error of different models in predicting the deformation mode of pure shear. Note: Only the top five models are presented in Figure 6b–d. $\bar{\delta}$ means the average relative error.

Isoprene Vulcanized Rubber

The isoprene vulcanized rubber obtained by Kawabata et al. [38] and the natural vulcanized rubber obtained by Treloar [37] are similar, both of which belong to unfilled rubber. Continuing to calibrate the constitutive model based on the dataset of the isoprene vulcanized rubber will further validate the performance of the modified constitutive model proposed in this study. Table 2 summarizes some quantitative results obtained based on this dataset. Obviously, compared with other constitutive models in the table, the modified model proposed in this study achieved the lowest total prediction error (1.5%). This means that the improved model has the best comprehensive description effect on the hyperelastic properties of this material. The corresponding more intuitive prediction curves for different deformation modes are shown in Figure 7a. From the graph, it can be seen that the prediction curves of this modified model for different deformation modes almost perfectly coincide with the corresponding experimental data points. In order to compare the prediction accuracy of different models for different deformation modes in more detail, the corresponding relative error curves are shown in Figure 7b–d. In order to present the results clearly, only the relative error curves of the models with the top five prediction accuracies are shown in the corresponding figures and the average relative error of the corresponding model is marked on the legend. It can be observed that almost all the models presented in these figures have relatively low relative error, while the modified model in this study has lower relative errors for predicting different deformation modes. The vast majority of its curves are below 3%, resulting in an average relative error for predicting different deformation modes ranging from 1.2% to 1.3%, which slightly outperforms the third-order Ogden model (0.9–1.9%) and the Alexander model (1.6–2.6%).

Table 2. Total error and goodness of fit of different models on isoprene vulcanized rubber [38].

Models	The Number of Coefficient	Error	UT	ET	PS
Ogden-N3	6	0.017	0.970	0.996	0.984
Alexander	5	0.025	0.986	0.986	0.954
Yeoh	3	0.092	0.964	0.815	0.944
Melly	4	0.032	0.979	0.976	0.950
Modified Yeoh	5	0.068	0.984	0.845	0.966
Generalized Yeoh	6	0.068	0.987	0.847	0.963
Anssari–Benam	3	0.081	0.987	0.800	0.976
Modified Anssari–Benam	5	0.019	0.980	0.997	0.965
This work	5	**0.015**	**0.981**	**0.997**	**0.977**

Notes: Error indicates the total error of different models; UT, ET and PS represent the goodness of fit of different models for uniaxial tension, equibiaxial tension and pure shear, respectively. The model parameters are shown in Table S10 in supporting information.

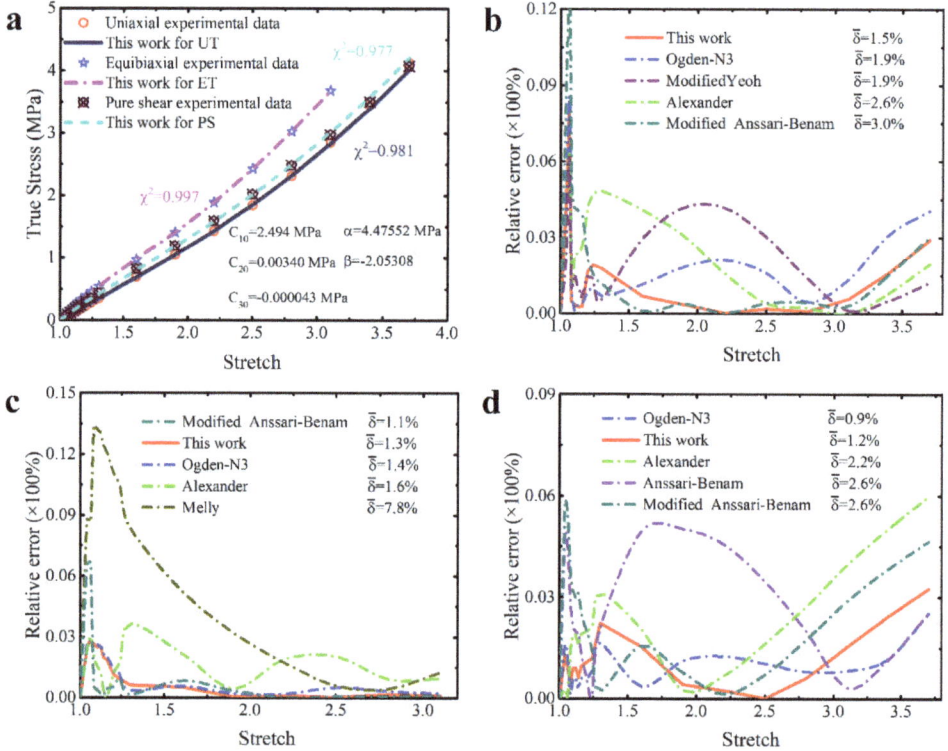

Figure 7. The prediction performance and corresponding prediction error curves of modified constitutive model for isoprene vulcanized rubber. (**a**) Prediction curves for different deformation modes; (**b**) relative error of different models in predicting uniaxial deformation mode; (**c**) relative error of different models in predicting equibiaxial deformation mode; (**d**) relative error of different models in predicting the deformation mode of pure shear. Note: Only the top five models are presented in Figure 7b–d. $\bar{\delta}$ means the average relative error.

Overall, these results once again confirm the excellent predictive performance of the modified model proposed in this study on the multiaxial deformation of unfilled rubber. Specifically, the performance of the modified model has a several-fold improvement compared to other improved Yeoh models, which once again demonstrates that the improvement of the Yeoh model based on the correction term of this study is worthwhile. In addition, the improved model outperforms the modified Anssari–Benam model, indicating that the improvement of the Yeoh model based on our correction term is superior to the improvement of the Anssari–Benam model based on the same correction term.

Unfilled Silicone Rubber

Silicone rubber has good mechanical properties and biocompatibility [17]. Completing more accurate modeling of the mechanical properties is of great significance for its fuller utilization in the field of biomedicine. Here, the modified constitutive model proposed in this study is applied to the unfilled silicone rubber [17] to verify its usability in more precisely modeling silicone rubber. Table 3 quantitatively shows the prediction performance of different models on the hyperelastic property of unfilled silicone rubber. It can be seen from these quantitative data that although the modified constitutive model proposed in this study is not as effective as the third-order Ogden model with six undetermined parameters in predicting the multiaxial stress state of silicone rubber, it ranks second alongside the Alexander model and the modified Anssari–Benam model (both with five undetermined parameters). The reason for ranking second is that the modified model in this study slightly overestimates the uniaxial tensile deformation when characterizing the multiaxial deformations. As shown in Figure 8a, the prediction curves of the modified model for equibiaxial tension, pure shear and uniaxial compression fit well with the corresponding experimental data points, while its prediction curve for uniaxial tension is slightly higher than the experimental data points. Nonetheless, its average relative errors for predicting different deformations are no more than 8% (as shown in Figure 8b–d). And, considering that it far outperforms the Yeoh model with a total prediction error of 4.4%, and slightly surpasses these existing improved models for the Yeoh model, we still believe that the modified constitutive model proposed in this study has potential for application for unfilled silicone rubber.

Table 3. Total error and goodness of fit of different models on unfilled silicone rubber [17].

Models	The Number of Coefficient	Error	UT	ET	PS
Ogden-N3	6	0.029	0.970	0.985	0.958
Alexander	5	0.044	0.936	0.988	0.943
Yeoh	3	0.233	0.606	0.897	0.800
Melly	4	0.048	0.939	0.963	0.955
Modified Yeoh	5	0.047	0.932	0.961	0.965
Generalized Yeoh	6	0.048	0.931	0.955	0.970
Anssari–Benam	3	0.046	0.922	0.983	0.956
Modified Anssari–Benam	5	0.044	0.934	0.986	0.947
This work	**5**	**0.044**	**0.920**	**0.972**	**0.978**

Notes: Error indicates the total error of different models; UT, ET and PS represent the goodness of fit of different models for uniaxial tension, equibiaxial tension and pure shear, respectively. The model parameters are shown in Table S11 in supporting information.

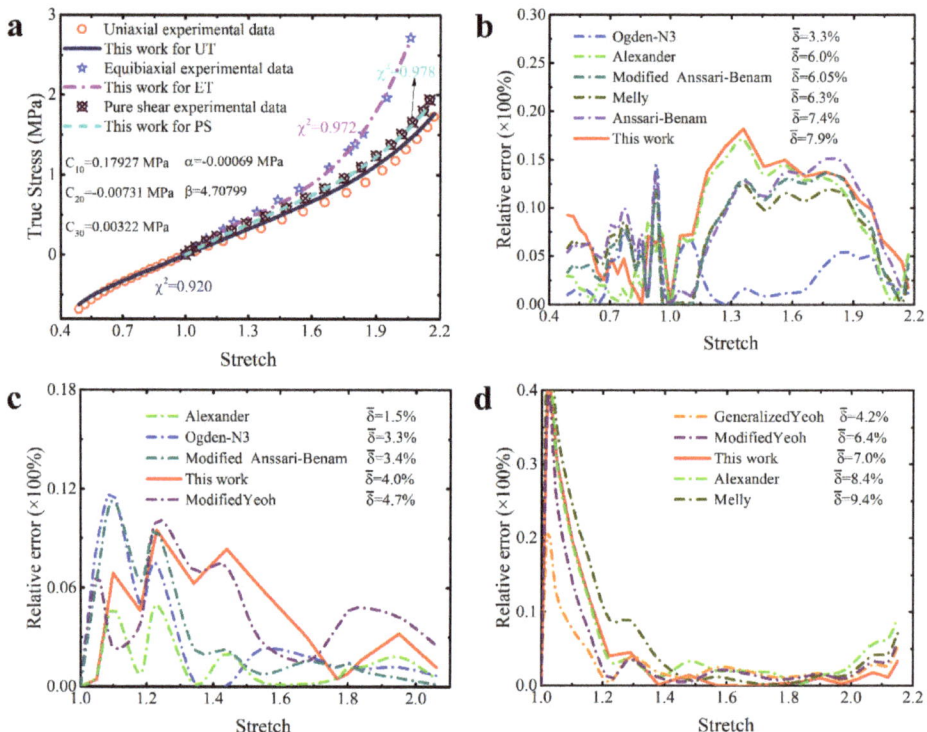

Figure 8. The prediction performance and corresponding prediction error curves of modified constitutive model for unfilled silicone rubber. (**a**) Prediction curves for different deformation modes; (**b**) relative error of different models in predicting uniaxial deformation mode; (**c**) relative error of different models in predicting equibiaxial deformation mode; (**d**) relative error of different models in predicting the deformation mode of pure shear. Note: Only the top five models are presented in Figure 8b–d. $\bar{\delta}$ means the average relative error.

Poly-Acrylamide Hydrogel

Like silicone rubber, hydrogels are also widely used in the biomedical field. But compared to silicone rubber, they undergo greater deformation due to their softness [25]. Currently, accurately characterizing their mechanical properties also has important practical significance. Here, we also apply the modified constitutive model proposed in this study to the poly-acrylamide hydrogel [39] to explore the ability of this model in describing the hyperelastic property of hydrogel. As shown in Table 4, the modified constitutive model proposed in this study has the best prediction effect on the multiaxial deformation of hydrogel. Its total prediction error for different deformation modes is only 1.1%. Figure 9 visually shows its actual prediction performance. It can be clearly seen from the figure that the prediction curves of the modified constitutive model proposed in this study for different deformations of hydrogels are in good agreement with the corresponding experimental data points, and the corresponding goodness of fit reaches 0.98. Moreover, the relative error curves of the modified model for different deformation modes are mostly below 4%, resulting in an average relative error of 1.8–3.1%. Looking carefully at the coefficients of the modified model at this time (see Figure 9a), it can be found that the coefficient of the cubic term of the first invariant is zero, which means that the modified constitutive model only needs four parameters to characterize the hydrogel. Based on the above results, we have reason to believe that the modified constitutive model also has the ability to accurately characterize the multiaxial deformation of hydrogels.

Table 4. Total error and goodness of fit of different models on poly-arcylamide hydrogel [39].

Models	The Number of Coefficient	Error	UT	ET	PS
Ogden-N3	6	0.013	0.982	0.992	0.988
Alexander	5	0.014	0.981	0.993	0.986
Yeoh	3	0.090	0.966	0.778	0.986
Melly	4	0.012	0.981	0.995	0.987
Modified Yeoh	5	0.089	0.967	0.777	0.991
Generalized Yeoh	6	0.089	0.965	0.780	0.988
Anssari–Benam	3	0.095	0.956	0.783	0.977
Modified Anssari–Benam	5	0.018	0.984	0.980	0.984
This work	**5**	**0.011**	**0.989**	**0.994**	**0.985**

Notes: Error indicates the total error of different models; UT, ET and PS represent the goodness of fit of different models for uniaxial tension, equibiaxial tension and pure shear, respectively. The model parameters are shown in Table S12 in supporting information.

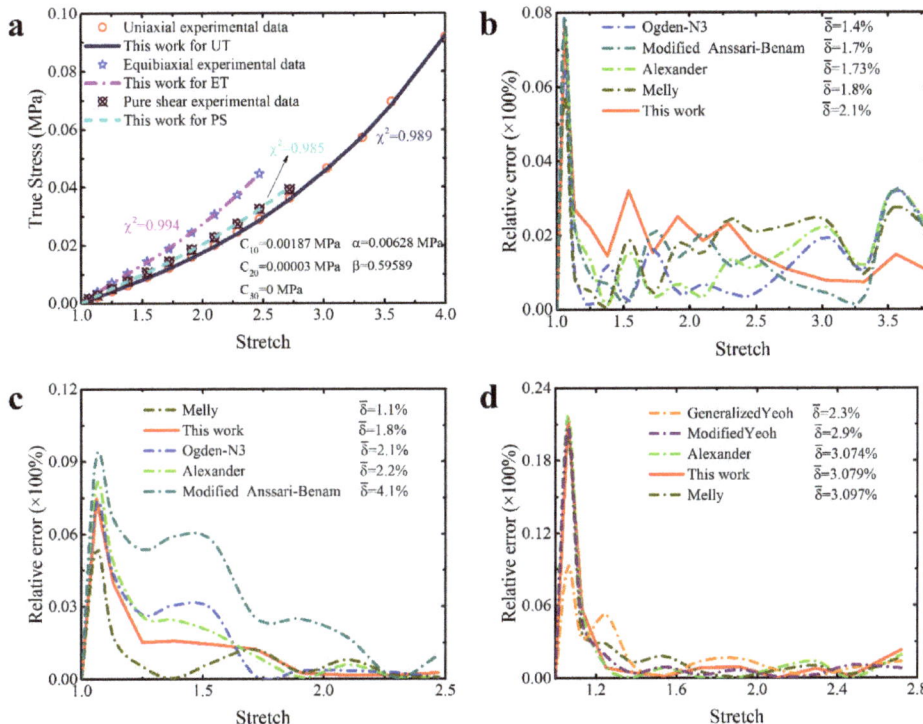

Figure 9. The prediction performance and corresponding prediction error curves of modified constitutive model for poly-arcylamide hydrogel. (**a**) Prediction curves for different deformation modes; (**b**) relative error of different models in predicting uniaxial deformation mode; (**c**) relative error of different models in predicting equibiaxial deformation mode; (**d**) relative error of different models in predicting the deformation mode of pure shear. Note: Only the top five models are presented in Figure 9b–d. $\bar{\delta}$ means the average relative error.

Carbon-Black-Filled Styrene Butadiene Rubber

Due to the filling of additives, the stress–strain curve of filled rubber is significantly different from that of unfilled rubber [40]. Based on this, there are few constitutive models that can properly capture the hyperelastic property of this material in the current literature [25]. If the modified constitutive model proposed in this study can also well characterize the multiaxial deformation characteristics of filled rubber, it will further confirm its adaptability to different materials. Therefore, we further apply the modified model proposed in this study to a dataset related to the carbon-black-filled styrene butadiene rubber [40]. As summarized in Table 5, the modified model proposed in this study also has good characterization ability for the multiaxial deformation of this filled rubber. Although its overall performance is not the best, it ranks behind the Alexander model and before the third-order Ogden model with a total prediction error of 3.4%. Moreover, the difference in goodness of fit between the modified model and the Alexander model for the two deformation modes of uniaxial and equibiaxial tension is no more than 1%, and the difference in goodness of fit between the two models for pure shear is only 2.2%. Furthermore, like predicting other materials, the modified model proposed in this study also outperforms other improved models of the Yeoh model in describing the multiaxial deformation of filled rubber. Figure 10 shows the intuitive prediction performance of the modified model for different deformation modes. Just as the case indicated by the aforementioned quantitative indicators, the prediction curves of the modified model for different deformation modes are in good agreement with the corresponding experimental data points (see Figure 10a) and the corresponding relative error is in a relatively low position (see Figure 10b–d). Although it can be seen from Figure 10d that the average relative error of the modified constitutive model on pure shear (7.1%) is slightly higher than that of the generalized Yeoh model and modified Yeoh model, most of the zones of its relative error curve are below 6%. Moreover, from the perspective of characterizing multiaxis deformations, there are trade-offs between the predictive performance of the model for different deformation modes during calibrating the parameters of the model, so as to optimize the overall characterization ability of the model. Based on this, the slight advantage presented by the modified Yeoh model and generalized Yeoh model in characterizing pure shear do not make them possess better overall prediction accuracy, while the result is that the modified model in this study is better than them in overall prediction accuracy. The prediction accuracy of the modified model proposed in this study for different deformation modes is relatively balanced, and each goodness of fit reaches 0.96.

Table 5. Total error and goodness of fit of different models on carbon-black-filled styrene butadiene rubber [40].

Models	The Number of Coefficient	Error	UT	ET	PS
Ogden-N3	6	0.037	0.968	0.966	0.954
Alexander	5	0.023	0.975	0.975	0.982
Yeoh	3	0.106	0.887	0.885	0.909
Melly	4	0.049	0.958	0.960	0.936
Modified Yeoh	5	0.080	0.916	0.887	0.956
Generalized Yeoh	6	0.079	0.914	0.883	0.965
Anssari–Benam	3	0.107	0.890	0.875	0.914
Modified Anssari–Benam	5	0.039	0.963	0.965	0.954
This work	**5**	**0.034**	**0.967**	**0.969**	**0.961**

Notes: Error indicates the total error of different models; UT, ET and PS represent the goodness of fit of different models for uniaxial tension, equibiaxial tension and pure shear, respectively. The model parameters are shown in Table S13 in supporting information.

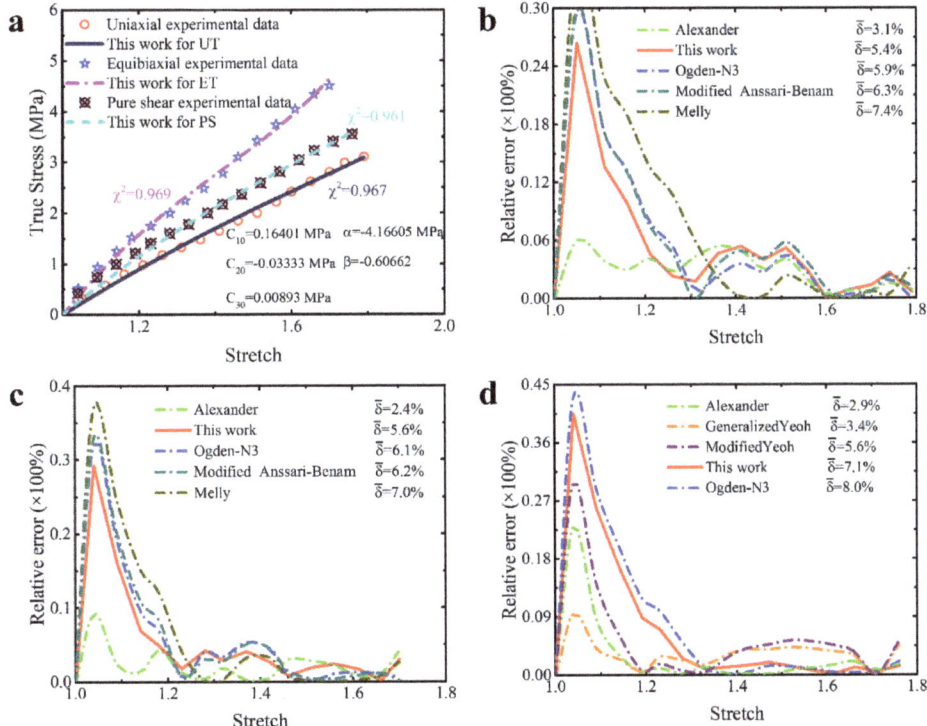

Figure 10. The prediction performance and corresponding prediction error curves of modified constitutive model for carbon-black-filled styrene butadiene rubber. (**a**) Prediction curves for different deformation modes; (**b**) relative error of different models in predicting uniaxial deformation mode; (**c**) relative error of different models in predicting equibiaxial deformation mode; (**d**) relative error of different models in predicting the deformation mode of pure shear. Note: Only the top five models are presented in Figure 10b–d. $\bar{\delta}$ means the average relative error.

Human Brain Cortex Tissue

Due to the unique asymmetry and nonlinearity presented by the deformation data from human brain tissue [41], there are currently few constitutive models that can accurately capture it without losing convexity [25]. Applying the modified model proposed in this study to a dataset from human brain tissue will further validate the potential of this modified model in the biomechanical modeling of brain tissue. Table 6 and Figure 11 show some of the results obtained by applying the modified constitutive model proposed in this study to a dataset reflecting the deformation of human brain cortex tissue [41]. From the quantitative indicators in Table 6, it can be found that the total prediction error of the modified constitutive model proposed in this study for the multiaxial deformation of human brain cortex tissue is only 5.6%. Although the error value is slightly higher than the 5.4% of the third-order Ogden model, the difference between the two is only about 3.6%. This difference is smaller when it comes to local indicator—goodness of fit. Specifically, the difference in goodness of fit between the proposed modified model and the third-order Ogden model for uniaxial tension and compression of brain tissue is only 0.1%, while the difference in goodness of fit between simple shear is only 0.4%. Based on these small differences, we believe that our modified model has a comparable ability to describe the multiaxis deformations of human brain cortex tissue to the third-order Ogden model. As shown in Figure 11a,b, our proposed modified model appropriately captures the asymmetry of brain cortex tissue during uniaxial tension and compression, as well as its high nonlinearity during simple shear. Moreover, the relative error of our modified model is also on par with that of the third-order Ogden model (see Figure 11c,d).

Table 6. Total error and goodness of fit of different models on human brain cortex tissue [41].

Models	The Number of Coefficient	Error	UT	SS
Ogden-N3	6	0.054	0.927	0.966
Generalized Anssari–Benam	4	0.083	0.922	0.913
Modified Anssari-Benam	5	0.060	0.915	0.965
This work	**5**	**0.056**	**0.926**	**0.962**

Notes: Error indicates the total error of different models; UT, ET and PS represent the goodness of fit of different models for uniaxial tension, equibiaxial tension and pure shear, respectively. The model parameters are shown in Table S14 in supporting information.

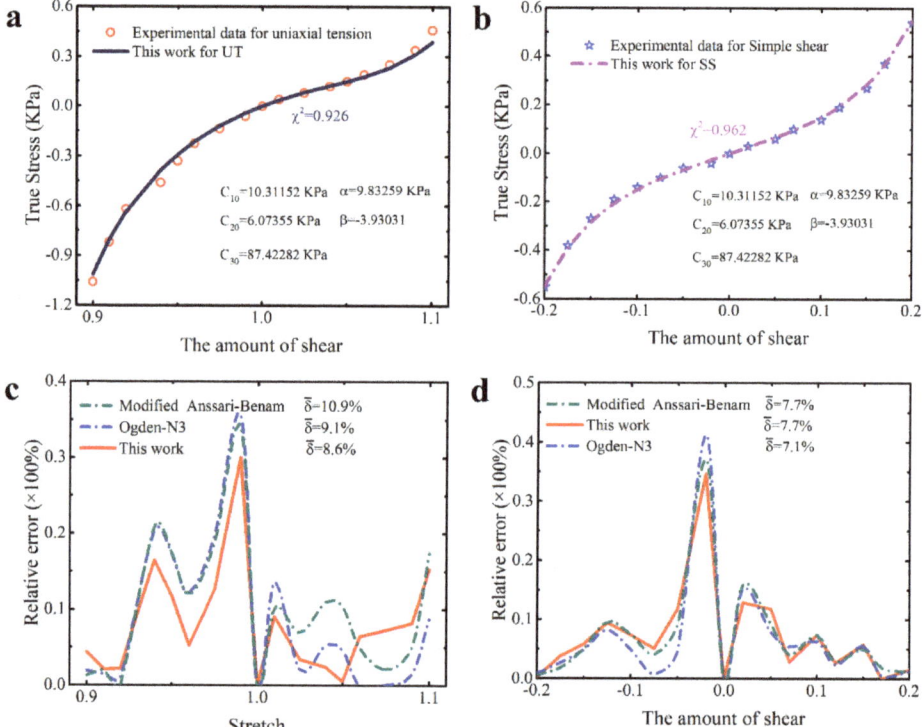

Figure 11. The prediction performance and corresponding prediction error curves of modified constitutive model for human brain cortex tissue. (a) Prediction curves for different deformation modes; (b) relative error of different models in predicting uniaxial deformation mode; (c) relative error of different models in predicting equibiaxial deformation mode; (d) relative error of different models in predicting the deformation mode of pure shear. Note: Only the top three models are presented in Figure 11b–d. $\bar{\delta}$ means the average relative error.

In [25], the researcher generalized the Anssari–Benam model, and the first-order form of the obtained model (we call it the generalized Anssari–Benam model) also has a good effect on describing the multiaxial deformation of brain cortex tissue without losing convexity. Because the dataset we used here is the same as that used by the aforementioned researcher (see Table S6 for details), we directly used the parameters of the model obtained by the aforementioned researcher to obtain the quantitative results of the generalized Anassari–Benam model in describing the deformation of brain cortex tissue. It is evident that our proposed modified model outperforms the generalized Anssari–Benam model in describing the multiaxis deformation of brain cortex tissue, both globally and locally (see Table 6). Here, we need to emphasize that when we reproduced the prediction performance of the generalized Anssari–Benam model based on the same parameters and dataset, we found that there is a deviation between its corresponding peak of relative error and that of [25]. The peaks of relative error obtained by us for both uniaxial tension and pure shear slightly exceed 25%. Anyway, our conclusion is consistent with [25], that is, the third-order Ogden model has improved prediction performance compared to the generalized Anssari–Benam model and the performance of our proposed modified model is comparable to the third-order Ogden model; so, our model should still be better than the generalized Anssari–Benam model proposed in [25]. In the next section, we also confirm that our proposed modified model remains stable and convex. In Table 6, we also provide the results of the modified Anssari–Benam model. Figure 11c,d also presents its relative error curves. Based on these results, it is clear that our modified model is still better.

3.1.3. Predictive Ability of the Modified Model to the Untested Deformation Mode

The Treloar's dataset on natural vulcanized rubber is the most commonly used dataset in verifying the ability of constitutive models to describe untested deformations [15,18,27,30–32]. Considering this, we first utilized this dataset to demonstrate the ability of our proposed modified model to predict untested deformations. But, to avoid controversy, here we will mainly use data from UT and PS to calibrate the model, and verify the predictive ability of the obtained model for the untested equibiaxial deformation mode. As introduced in the introduction, this is also a common situation in practice, where the experiments of uniaxial tension and pure shear are easier for researchers compared to the experiments of equibiaxial tension.

Maintaining consistency with Section 3.1.2 the experimental data of UT and PS from different deformation ranges of this dataset are applied to complete the parameter identification of constitutive models, respectively. The corresponding results are shown in Table 7 and Figure 12 below. Combining these quantitative and qualitative results, it can be seen that the proposed modified constitutive model not only has a good, predicted capability for the modes of deformation which have been tested such as uniaxial tension and pure shear, but can also effectively predict the equibiaxial mode of deformation which is untested. Specifically, the modified model has better predictive ability for the untested equibiaxial deformation mode in different deformation ranges than all models except for the third-order Ogden model in Table 7. In fact, like the other studies mentioned above [15,18,27,30–32], if only Treloar's data from uniaxial tension were used to calibrate these models, the effectively predictive ability of the proposed model for the other two untested deformation modes (ET and PS) would be superior to all models in the comparative list, including the third-order Ogden model (as shown in Table S15 and Figure S1 in the supporting information). This shows that the addition of the correction term not only does not affect the ability of the original Yeoh model to effectively predict untested deformations, but further improves its ability in the modified model.

Table 8 further summarizes the quantitative prediction results of our modified model calibrated by using data of UT and PS from the other four datasets (the corresponding prediction curve is shown in Figure S2). Obviously, the proposed modified constitutive model can also effectively predict the untested equibiaxial deformation of isoprene vulcanized rubber, unfilled silicone rubber, poly-arcylamide hydrogel and carbon-black-filled styrene butadiene rubber. Except for a higher total predictive error for the unfilled silicone rubber, the proposed modified model has a total predictive error of less than 5% for the other three types of rubber. Moreover, compared to other models, these results are also superior to most other models (as shown in Tables S19–S22 in the supporting information), especially the several existing improved models of the Yeoh model. It should be clarified that our modified model is not the only one with a relatively low effective prediction accuracy for untested equibiaxial deformation of the unfilled silicone rubber. As shown in Table S20 (see the supporting information), compared to our proposed modified model, the widely recognized third-order Ogden model and Alexander model with good performance have lower effective prediction accuracy for untested equibiaxial deformation modes of the unfilled silicone rubber. And, the Anssari–Benam model and the modified Anssari–Benam model are directly unable to effectively predict the untested equibiaxial deformation of the unfilled silicone rubber. Although the results show that the modified Yeoh model and the generalized Yeoh model have slightly high prediction accuracy for the untested equibiaxial deformation of the unfilled silicone rubber, their prediction accuracy for UT is not high such that their overall prediction accuracy is not excellent (with a total error of over 8%).

Table 7. The predictive results of different models calibrated by the data from UT and PS of natural vulcanized rubber [37].

Model	Large Deformation				Medium Deformation				Small Deformation			
	Error	UT	ET	PS	Error	UT	ET	PS	Error	UT	ET	PS
Ogden-N3	0.043	0.975	0.907	0.989	0.018	0.991	0.960	0.995	0.014	0.988	0.971	0.997
Alexander	0.175	0.983	0.501	0.991	0.111	0.983	0.689	0.993	0.027	0.988	0.934	0.998
Yeoh	0.086	0.933	0.839	0.971	0.095	0.952	0.794	0.968	0.128	0.941	0.756	0.918
Melly	0.069	0.957	0.844	0.993	0.039	0.988	0.905	0.989	0.022	0.987	0.957	0.989
Modified Yeoh	0.079	0.949	0.829	0.986	0.086	0.948	0.802	0.993	0.098	0.936	0.774	0.995
Generalized Yeoh	0.086	0.970	0.800	0.973	0.092	0.954	0.787	0.983	0.108	0.961	0.757	0.959
Anssari–Benam	0.082	0.972	0.813	0.971	0.096	0.952	0.790	0.970	0.100	0.932	0.773	0.995
Modified Anssari–Benam	0.225	0.626	0.757	0.942	0.029	0.988	0.936	0.991	0.015	0.986	0.971	0.998
This work	0.066	0.953	0.855	0.993	0.027	0.992	0.933	0.994	0.013	0.988	0.976	0.997

Notes: "Error" indicates the total error in predicting the three modes of deformation. UT, ET and PS represent the goodness of fit of different models for uniaxial tension, equibiaxial tension and pure shear, respectively. The model parameters are shown in Tables S16–S18 in supporting information.

Table 8. The predictive results of the modified model calibrated by the data from UT and PS of others rubber.

Different Materials	Error	UT	ET	PS
Isoprene vulcanized rubber	0.041	0.974	0.911	0.993
Unfilled silicone rubber	0.134	0.982	0.642	0.974
Poly-arcylamide hydrogel	0.035	0.992	0.914	0.990
Carbon-black-filled styrene butadiene rubber	0.049	0.968	0.938	0.949

Notes: "Error" indicates the total error in predicting the three modes of deformation. UT, ET and PS represent the goodness of fit of different models for uniaxial tension, equibiaxial tension and pure shear, respectively. The model parameters are shown in Tables S23–S26 in supporting information.

Anyway, in view of the above analysis, it can be believed that it is worthwhile to modify the Yeoh model by the proposed correction term in this study. The correction term expressed in the form of power function based on the principal stretches gives the modified model improved ability in predicting the untested equibiaxial deformation, which means that the proposed modified constitutive model can be applied to effectively predict the equibiaxial mode of deformation when the equibiaxial tension cannot be completed under limited hardware conditions.

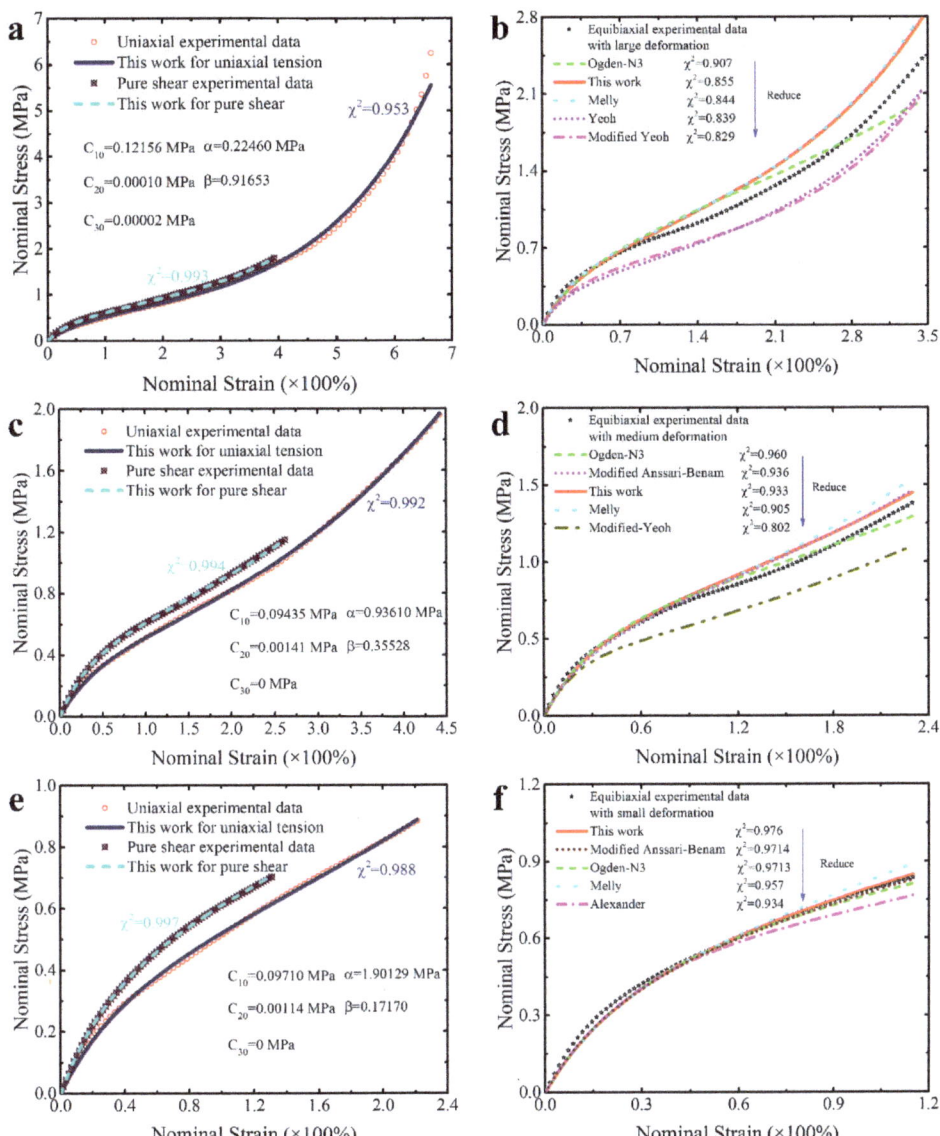

Figure 12. The prediction effects of the modified model for different deformation modes with different deformation ranges: (**a**,**b**) in large deformation; (**c**,**d**) in medium deformation; (**e**,**f**) in small deformation. Here, the parameters of models are calibrated by the experimental data of uniaxial tension and pure shear with different deformation ranges. Only the top five constitutive models are listed on the right side of the figure.

3.2. The A Posteriori Check of the Modified Model

Possessing the excellent ability to characterize the hyperelastic behavior is only a basic requirement for a constitutive model. If the constitutive model can be further used, it must obey the laws of thermodynamics, and it must also ensure that it has a solution to the boundary-value problem (as explained in Section 2.2). Accordingly, the a posteriori check on the proposed modified model is performed. The a posteriori check is conducted within the range of minimum to maximum principal stretch of the corresponding material datasets ($\lambda_{min} \leq \lambda \leq \lambda_{max}$). It is verified that the proposed modified constitutive model calibrated in the above subsection satisfies the inequalities of (12) and can make the matrix of D positive definite. This means that the modified model with calibrated parameters has a posteriori poly-convexity and Drucker stability. For a more intuitive display, Figure 13 demonstratively presents the iso-energy contour plots of the modified models with different calibrated parameters obtained in Section 3.1.2. It is clear that the stationary points are positioned in the undeformed state ($\lambda_1 = \lambda_2 = 1$), namely, the modified models with different calibrated parameters obtained based on different material datasets are stable and possess the convexity [4,52].

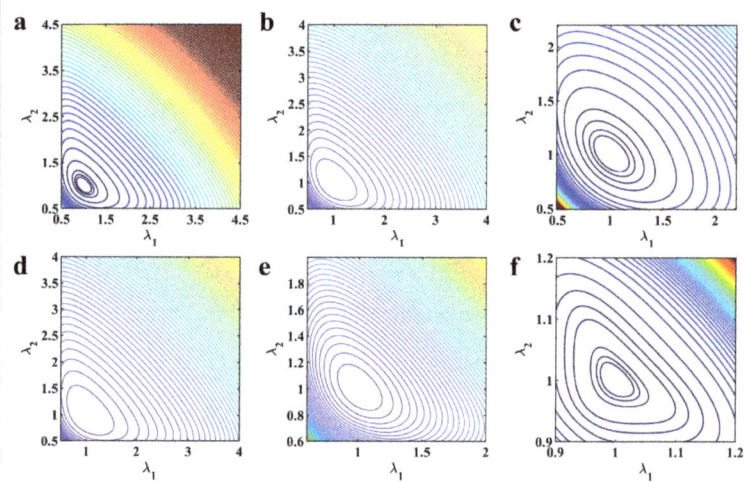

Figure 13. Iso-energy contour plots of the modified models with different calibrated parameters obtained based on different material datasets: (**a**) the dataset of natural vulcanized rubber containing medium deformation; (**b**) the dataset of isoprene vulcanized rubber; (**c**) the dataset of unfilled silicone rubber; (**d**) the dataset of poly-acrylamide hydrogel; (**e**) the dataset of carbon-black-filled styrene butadiene rubber; (**f**) the dataset of human brain cortex tissue.

3.3. Validation of the Parameter Identification Method

In order to validate the validity of the parameter identification method introduced in this study, several algorithms that have been used (including the damped least squares algorithm (L-M) [48], genetic algorithm (GA) [50], pattern search algorithm (PS) and the hybrid algorithm (PS-LM) consisting of pattern search and damped least squares [51]) are used to compare with the cyclic genetic-pattern search algorithm (CGA-PS) introduced in this study. These algorithms are used to identify the parameters of the third-order Ogden model and the modified constitutive model, respectively. The dataset used in the identification process is the dataset from the natural vulcanized rubber containing the large deformation. Considering that the L-M algorithm and the PS algorithm are affected by the initial value, the initial values of the L-M algorithm, the PS algorithm, and the PS-LM algorithm will be randomly generated by a random function within five initial intervals of 0–1, 0–0.5, 0–0.25, 0–0.1 and 0–0.05, respectively. And two different initial values are generated in each interval, which means that each algorithm needs to run independently twice with two different initial values within the corresponding interval. Therefore, each algorithm will run independently ten times due to being randomly assigned ten initial values separately. Owing to the inherent random search property, the initial population interval of the GA algorithm and the CGA-PS algorithm is fixed at 0–1, and they will also run ten times independently. Finally, for the ten sets of running results of each algorithm, only the set of results with the minimum total error ($Error_{total}$, see Equation (17)) will be recorded.

According to the minimum total error shown in Figure 14a, it can be seen that the parameter identification method based on the cyclic genetic-pattern search algorithm introduced in this study significantly improves the performance of constitutive models for characterizing hyperelastic properties. In view of the standard deviation of the total error shown in Figure 14b, the parameter identification method introduced in this study has the minimum standard deviation compared with other methods. So, there is reason to believe that the introduced method of parameter identification performs better in terms of uniform convergence. Figure 15 presents the prediction performance of different modes of deformation by the modified model based on different methods of parameter identification (the relevant prediction curve of the third-order Ogden model is shown in Figure S3). It is clear that the introduced method of parameter identification based on the cyclic genetic-pattern search algorithm (CGA-PS) in this study has better performance, which can make the final calibrated models have better characterization capability for different modes of deformation. In addition, we also compared the identification ability of the introduced parameter identification method and the parameter identification method integrated in ABAQUS by the three models of the Arruda–Boyce model, Ogden model and Yeoh model. This comparison procedure uses the experimental data consistent with the above. From this result (as shown in Table 9), it can be found that the introduced identification method also slightly outperforms the parameter identification method integrated in ABAQUS. Hence, one can see that the introduced method of CGA-PS is not only effective, but also has better performance in the process of parameter identification for the hyperelastic constitutive model.

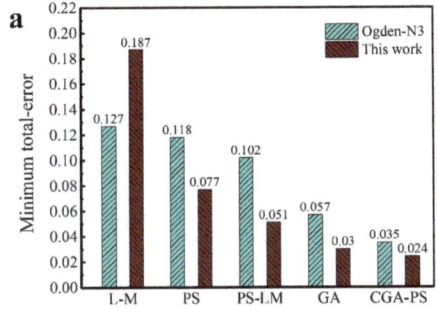
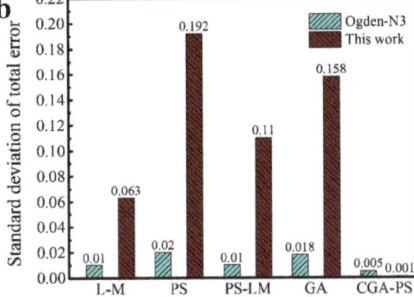

Figure 14. The comparison of different parameter identification methods: (**a**) the minimum total-error after ten runs of different parameter identification methods; (**b**) the standard deviation of total error after ten runs of different parameter identification methods.

Table 9. The comparison between the ability of the introduced parameter identification method and the ability of parameter identification method integrated in ABAQUS.

Method	Arruda–Boyce Model	Third-Order Ogden Model	Yeoh Model
ABAQUS	Error = 0.103 $\chi^2_{UT} = 0.894$ $\chi^2_{ET} = 0.877$ $\chi^2_{PS} = 0.920$	Error = 0.041 $\chi^2_{UT} = 0.925$ $\chi^2_{ET} = 0.978$ $\chi^2_{PS} = 0.974$	Error = 0.083 $\chi^2_{UT} = 0.932$ $\chi^2_{ET} = 0.858$ $\chi^2_{PS} = 0.962$
This work	Error = 0.087 $\chi^2_{UT} = 0.934$ $\chi^2_{ET} = 0.852$ $\chi^2_{PS} = 0.955$	Error = 0.035 $\chi^2_{UT} = 0.927$ $\chi^2_{ET} = 0.987$ $\chi^2_{PS} = 0.981$	Error = 0.077 $\chi^2_{UT} = 0.943$ $\chi^2_{ET} = 0.839$ $\chi^2_{PS} = 0.988$

Note: The $Error$ is the total prediction error, χ^2_{UT} is the goodness of fit for the uniaxial tension, χ^2_{ET} is the goodness of fit for the equibiaxial tension, and χ^2_{PS} is the goodness of fit for the pure shear. The model parameters are shown in Table S27 in supporting information.

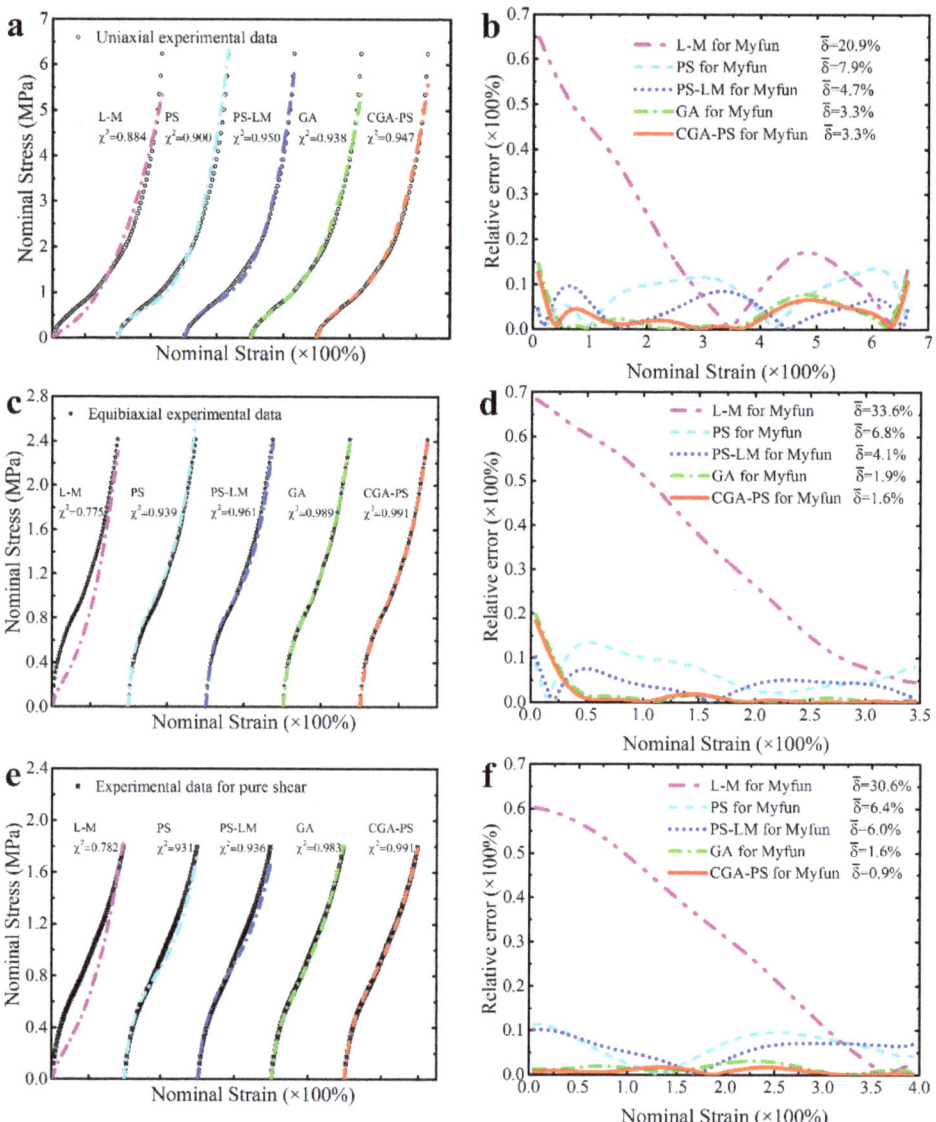

Figure 15. The comparison of prediction performance by the modified model based on different methods of parameter identification: (**a**,**b**) the prediction performance of uniaxial tension by the modified model based on different methods of parameter identification; (**c**,**d**) the prediction performance of equibiaxial tension by the modified model based on different methods of parameter identification; (**e**,**f**) the prediction performance of pure shear by the modified model based on different methods of parameter identification.

4. Conclusions

This paper presents a modified hyperelastic constitutive model based on the first strain invariant and the principal stretches. In order to more accurately identify the parameters of model, a special method of parameter identification based on the cyclic genetic-pattern search algorithm has also been introduced. Combining the experiment data of different rubber materials with the introduced parameter identification method, the performance of the proposed modified constitutive model is fully evaluated. The results show that the proposed modified model not only possesses a significantly improved ability to predict multiaxial deformation, but also has a wider range of material applicability. This advantage is not only reflected in the comparison with the original Yeoh model, but also in the comparison with other improved Yeoh models (such as the modified Yeoh model, the generalized Yeoh model and the Melly model). Even compared with the excellent third-order Ogden model and Alexander model, our proposed modified model with only five undetermined parameters is not necessarily inferior in characterizing the multiaxial deformation of rubber materials. For example, the overall prediction accuracy of the modified model for isoprene vulcanized rubber and poly-acrylamide hydrogel is slightly better than that of the third-order Ogden model and Alexander model. Anyway, for the five rubber materials used in this study, our modified model has a similar level of total prediction accuracy as the two models of the third-order Ogden model and Alexander model (total error is less than 5%). Furthermore, our modified model is also proven to have a good ability to characterize the deformation of human brain cortex tissue. Its prediction accuracy for multiaxial deformations of human brain cortex tissue is not only similar to that of the third-order Ogden model (the third-order Ogden model is considered to be the best model currently characterizing the deformation of the human brain cortex tissue.), but also has no loss of convexity compared to the third-order Ogden model. In addition, the proposed modified model in this study is also proven to hold the improved ability to predict the untested equibiaxial deformation of most rubber materials used in this study. The modified constitutive model calibrated based on different datasets is also verified to have a posteriori Drucker stability and polyconvexity, which lays a foundation for the modified constitutive model to be applied to finite element analysis.

Based on the above conclusions, we believe that the main advantages of the modified constitutive model proposed by us are as follows:

Firstly, although the modified model has five undetermined parameters, it still maintains a relatively simple functional form compared to those models containing logarithmic, exponential or integral expressions, making it easier to perform related mathematical calculations. Having five parameters also ensures that it is neither incompatible with fourth-order weak nonlinear elasticity theory due to having too few parameters, nor does it face difficulties in parameter identification due to having too many parameters.

Secondly, thanks to our proposed correction term based on the principal stretch, this modified constitutive model has significantly improved ability for predicting multiaxial deformation. Moreover, this improved predictive ability has certain universality, that is, the modified model can be applied to various different types of rubber materials (including natural unfilled or filled rubber, silicone rubber and hydrogel). There are currently few studies that apply their model to so many types of material.

Thirdly, this modified constitutive model can, with relative accuracy, predict the multiaxial deformation of human brain cortex tissue (including uniaxial tension, uniaxial compression and simple shear) while ensuring convexity. This gives it the potentiality to characterize the biomechanics of soft biological tissues.

Finally, the modified constitutive model also has an improved capacity to predict untested equibiaxial deformation. This advantage is very useful in the situation where equibiaxial tension cannot be completed under limited hardware conditions.

Compared with other parameter identification methods, the introduced method of parameter identification has been proved not only to take into account different modes of deformation, but also makes models have better performance. The excellent capability of the introduced method of parameter identification benefits from the strong ability of global search of the genetic algorithm and the strong ability of local search of the pattern search algorithm, and the addition of the cyclic structure further weakens its dependence on the initial value, thereby making its uniform convergence better.

Supplementary Materials: The following supporting information can be downloaded at https://www.mdpi.com/article/10.3390/polym15153172/s1.

Author Contributions: Conceptualization, W.W. and Y.L.; data curation, W.W., Y.L. and Z.X.; formal analysis, W.W. and Y.L.; funding acquisition, Y.L. and Z.X.; investigation, W.W.; methodology, W.W.; project administration, Y.L. and Z.X.; resources, W.W. and Y.L.; software, W.W.; supervision, Y.L. and Z.X.; validation, W.W.; visualization, W.W.; writing—original draft, W.W.; writing—review and editing, Y.L. and Z.X. All authors have read and agreed to the published version of the manuscript.

Funding: This work was supported by the Major Research Plan of National Natural Science Foundation of China (grant number 91848202). This work was also supported by the Self-Planned Task of State Key Laboratory of Robotics and System (HIT) (grant number SKLRS202106B) and the Special Foundation (Pre-Station) of China Postdoctoral Science (grant number 2021TQ0089).

Institutional Review Board Statement: Not applicable.

Data Availability Statement: All data from this study are presented in the paper.

Conflicts of Interest: The authors declare no conflict of interest.

References

1. Kanyanta, V.; Ivankovic, A. Mechanical characterisation of polyurethane elastomer for biomedical applications. *J. Mech. Behav. Biomed.* **2010**, *3*, 51–62. [CrossRef] [PubMed]
2. Rosset, S.; Shea, H.R. Flexible and stretchable electrodes for dielectric elastomer actuators. *Appl. Phys. A-Mater.* **2013**, *110*, 281–307. [CrossRef]
3. Wang, H.B.; Totaro, M.; Beccai, L. Toward perceptive soft robots: Progress and challenges. *Adv. Sci.* **2018**, *5*, 1800541. [CrossRef] [PubMed]
4. Osterlof, R.; Wentzel, H.; Kari, L. An efficient method for obtaining the hyperelastic properties of filled elastomers in finite strain applications. *Polym. Test.* **2015**, *41*, 44–54.
5. Boyce, M.C.; Arruda, E.M. Constitutive models of rubber elasticity: A review. *Rubber Chem. Technol.* **2000**, *73*, 504–523. [CrossRef]
6. Mansouri, M.R.; Darijani, H. Constitutive modeling of isotropic hyperelastic materials in an exponential framework using a self-contained approach. *Int. J. Solids Struct.* **2014**, *51*, 4316–4326. [CrossRef]
7. Marckmann, G.; Verron, E. Comparison of hyperelastic models for rubber-like materials. *Rubber Chem. Technol.* **2006**, *79*, 835–858. [CrossRef]
8. Gernay, T.; Millard, A.; Franssen, J.M. A multiaxial constitutive model for concrete in the fire situation: Theoretical formulation. *Int. J. Solids Struct.* **2013**, *50*, 3659–3673. [CrossRef]
9. Mooney, M. A theory of large elastic deformation. *J. Appl. Phys.* **1940**, *11*, 582–592. [CrossRef]
10. Rivlin, R.S. Large elastic deformations of isotropic materials iv further developments of the general theory. *Philos. T R Soc.-A Math. Phys. Sci.* **1948**, *241*, 379–397.
11. Attard, M.M.; Hunt, G.W. Hyperelastic constitutive modeling under finite strain. *Int. J. Solids Struct.* **2004**, *41*, 5327–5350. [CrossRef]
12. Haines, D.W.; Wilson, W.D. Strain-energy density function for rubberlike materials. *J. Mech. Phys. Solids* **1979**, *27*, 345–360. [CrossRef]
13. Hartmann, S.; Neff, P. Polyconvexity of generalized polynomial-type hyperelastic strain energy functions for nearincompressibility. *Int. J. Solids Struct.* **2003**, *40*, 2767–2791. [CrossRef]
14. Lopez-Pamies, O. A new I1-based hyperelastic model for rubber elastic materials. *Comptes Rendus Mec.* **2010**, *338*, 3–11. [CrossRef]
15. Seibert, D.J.; Schoche, N. Direct comparison of some recent rubber elasticity models. *Rubber Chem. Technol.* **2000**, *73*, 366–384. [CrossRef]
16. Tschoegl, N.W. Constitutive equations for elastomers. *J. Polym. Sci. A-Polym. Chem.* **1971**, *9*, 1959–1970. [CrossRef]
17. Meunier, L.; Chagnon, G.; Favier, D.; Orgeas, L.; Vacher, P. Mechanical experimental characterization and numerical modelling of an unfilled silicone rubber. *Polym. Test* **2008**, *27*, 765–777. [CrossRef]
18. Khajehsaeid, H.; Arghavani, J.; Naghdabadi, R. A hyperelastic constitutive model for rubber-like materials. *Eur. J. Mech. A-Solid* **2013**, *38*, 144–151. [CrossRef]
19. Kulcu, I.D. A hyperelastic constitutive model for rubber-like materials. *Arch. Appl. Mech.* **2020**, *90*, 615–622. [CrossRef]
20. Darijani, H.; Naghdabadi, R. Hyperelastic materials behavior modeling using consistent strain energy density functions. *Acta Mech.* **2010**, *213*, 235–254. [CrossRef]
21. Ogden, R.W. Large deformation isotropic elasticity–on the correlation of theory and experiment for incompressible rubberlike solids. *Proc. R. Soc. London. A. Math. Phys. Sci.* **1972**, *326*, 565–584. [CrossRef]
22. Rubin, M.B.; Ehret, A.E. An invariant-based Ogden-type model for incompressible isotropic hyperelastic materials. *J. Elasticity* **2016**, *125*, 63–71. [CrossRef]

23. Melly, S.K.; Liu, L.W.; Liu, Y.J.; Leng, J.S. A phenomenological constitutive model for predicting both the moderate and large deformation behavior of elastomeric materials. *Mech. Mater.* **2022**, *165*, 104179. [CrossRef]
24. Anssari-Benam, A.; Horgan, C.O. A three-parameter structurally motivated robust constitutive model for isotropic incompressible unfilled and filled rubber-like materials. *Eur. J. Mech. A-Solid* **2022**, *95*, 104605. [CrossRef]
25. Anssari-Benam, A. Large isotropic elastic deformations: On a comprehensive model to correlate the theory and experiments for incompressible rubber-like materials. *J. Elasticity* **2023**, *153*, 219–244. [CrossRef]
26. Dal, H.; Açıkgöz, K.; Badienia, Y. On the performance of isotropic hyperelastic constitutive models for rubber-like materials: A state of the art review. *Appl. Mech. Rev.* **2021**, *73*, 020802. [CrossRef]
27. Steinmann, P.; Hossain, M.; Possart, G. Hyperelastic models for rubber-like materials: Consistent tangent operators and suitability for Treloar's data. *Arch. Appl. Mech.* **2012**, *82*, 1183–1217. [CrossRef]
28. Yeoh, O.H. Characterization of elastic properties of carbon-black-filled rubber vulcanizates. *Rubber Chem. Technol.* **1990**, *63*, 792–805. [CrossRef]
29. Urayama, K. An experimentalist's view of the physics of rubber elasticity. *J. Polym. Sci. Pol. Phys.* **2006**, *44*, 3440–3444. [CrossRef]
30. Melly, S.K.; Liu, L.W.; Liu, Y.J.; Leng, J.S. Modified yeoh model with improved equibiaxial loading predictions. *Acta. Mech.* **2022**, *233*, 437–453. [CrossRef]
31. Melly, S.K.; Liu, L.W.; Liu, Y.J.; Leng, J.S. Improved Carroll's hyperelastic model considering compressibility and its finite element implementation. *Acta Mech. Sin.* **2021**, *37*, 785–796. [CrossRef]
32. Fujikawa, M.; Maeda, N.; Koishi, M. Performance evaluation of hyperelastic models for carbon-black-filled SBR vulcanizates. *Rubber Chem. Technol.* **2020**, *93*, 142–156. [CrossRef]
33. Yeoh, O.H. Some forms of the strain energy function for rubber. *Rubber Chem. Technol.* **1993**, *66*, 754–771. [CrossRef]
34. Hoss, L.; Marczak, R.J. A new constitutive model for rubber-like materials. In *20th International Congress of Mechanical Engineering*; ABCM: Gramado, Brazil, 2010.
35. Benam, A.; Bucchi, A.; Saccomandi, G. On the central role of the invariant I2 in nonlinear elasticity. *Int. J. Eng. Sci.* **2021**, *163*, 103486. [CrossRef]
36. Hohenberger, T.W.; Windslow, R.J.; Pugno, N.M.; Busfield, J.J. A constitutive model for both low and high strain nonlinearities in highly filled elastomers and implementation with user-defined material subroutines in ABAQUS. *Rubber Chem. Technol.* **2019**, *92*, 653–686. [CrossRef]
37. Treloar, L.R.G. Stress-strain data for vulcanized rubber under various type of deformation. *Trans. Faraday Soc.* **1944**, *40*, 59–70. [CrossRef]
38. Kawabata, S.; Matsuda, M.; Tei, K.; Kawai, H. Experimental survey of the strain energy density function of isoprene rubber vulcanizate. *Macromolecules* **1981**, *14*, 154–162. [CrossRef]
39. Yohsuke, B.; Urayama, K.; Takigawa, T.; Ito, K. Biaxial strain testing of extremely soft polymer gels. *Soft Matter* **2011**, *7*, 2632–2638. [CrossRef]
40. Fujikawa, M.; Maeda, N.; Yamabe, J.; Kodama, Y.; Koishi, M. Determining stress-strain in rubber with in-plane biaxial tensile tester. *Exp. Mech.* **2014**, *54*, 1639–1649. [CrossRef]
41. Budday, S.; Sommer, G.; Birkl, C.; Langkammer, C.; Haybaeck, J.; Kohnert, J.; Bauer, M.; Paulsen, F.; Steinmann, P.; Kuhl, E.; et al. Mechanical characterization of human brain tissue. *Acta Biomate* **2017**, *15*, 319–340. [CrossRef]
42. Anssari-Benam, A.; Bucchi, A. A generalised neo-hookean strain energy function for application to the finite deformation of elastomers. *Int. J. Non-Lin. Mech.* **2021**, *128*, 103626. [CrossRef]
43. Alexander, H. A constitutive relation for rubber-like materials. *Int. J. Eng. Sci.* **1968**, *6*, 549–563. [CrossRef]
44. Li, X.B.; Wei, Y.T. An improved yeoh constitutive model for hyperelstic material. *Eng. Mech.* **2016**, *12*, 38–43. [CrossRef]
45. Ogden, R.W. *Non-Linear Elastic Deformations*; Courier Corporation: Toronto, ON, Canada, 1997.
46. Bonet, J.; Gil, A.J.; Ortigosa, R. A computational framework for polyconvex large strain elasticity. *Comput. Method Appl. M* **2015**, *283*, 1061–1094. [CrossRef]
47. Johnson, A.; Quigley, C.; Mead, J. Large strain viscoelastic constitutive models for rubber, Part I: Formulations. *Rubber Chem. Technol.* **1994**, *67*, 904–917. [CrossRef]
48. Ogden, R.W.; Saccomandi, G.; Sgura, I. Fitting hyperelastic models to experimental data. *Comput. Mech.* **2004**, *34*, 484–502. [CrossRef]
49. Gendy, A.S.; Saleeb, A.F. Nonlinear material parameter estimation for characterizing hyper elastic large strain models. *Comput. Mech.* **2000**, *25*, 66–77. [CrossRef]
50. Dal, H.; Badienia, Y.; Acıkgoz, K.; Denli, F.A. A novel parameter identification toolbox for the selection of hyperelastic constitutive models from experimental data. In Proceedings of the 7th GACM Colloquium on Computational Mechanics for Young Scientists from Academia and Industry, Stuttgart, Germany, 11–13 October 2017; ACM: Stuttgart, Germany, 2017.

51. Wu, Y.F.; Wang, H.; Li, A.Q. Parameter identification methods for hyperelastic and hyper-viscoelastic models. *Appl. Sci.* **2016**, *6*, 386. [CrossRef]
52. Anssari-Benam, A.; Bucchi, A.; Horgan, C.O.; Saccomandi, G. Assessment of a new isotropic hyperelastic constitutive model for a range of rubber-like materials and deformations. *Rubber Chem. Technol.* **2022**, *95*, 200–217. [CrossRef]

Disclaimer/Publisher's Note: The statements, opinions and data contained in all publications are solely those of the individual author(s) and contributor(s) and not of MDPI and/or the editor(s). MDPI and/or the editor(s) disclaim responsibility for any injury to people or property resulting from any ideas, methods, instructions or products referred to in the content.

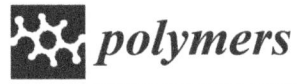

Article

Phase Change Microcapsule Composite Material with Intelligent Thermoregulation Function for Infrared Camouflage

Ying Su [1], Xiaoming Zhao [1,2,3,*] and Yue Han [1]

1. School of Textile Science and Engineering, Tiangong University, Tianjin 300387, China
2. Tianjin Key Laboratory of Advanced Textile Composites, Tiangong University, Tianjin 300387, China
3. Tianjin Municipal Key Laboratory of Advanced Fiber and Energy Storage, Tiangong University, Tianjin 300387, China
* Correspondence: texzhao@163.com

Abstract: The infrared camouflage textile materials with soft and wear-resistant properties can effectively reduce the possibility of soldiers and military equipment being exposed to infrared detectors. In this paper, the infrared camouflage textile composites with intelligent temperature adjustment ability were prepared by different methods, using phase change microcapsule as the main raw material and high polymer polyurethane as the matrix, combining the two factors of temperature control and emissivity reduction. It was tested by differential scanning calorimeter, temperature change tester, infrared emissivity tester, and infrared imager. The results show that the temperature regulation effect of textile materials finished by coating method is better than dip rolling method, the temperature regulation ability and presentation effect are the best when the microcapsule content is 27%. When the bottom layer of infrared camouflage textile composite is 27% phase change microcapsule and the surface layer is 20% copper powder, its infrared emissivity in the band of 2–22 μm is 0.656, and the rate of heating and cooling is obviously slowed down. It has excellent heat storage and temperature regulation function, which can reduce the skin surface temperature by more than 6 °C and effectively reduce the infrared radiation. This study can provide reference for laboratory preparation and industrial production of infrared camouflage composite material. The infrared camouflage textile composite prepared are expected to be used in the field of military textiles.

Keywords: phase change microcapsules; infrared camouflage; emissivity; intelligent temperature regulation; textile composites

Citation: Su, Y.; Zhao, X.; Han, Y. Phase Change Microcapsule Composite Material with Intelligent Thermoregulation Function for Infrared Camouflage. *Polymers* **2023**, *15*, 3055. https://doi.org/10.3390/polym15143055

Academic Editors: Jiangtao Xu and Sihang Zhang

Received: 19 June 2023
Revised: 9 July 2023
Accepted: 14 July 2023
Published: 15 July 2023

Copyright: © 2023 by the authors. Licensee MDPI, Basel, Switzerland. This article is an open access article distributed under the terms and conditions of the Creative Commons Attribution (CC BY) license (https:// creativecommons.org/licenses/by/ 4.0/).

1. Introduction

Matter is generally divided into three phases: solid, liquid, and gas. The transition between different phase states of the same material is called phase transition [1]. The substance whose state can be changed is called phase change material (PCM). When the phase change occurs, there is a significant energy exchange between the material and the environment, which will be strongly coupled with the heat transfer, so that the material has a certain temperature control and heat release function [2,3]. With this capability of phase change materials, the temperature around the working source or materials can be adjusted and controlled to reduce the mismatch between energy supply and demand in time and speed [4]. Therefore, phase change materials are applied broadly in the field of energy storage and temperature regulation [5,6]. However, phase change materials have problems such as large volume changes, easy leakage, and low thermal conductivity. Microencapsulation of phase change materials is an advanced application method. Microcapsule phase change material (MPCM), also known as phase change microcapsule, is a new type of composite phase change material with core-shell structure. It is coated with a stable polymer film on the surface of solid-liquid phase change material particles.

The shell structure of microcapsules can provide good protection for phase change core materials, improve the stability of phase change materials, prevent chemical reactions with the outside world and leakage during long-term cyclic use, and significantly increase the contact area with the matrix material to improve thermal conductivity, thereby improving the working performance of phase change materials [7–9]. When the external temperature changes, the core material in the microcapsule will undergo phase change. The phase change material absorbs or releases a large amount of heat, and the temperature of the microcapsule itself remains constant, to achieve the effect of intelligent temperature regulation [10–13]. Phase change microcapsules with temperature regulation ability are widely used in construction [14–18], solar energy [19], food industry [20], textile [21,22], and other fields.

In the modern battlefield, the infrared radiation energy of general military targets is higher than the background, so it is easy to find out by using infrared detectors. According to Stefan Boltzmann's Law (1) [23], the total infrared radiation energy of an object is directly proportional to the fourth power of its emissivity and absolute temperature. Therefore, the possibility of the target being discovered by the infrared detector can be reduced by reducing the emissivity and controlling the temperature, to achieve its camouflage effect in the infrared band. Therefore, infrared camouflage materials that protect military targets without changing the shape and structure of the target have attracted extensive attention in the national defense and military industry [24]. Various infrared camouflage materials developed around fibers and fiber products are called infrared camouflage textile materials. Infrared camouflage textile materials are soft, portable, and wearable. They are the main raw materials of infrared camouflage clothing, backpacks, camouflage nets and tents. They can provide guarantee for the survival of soldiers and weapons and equipment, and plays an extremely important role in the battlefield [25,26]. In terms of reducing infrared emissivity, it mainly includes developing new low emissivity fibers [27–30], modifying existing fibers [31–33] or coating low infrared emissivity coatings [34–36]. In terms of temperature control, in addition to thermal insulation and structural design [37–40], the temperature regulating textile [41–44] and infrared camouflage textile [45–47] are prepared by combining phase change microcapsules with textile, which can effectively reduce the infrared radiation energy of the target [48–50]. However, it is difficult to use phase change microcapsules for infrared camouflage alone. Its phase change temperature, latent heat of phase change, and thermal conductivity can hardly meet the requirements of thermal camouflage. Only by combining with other functional materials can infrared camouflage be better realized [51]. According to Kirchhoff's law, opaque objects with high reflectivity generally have low emissivity. Metal is a typical low-emissivity material, which is generally used in the field of infrared camouflage in the form of coating. Among them, copper and aluminum have become the main force of metal fillers due to their low cost and easy availability, excellent performance, and wide application. In this paper, phase change microcapsules are finished on the fabric, and the temperature-regulated fabric is obtained by changing different parameters. On this basis, the infrared camouflage fabric was prepared by adding low emissivity materials. Then its temperature adjustment ability, infrared camouflage effect, and mechanism are analyzed systematically. Compared with the untreated fabric, the prepared textile has a certain degree of infrared camouflage ability, which can delay the speed of temperature rise and effectively reduce the infrared thermal radiation.

$$E = \sigma \varepsilon T^4 \qquad (1)$$

where E is the infrared radiation (J/(s·m^2)), σ is Stefan Boltzmann constant, ε is the emissivity of the target surface, and T is the thermodynamic temperature of the target surface (K).

2. Materials and Methods

2.1. Materials

The fabric used in the experiment was cotton fabric, which was purchased from Hongfei Textile Manufacturing, Baoding, China. Phase change microcapsules are prepared by in-situ polymerization with paraffin as core material and urea-formaldehyde resin as wall material. Urea and formaldehyde aqueous solutions were purchased from Meryer (Shanghai) Chemical Technology, Shanghai, China. Paraffin, OP emulsifier, triethanolamine, citric acid and petroleum ether were purchased from Beijing enokai Technology, Beijing, China. Polyurethane resin was purchased from Guangzhou Yuheng environmental protection materials, China. Hollow glass beads were purchased from Henan Bairun casting materials, Zhengzhou, China. Silicon dioxide was purchased from Jiangsu Tianxing's new materials, Huaian, China. The copper powder was purchased from Nangong Xindun alloy welding material spraying, Xingtai, China. Aluminum powder was purchased from Shanghai Aladdin Biochemical Technology, Shanghai, China. All reagents were of analytical grade and used directly without further purification. The details are shown in Table 1.

Table 1. Material Information.

Material	Particle Size/Model	Manufacturer
Hollow glass beads	30–100 μm	Henan Bairun casting materials, China
Silicon dioxide	20 nm	Jiangsu Tian xing's new materials, China
Copper powder	38 μm	Nangong Xindun alloy welding mate-rial spraying, China
Aluminum powder	25 μm	Shanghai Aladdin Bio-chemical Technology, China
Polyurethane	PU2540	Guangzhou Yuheng environmental protection materials, China
Defoamer	AFE-1410	Shandong Yousuo Chemical Technology, China
Thickener	7011	Guangzhou Dianmu Composite Materials Business Department, China
Dispersant	5040	Shandong Yousuo Chemical Technology, China
Urea, Formaldehyde aqueous solutions		Meryer (Shanghai) Chemical Technology, China
Paraffin, OP emulsifier, Triethanolamine, Citric acid, Petroleum ether		Beijing enokai Technology, China

2.2. Methods

2.2.1. Preparation of Phase Change Microcapsules

Mix urea and formaldehyde aqueous solution in a certain proportion, drop triethanolamine to adjust the pH value of the solution to be weakly alkaline, react at 70 °C for 1 h, and add deionized water to form a stable urea/formaldehyde prepolymer solution. Add a certain amount of OP emulsifier and paraffin into deionized water, heat and melt, emulsify and disperse for 30 min (3000 r/min) at 60 °C, and form a stable emulsion. Drop the prepolymer solution into the emulsion and stir for 20 min after dropping. Then slowly add citric acid solution, adjust the final pH value of the solution to be acidic, keep the temperature at 60 °C for reaction for 1 h, and then raise the temperature to 90 °C, and keep the temperature for reaction for 2 h. After the reaction, the microcapsule lotion was poured out, cooled, separated and filtered. The obtained microcapsules were washed twice with petroleum ether and deionized water, and dried to obtain white powder microcapsules.

2.2.2. Preparation of Phase Change Microcapsule Temperature Regulating Textile by Dip Rolling

Disperse the phase change microcapsule particles in water, add dispersant, adhesive, and penetrant, mix evenly to obtain the phase change microcapsule solution. The cotton fabric with 10×10 cm^2 was washed and dried. Put it into a beaker containing phase change microcapsule solution and fully wet it, with a bath ratio of 1:20. The phase change microcapsule fabric was obtained by two dipping and two rolling processes, drying at 80 °C for 5 min, and then baking at 120 °C for 2 min. Two groups of phase change microcapsule fabrics were prepared by changing the microcapsule content and adhesive content respectively.

2.2.3. Preparation of Phase Change Microcapsule Temperature Regulating Textile by Coating Method

Disperse the phase change microcapsule particles in water, add adhesive, thickener, dispersant and defoamer, and stir evenly to obtain phase change microcapsule coating.

Take 20 × 20 cm² cotton fabric, washed, dried, and ironed flat. The coating was evenly coated on the cotton fabric by a small sample coating machine, dried at 80 °C for 10 min, and dried to obtain the phase change microcapsule fabric. Three groups of phase change microcapsule fabrics were prepared by changing the content of phase change microcapsule, the thickness of coating and the type of thermally conductive materials.

2.2.4. Preparation of Infrared Camouflage Textile

In this paper, aluminum powder and copper powder are selected as coating materials with low emissivity, combined with phase change microcapsules to achieve a better-infrared camouflage effect on the fabric. Firstly, aluminum powder or copper powder is directly added into the phase change microcapsule solution, mixed evenly to obtain the infrared camouflage coating, and according to Section 2.2.3 to prepare infrared camouflage fabric. In addition, aluminum powder or copper powder shall be directly mixed with the adhesive to obtain low emissivity coating, and then phase change microcapsule coating can be obtained according to Section 2.2.3. The infrared camouflage fabric was prepared by coating phase change microcapsule coating and low emissivity coating on cotton fabric in turn.

2.3. Characterizations and Measurements

The characteristic functional groups of samples were measured by infrared spectrometer (FRONTIER, FT-IR, made by PerkinElmer, Waltham, MA, USA). The phase change microcapsules and fabrics were observed by scanning electron microscope (HITACHI, SEM, made by Hitachi Limited, Tokyo, Japan), and the distribution of microcapsules and metal particles on the fabric surface was compared. The enthalpy values of phase change microcapsules, unfinished fabrics and phase change microcapsule temperature regulating fabrics were measured by differential scanning calorimetry (NETZSCH, DSC200F3, made by Netzsch Group, Bavaria, Germany). The heating temperature range was 20–50 °C, the cooling temperature range was 50–20 °C, and the heating and cooling rates were 5K/min. The time when the sample rises to a specific temperature is measured by a self-made temperature rising instrument to evaluate its temperature regulation ability. The infrared thermal source is an infrared lamp (Philips 175R, made by Philips, Amsterdam, The Netherlands). The temperature rising range is 20–60 °C. The unfinished fabrics and infrared camouflage fabrics with different parameters were heated on a constant temperature (simulating human body surface temperature) test bench, and the infrared thermal image was taken with an infrared thermal imager (FlIR®TG165, made by FLIR Systems, Wilsonville, OR, USA) to test their thermal insulation and infrared camouflage properties. The emissivity of infrared camouflage fabric was measured by far infrared emissivity tester (TSS-5X, made by Japan sensor corporation, Tokyo, Japan).

3. Results

3.1. Characterization and Analysis of Phase Change Microcapsules

The phase change microcapsules used in this experiment are prepared with paraffin as the core material and urea formaldehyde resin as the wall material. The structural diagram is shown in Figure 1a. From the SEM Figure 1b,c, it can be seen that the phase change microcapsule is spherical with a particle size of 10 μm or so. According to the infrared spectrum in Figure 1d, the phase change microcapsule has 7 absorption peaks. The absorption peaks at 2926 cm^{-1} and 2855 cm^{-1} are related to the asymmetric and symmetric stretching vibration of C-H, the absorption peak at 1467 cm^{-1} is caused by the bending vibration of C-H$_2$, the absorption peak at 721 cm^{-1} is caused by the rocking vibration of C-H$_2$, and the absorption peak at 1745 cm^{-1} is caused by the stretching vibration of C=O, the absorption peaks at 1171 cm^{-1} and 1111 cm^{-1} are caused by the stretching vibration of C-N, and the band at 804 cm^{-1} corresponds to the bending vibration of N-H. It can be seen that there are both characteristic peaks of paraffin and urea formaldehyde resin in the infrared spectrum, which also proves that the microcapsule is composed of wall urea formaldehyde resin and core paraffin. Figure 1e shows the DSC curve of phase change microcapsules. The heat storage and temperature regulation performance of phase change microcapsules are mainly determined by the solid-liquid phase change of its core paraffin. During heating up, the phase change microcapsules began to melt and absorb heat at 24.4 °C, and the phase change temperature range was 24.4–34.7 °C. During cooling, the phase change microcapsules began to solidify and release heat at 25.6 °C, and the phase change temperature range was 25.6–19.0 °C. According to the indicators formulated by outlast, the body surface contact air layer is 18.3–29.4 °C, belonging to the cold climate temperature area, the body surface contact air layer is 26.7–37.8 °C, belonging to the mild or comfortable temperature area, and the body surface contact air layer is 32.2–43.3 °C, belonging to the temperature area during hot or intense exercise [52]. The phase change microcapsule used in this paper can be combined with fabric to adjust the temperature and infrared radiation energy of the body surface air layer in cold climates.

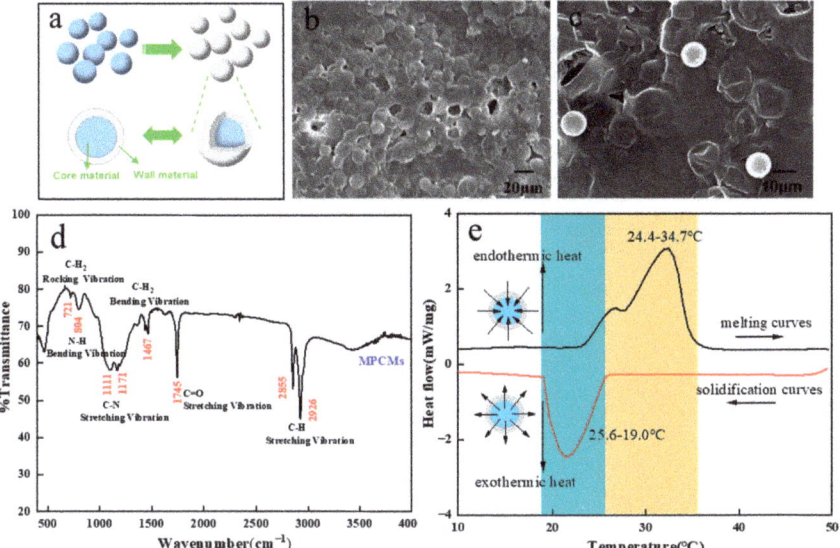

Figure 1. (**a**) Structure diagram of phase change microcapsule, (**b**,**c**) SEM of phase change microcapsules under different magnification of electron microscope, (**d**) Fourier transform infrared spectroscopy of phase change microcapsules, (**e**) DSC curve of phase change microcapsules.

3.2. Preparation and Performance Analysis of Phase Change Microcapsule Temperature Regulating Fabric

According to the steps described in Sections 2.2.2 and 2.2.3, phase change microcapsules are combined with cotton fabric by dip rolling and coating (Figure 2a,b). Before finishing, the surface of cotton fiber is smooth and tidy, flat and longitudinally twisted (Figure 2c). The phase change microcapsule solution was treated on the cotton fabric by dip rolling. The observation of the SEM (Figure 2d) showed that the phase change microcapsule particles were attached to the bending and depression of the cotton fiber through the adhesive. The phase change microcapsule coating was applied to the cotton fabric. It can be seen from the SEM image (Figure 2e) that the adhesive wrapped the phase change microcapsule and covered the surface of the cotton fabric to form a complete coating.

When the phase change microcapsule temperature regulating fabric is prepared by dip rolling, the process parameters shall be consistent, the content of phase change microcapsules and adhesives is the main factor affecting the related properties of fabrics. Therefore, we analyzed the effects of different content of phase change microcapsules and adhesives on the temperature regulation ability of the fabric. The reference sample in the figure is untreated cotton fabric. It can be seen from Figure 3a that the content of phase change microcapsules basically does not affect the phase change initial temperature of the sample. The initial temperature of exothermic and endothermic is around 28 °C. The latent heat of phase transformation of the sample increases with the increase of the content of microcapsules, and the latent heat of phase transformation of the sample is the largest when the percentage content is 36%, because the more the content, the more phase change microcapsules attached to the fabric after dip rolling treatment, and the greater the overall latent heat of phase change. The fabric treated by phase change microcapsule can absorb or release a certain amount of heat, so as to achieve the purpose of temperature regulation. As can be seen from Figure 3b, the phase transition latent heat of the sample increases with the increase of the binder content. Because the larger the binder content, the more phase change microcapsules adhere to the fabric surface. The greater the latent heat of phase change of fabric, the more obvious the effect of heat absorption and release. With the increase of binder content, the phase transition temperature of the sample will also change. The reason is that the added adhesive can block the heat transfer and increase the phase transition temperature of the sample. Figure 3c shows the temperature rise curve of samples with different content of phase change microcapsules. When the content of the phase change microcapsule is 9%, the heating rate of the sample is the fastest, and it takes 65 s to rise from 20 to 60 °C. When the content of the phase change microcapsule is 27%, the heating rate of the sample is the slowest, and it takes 83 s to rise from 20 to 60 °C. That is when the content of phase change microcapsule is 27%, the temperature regulation and heat storage effect of the sample is the best. The reason is that the existence of phase change microcapsules will delay the rate of fabric temperature change, that is, the heat released and absorbed will be temporarily supplemented and stored through phase change materials. When the content of phase change microcapsules is 36%, the heating rate of the sample is not the slowest. The reason is that when the content of phase change microcapsules in the solution is too high, agglomeration will occur in the mixing process, resulting in poor dispersion effect of phase change microcapsules, which affects the overall heat storage capacity of the sample. Figure 3d is the temperature rise curve of the sample with different adhesive content. When the adhesive content is 50%, the heating rate of the sample is the slowest, and the time is 95 s. When the content of adhesive is 20%, the heating rate of the sample is the fastest, and the time is 60 s. This may be because when the binder content is low, the content of phase change microcapsules entering the fabric interior and attached to the fabric surface after dip rolling treatment is relatively small. When the binder content is high, the proportion of the binder solidified on the fabric surface will increase, and a layer of film will be formed on the fabric surface, which will inhibit the heat transfer and slow down the heating rate of the sample. At the same time, the high content of adhesive will also increase the phase change microcapsules adhered to the fabric surface and immersed into the fabric, and further reduce the heating rate of the sample.

When the phase change microcapsule temperature regulating fabric is prepared by coating method, the process parameters shall be consistent, the content of phase change microcapsules and coating thickness are the main factors affecting the related properties of fabrics, and the thermal conductivity also has a great influence on the temperature regulation ability of materials. Therefore, we analyzed the effects of phase change microcapsule content, coating thickness and thermal conductivity on the temperature regulation ability of the fabric. The reference sample in the figure is untreated cotton fabric.

Figure 4a,c show the DSC curves of samples with different content of phase change microcapsule, different coating thickness and different thermal conductivity materials. It can be seen from Figure 4a that the adhesives and other additives have no influence on the phase change temperature and latent heat of the phase change microcapsule fabric. With the increase of the content of phase change microcapsules in the coating, more and more microcapsule phase change materials are attached to the surface of the fabric, which improves the heat storage and temperature adjustment ability of the prepared sample. However, in the experiment, it is found that when the content of phase change microcapsules in the coating is too high, the phenomenon of agglomeration and uneven dispersion will appear, and cracks will appear on the surface of the prepared coating. Therefore, the content of phase change microcapsules in the coating was fixed at 27% in the subsequent experimental study. It can be seen from Figure 4b that the coating thickness also has a certain effect on the phase transformation latent heat of the sample. When the phase change microcapsules melt endothermically, the phase change latent heat of the samples with coating thickness of 1.5 mm and 2.0 mm is close and relatively large. When the phase change microcapsules solidify exothermically, the phase change latent heat of the samples with coating thicknesses of 1.0 mm and 1.5 mm is close and relatively large. There is no positive correlation between thickness and latent heat of phase change. The main reason is that the sample taken in the DSC test is very small, and the larger the thickness, the smaller the sampling area, which cannot completely guarantee the content of dispersed phase change microcapsules in the test sample. Considering comprehensively, in the subsequent experimental research, the thickness of the coating is determined as 1.5 mm. When the coating thickness is 1.5 mm and the content of phase change microcapsules is 27%, the addition of materials with different thermal conductivity will also have a great influence on the heat storage and temperature adjustment ability of the sample. It can be seen from Figure 4c that adding different materials to the coating will have a certain influence on the initial temperature of the phase transition of the sample. When adding hollow glass beads and silicon dioxide thermal insulation materials with low thermal conductivity, the phase transition temperature of the sample will be reduced. When copper powder and aluminum powder with high thermal conductivity are added, the phase change latent heat of the sample is similar. The latent heat of melting phase transformation of samples containing silicon dioxide is the largest. This is because the silicon dioxide used in the experiment is nano-scale, and has the advantages of large specific surface area and high porosity. It can absorb a certain calorific value when heating up.

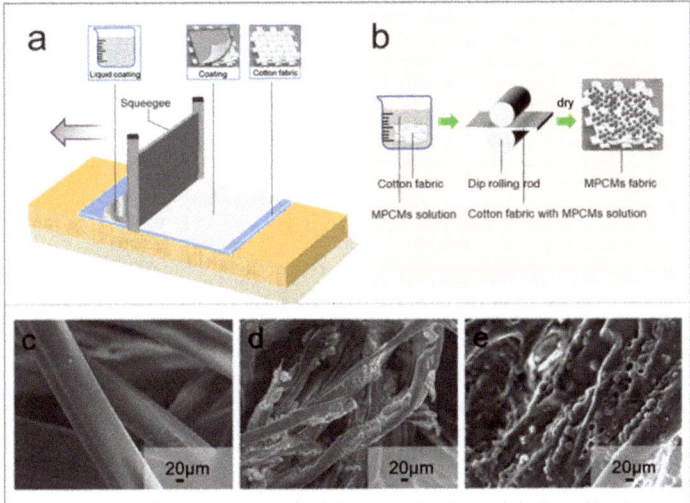

Figure 2. (**a**) Preparation of phase change microcapsule temperature control fabric by coating method, (**b**) Preparation of phase change microcapsule temperature control fabric by dip rolling, (**c**) SEM image of untreated cotton fabric, (**d**) SEM image of samples prepared by dip rolling method, (**e**) SEM image of samples prepared by coating method.

Figure 4d–f show the temperature rise curves of samples prepared by coating method with different content of phase change microcapsules, different coating thicknesses, and different thermal conductivity materials. It can be seen from the figure that the higher the content of phase change microcapsules, the longer the time required for the sample to rise from 20 to 60 °C, and the heating rate of the sample before 30 °C is slow. When one side of the fabric is heated, most of the heat is absorbed by the phase change microcapsules in the process of transferring the heat radiation to the other side of the fabric. Therefore, the higher the content of the phase change microcapsule, the more heat the sample and the slower the heating rate. When the coating thickness is 2.0 mm, the heating rate of the sample is the slowest, and the time for the sample to rise from 20 to 60 °C is 180 s. Materials with different thermal conductivity are added to the coating. When the materials are hollow glass beads and silicon dioxide, the heating speed of the sample is slow because of its low thermal conductivity. The sample with hollow glass beads has the slowest heating rate, mainly because the hollow structure of hollow glass beads is closed and contains more still air, which further enhances its thermal insulation performance. Figure 4g is the schematic diagram of sample heating rate test. To compare the thermal insulation capacity of the sample in a short time, the distance between the light source and the sample is close, so the heating speed will be much faster than in the actual situation.

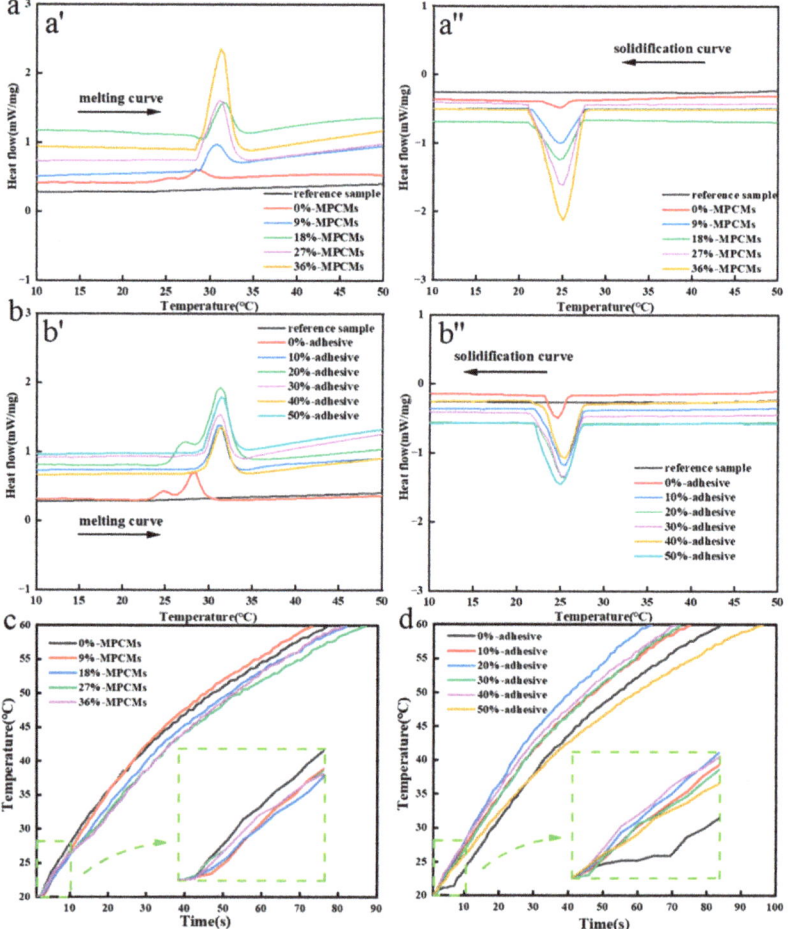

Figure 3. Dip rolling method (**a**) DSC curve of samples with different content of phase change microcapsules, (**b**) DSC curve of samples with different binder content, (**c**) Temperature rise curve of samples with different content of phase change microcapsules, (**d**) Temperature rise curve of samples with different binder content.

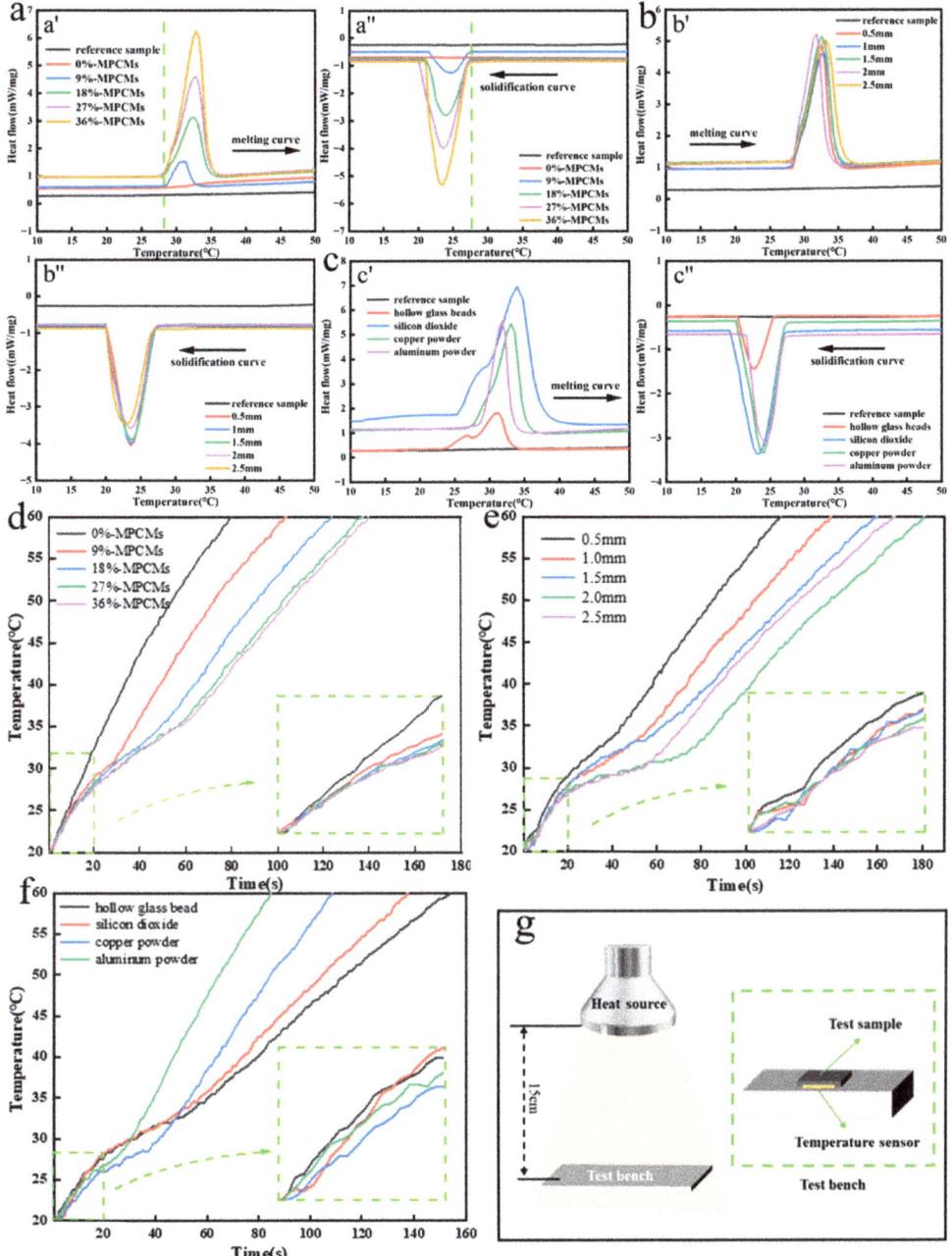

Figure 4. Coating method (**a**) DSC curve of samples with different content of phase change microcapsules, (**b**) DSC curves of samples with different coating thickness, (**c**) DSC curves of samples with different thermal conductivity, (**d**) Temperature rise curve of samples with different content of phase change microcapsules, (**e**) Temperature rise curve of samples with different coating thickness, (**f**) Temperature rise curve of samples with different thermal conductivity, (**g**) Schematic diagram of heating rate test.

Select the best samples prepared by the dip rolling method and coating method to compare and analyze. The experimental results show that, compared with the dip rolling method, the temperature-regulating fabric with phase change microcapsules prepared by the coating method has a greater latent heat of phase change (Figure 5a) and better temperature control ability (Figure 5b). Because the amount of phase change microcapsules on the sample in the coating method is more and the adhesion is firmer. Through the observation of the sample by the infrared thermal imager, it is also confirmed that the sample prepared by the coating method has better infrared camouflage effect (Figure 5c–f). Considering the performance, cost, thickness and other factors, we think that the content of phase change microcapsules is 27% and the coating thickness is 1.5 mm, which is the best choice for this experiment.

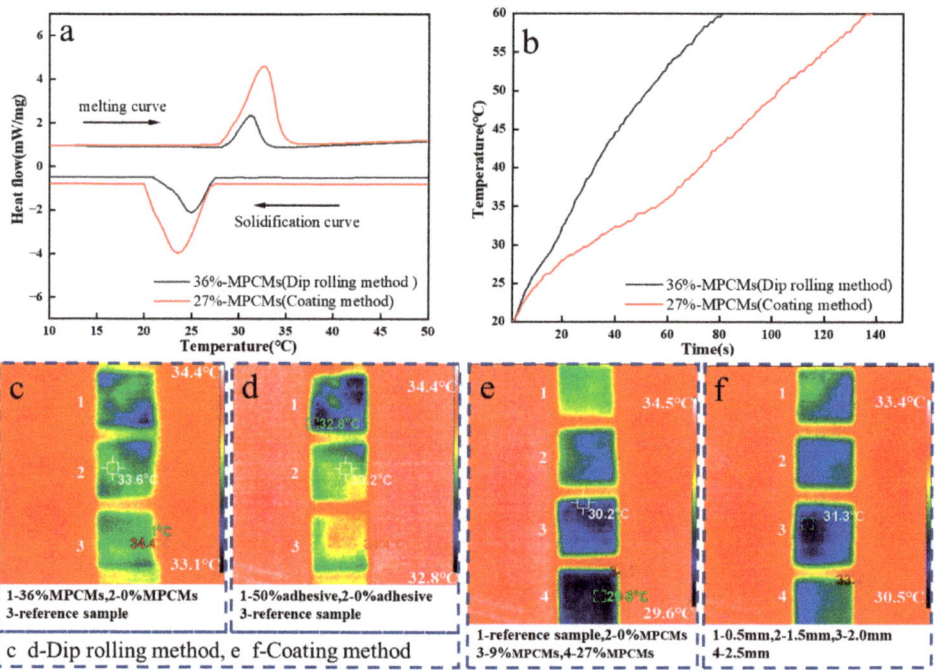

Figure 5. (**a**) Comparison of DSC curves between samples selected from dip rolling method and coating method, (**b**) Comparison of temperature rise curves between samples selected from dip rolling method and coating method, (**c**) Infrared thermogram of samples with different content of phase change microcapsules, (**d**) Infrared thermogram of samples with different binder content, (**e**) Infrared thermogram of samples with different content of phase change microcapsules, (**f**) Infrared thermogram of samples with different coating thickness.

3.3. Preparation and Performance Analysis of Phase Change Microcapsule Infrared Camouflage Fabric

According to the principle and preparation process described in Section 2.2.4, the infrared camouflage fabric is prepared by combining temperature-regulating phase change microcapsules with low emissivity metal materials. The SEM of low emissivity metallic copper powder and aluminum powder is shown in Figure 6a,b. According to the experimental results in Section 3.2, the content of phase change microcapsule is 27% and the coating thickness is 1.5 mm. Figure 6c shows the temperature rise curve of single-layer infrared camouflage coating sample after adding different kinds and contents of low emissivity materials. Figure 6d shows the infrared emissivity of the sample.

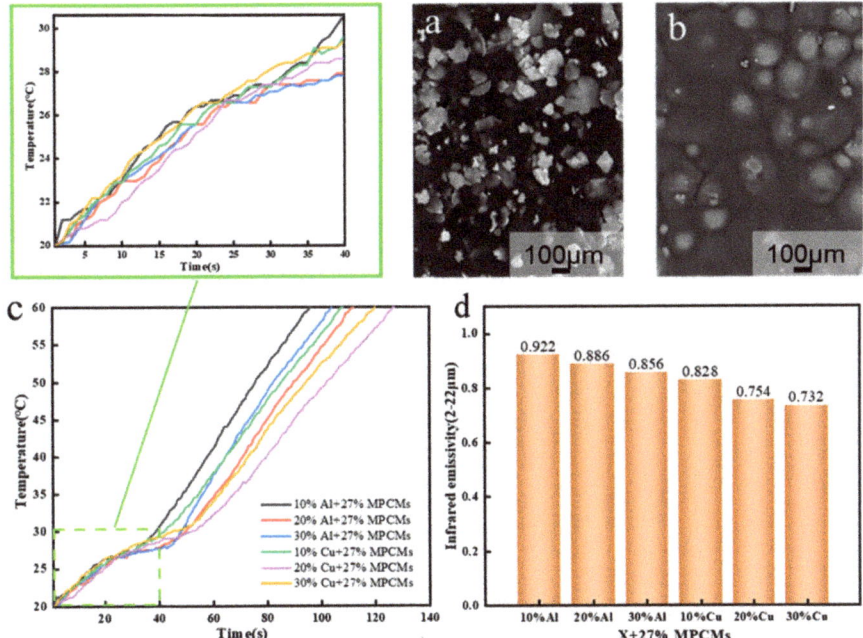

Figure 6. (**a**) SEM image of copper powder, (**b**) SEM image of aluminum powder, (**c**) Temperature rise curve of single-layer infrared camouflage coating samples after adding different kinds and contents of low emissivity materials, (**d**) Infrared emissivity of single layer infrared camouflage coating sample.

By analyzing the heating curve of the sample, it can be seen that it takes a long time for the sample to rise from 20 to 60 °C when the copper powder is added to the phase change microcapsule solution, especially when the copper powder content is 20% and 30%. At the beginning of heating, when the content of copper powder is 20%, the heating rate of the sample is the lowest. When the aluminum powder is added to the phase change microcapsule solution, the temperature rise of the sample is relatively fast. The reason is that when the content of metal particles is the same, the spherical aluminum powder is isotropic, the particle size is small, the distribution is relatively uniform, and the surface morphology is regular, which makes it easier to form mutually contacted heat conduction network chains, so as to improve the heat conduction efficiency of the material. Therefore, the heat transfer of the aluminum powder sample is faster when heating up, which reduces the temperature control performance of the sample. The test results of the two groups of samples show that the temperature control performance of the samples is better when the addition amount of low emissivity material is 20%. As can be seen from Figure 6d, the infrared emissivity of the sample with copper powder is less than that of the sample with aluminum powder. The reason can be explained by the microscopic morphology of the two kinds of materials. It can be seen that the copper powder has a flaky structure (Figure 6a), and the aluminum powder has a spherical structure (Figure 6b). The particle thickness of the flaky structure is small, and at the same particle concentration (mass-specific gravity), the content of flaky particles in the coating is more [53]. The flake particles can be arranged horizontally in the coating to form a compact reflective layer. This arrangement can effectively reduce the emissivity of the coated fabric [54]. The higher the content of copper powder, the smaller the infrared emissivity of the sample, but when the content of copper powder is 20% and 30%, the emissivity of the sample is close. Considering the material cost and the flexibility of the coating, the content of copper powder can be determined as 20%.

According to the theoretical analysis, two factors should be considered in infrared camouflage: temperature and infrared emissivity. It is considered that in this article, the infrared camouflage performance of the sample is the best when copper powder with low emissivity is added to the phase change microcapsule solution and the content of copper powder is 20%. At present, the prepared samples are all single-layer coatings. In order to ensure the ability of temperature regulation and reduce the infrared emissivity of the sample as much as possible, the double-layer coating can be prepared with copper powder and phase change microcapsule. The bottom layer is a phase change microcapsule (content: 27%) coating to ensure the temperature regulation performance of the sample, and the surface layer is a copper powder (content: 20%) coating to reduce the infrared emissivity of the sample (Figure 7a), while other factors remain unchanged.

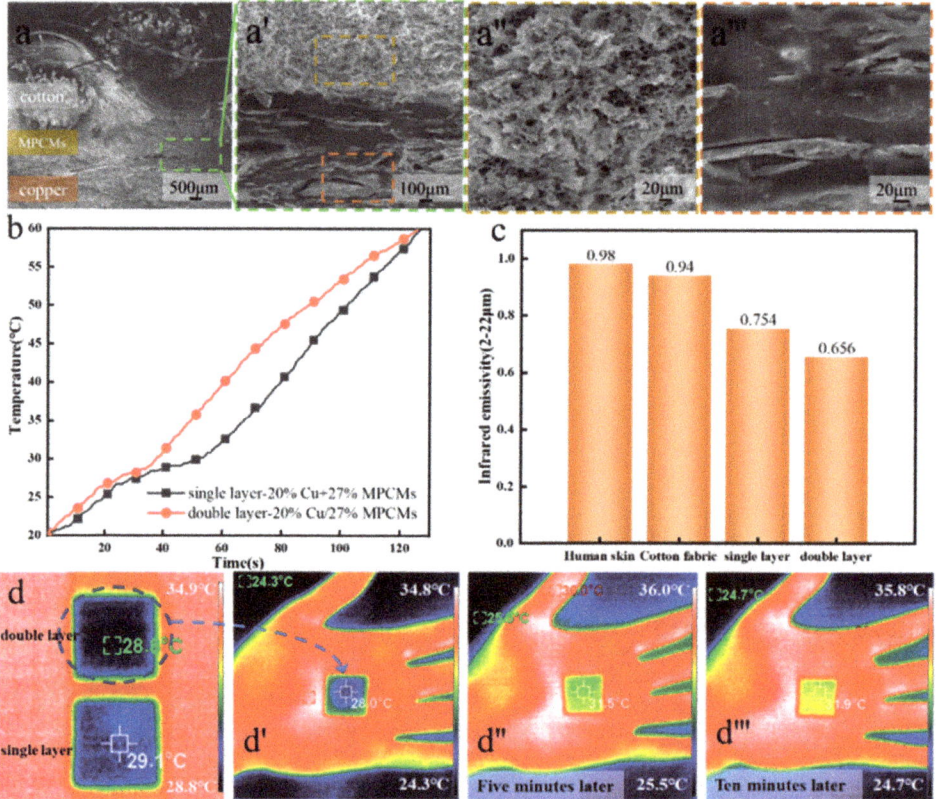

Figure 7. (**a**) Cross section SEM image of double-layer infrared camouflage fabric, (**b**) Comparison of temperature rise curves of single-layer and double-layer infrared camouflage fabrics, (**c**) Infrared emissivity of human skin, cotton fabric, single-layer and double-layer infrared camouflage fabrics, (**d**) Infrared thermogram of double-layer infrared camouflage fabric.

According to the temperature rise data in Figure 7b, the time required for the double-layer and single-layer samples to rise from 20 to 60 °C is very close, that is, the two samples have the same temperature adjustment ability. The infrared emissivity of the double-layer sample is significantly lower than that of the single-layer sample (Figure 7c). Therefore, the infrared camouflage performance of the double-layer coating sample is better. This conclusion is also confirmed in the infrared thermogram (Figure 7d). The double-layer infrared camouflage fabric is close to the environment in the infrared thermal image, which can reduce the detected surface temperature of human skin by 6.8 °C, and can still reduce 3.9 °C after covering for 10 min.

3.4. Analysis of Mechanical Properties of Phase Change Microcapsule Infrared Camouflage Fabric

The tensile and tear properties of infrared camouflage fabrics are tested by universal strength machine (Figure 8c), with reference to standard ISO34-1 [55]. The sample size is 200 × 50 mm, the clamping distance is 100 mm, and the tensile speed is 100 mm/min. Each sample shall be tested for 5 times, and the average value shall be taken as the final test result. The test results are shown in Figure 8a,b and Table 2. According to standard ISO 4604: 2011 [56], the coated fabric's extension length was measured using a fixed angle bending machine (Figure 8d), and the bending stiffness G of the coated fabric was about 3.36 mN · m according to Formula (2). YG (B) 401E Martindale wear tester is used to test the wear resistance of infrared camouflage fabric, with reference to standard ISO12947-3 [57]. The diameter of the sample is 38 mm. Record the mass loss of the fabric when rubbing 100 times, 250 times, 500 times, 750 times, and 1000 times respectively. Calculate the wear resistance index according to Formula (3). The results are shown in Table 3.

$$G = 9.81\rho_A (L/2)^3 \qquad (2)$$

where: G is the ordinary bending stiffness (mN · m), ρ_A is the mass per unit area (g/m^2), referring to standard ISO3374, and L is the average extension length (m).

$$Ai = n/\Delta m \qquad (3)$$

where Ai is the wear resistance index, the unit is times per milligram (times/mg); n is the total friction times, unit: times; Δm is the mass loss of the sample under the total friction times, the unit is mg.

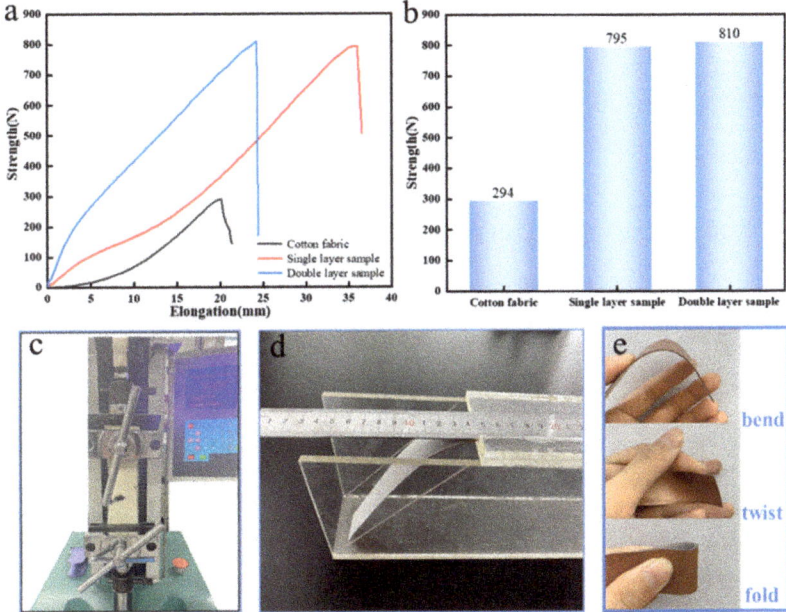

Figure 8. (**a**) Tensile strength displacement curve of infrared camouflage fabric, (**b**) Maximum tensile strength of infrared camouflage fabric, (**c**) Tensile strength test of infrared camouflage fabric, (**d**) Measurement of bending length of infrared camouflage fabric, (**e**) Soft performance display of infrared camouflage fabric.

Table 2. Tear strength of infrared camouflage fabric.

	Load (N)	Displacement (mm)	Tear Strength (N)
Cotton fabric	5.4	50.05	6.63
Single layer infrared camouflage fabric	7.5	50.05	21.31
Double layer infrared camouflage fabric	14.5	50.05	14.67

Table 3. Wear resistance index of infrared camouflage fabric with different friction times.

Friction Times	100	250	500	750	1000
Single layer infrared camouflage fabric	4.35	6.02	7.58	8.82	9.66
Double layer infrared camouflage fabric	3.85	4.90	6.71	7.77	8.77

The test shows that the tensile and tear strength of infrared camouflage fabric is much higher than that of cotton fabric. The tensile strength of double-layer infrared camouflage fabric is the largest (810 N), but the tensile displacement is less than that of single-layer infrared camouflage fabric. The calculation results of the fabric wear index show that the wear indexes of the two infrared camouflage fabrics are similar, and the more times of wear, the better the wear resistance. According to the actual picture of the fabric (Figure 8e) and the bending length test of the material, the infrared camouflage fabric has good softness and crimp ability. Table 4 shows the performance comparison between this material and other infrared camouflage materials. The results show that the infrared camouflage fabric has good mechanical properties and softness.

Table 4. Performance comparison between this material and other infrared camouflage materials.

Infrared Camouflage Materials	Infrared Emissivity (8–14 μm)	Temperature Regulation Range (°C)	Tensile Strength (N)
This material	0.507 0.656 (2–22 μm)	6.8	810
Material 1 [45]	0.575	5–10	
Material 2 [23]	0.795		398.4

4. Conclusions

Based on the principle of infrared camouflage and the characteristics of textile materials, from the point of view of controlling the temperature of materials to achieve the purpose of camouflage, phase change microcapsule temperature-regulating composite material with different parameters were prepared by padding method and coating method, and their infrared radiation performance was analyzed. The infrared camouflage textile composite was prepared by combining phase change microcapsule material with low emissivity metal material and using high molecular polyurethane as the matrix. The results of differential scanning calorimetry show that the phase change microcapsules melt and absorb heat at about 28 °C when heating up, and solidify and release heat at about 35 °C when cooling down. The temperature rises test shows that the content of adhesive, the content of phase change microcapsule, the thickness of the coating, and thermal conductivity have a great influence on the temperature adjustment ability of the sample. The temperature-regulating textile materials with phase change microcapsules were prepared by the coating method. When the content of phase change microcapsules is 27% and the coating thickness is 1.5 mm, the performance of the sample is the best. The results of the infrared thermogram and infrared emissivity test show that when the content of phase change microcapsule in the bottom layer is 27% and the content of flake copper powder in the surface layer is 20%, the double-layer coating sample has a good infrared camouflage effect. Its infrared emissivity in the band of 2–22 µm is 0.656, covering it on the surface of the human body can reduce the temperature to 6.8 °C, and effectively reduce the infrared radiation. Based on these results, we believe that this study can provide a reference for the preparation of infrared camouflage composite material in the laboratory and industry. The infrared camouflage textile composite is expected to be used in military textiles such as individual protective clothing, military tents, and equipment tarpaulins.

Author Contributions: All authors listed on this paper have contributed to this study. Conceptualization, X.Z. and Y.S.; methodology, Y.S. and X.Z.; investigation, Y.H. and Y.S.; writing—original draft preparation, Y.S.; writing—review and editing, X.Z. and Y.H. All authors have read and agreed to the published version of the manuscript.

Funding: This work was supported by the research project of "Development Strategy of New Materials Industry for Protective Textile" (grant numbers:2021DFZD1).

Institutional Review Board Statement: Not applicable.

Data Availability Statement: The data that support the findings of this study are available within the article.

Conflicts of Interest: The authors declare no conflict of interest.

References

1. Teggar, M.; Arici, M.; Mert, M.S.; Ajarostaghi, S.S.M.; Niyas, H.; Tuncbilek, E.; Ismail, K.A.R.; Younsi, Z.; Benhouia, A.T.; Mezaache, E. A comprehensive review of micro/nano enhanced phase change materials. *J. Therm. Anal. Calorim.* **2022**, *147*, 3989–4016. [CrossRef]
2. Chen, H.; Yue, Z.; Ren, D.; Zeng, H.; Wei, T.; Zhao, K.; Yang, R.; Qiu, P.; Chen, L.; Shi, X. Thermal Conductivity during Phase Transitions. *Adv. Mater.* **2019**, *31*, e1806518. [CrossRef]
3. Zhao, Y.X.; Zhang, X.L.; Hua, W.S. Review of preparation technologies of organic composite phase change materials in energy storage. *J. Mol. Liq.* **2021**, *336*, 115923. [CrossRef]
4. Fan, X.M.; Guan, Y.; Li, Y.Z.; Yu, H.Y.; Marek, J.; Wang, D.C.; Militky, J.; Zou, Z.Y.; Yao, J.M. Shape-Stabilized Cellulose Nanocrystal-Based Phase-Change Materials for Energy Storage. *Acs Appl. Nano Mater.* **2020**, *3*, 1741–1748. [CrossRef]
5. Yan, Y.; Li, W.; Zhu, R.; Lin, C.; Hufenus, R. Flexible Phase Change Material Fiber: A Simple Route to Thermal Energy Control Textiles. *Materials* **2021**, *14*, 401–418. [CrossRef] [PubMed]
6. Shaid, A.; Wang, L.J.; Islam, S.; Cai, J.Y.; Padhye, R. Preparation of aerogel-eicosane microparticles for thermoregulatory coating on textile. *Appl. Therm. Eng.* **2016**, *107*, 602–611. [CrossRef]
7. Tyagi, V.V.; Kaushik, S.C.; Tyagi, S.K.; Akiyama, T. Development of phase change materials based microencapsulated technology for buildings: A review. *Renew. Sustain. Energy Rev.* **2011**, *15*, 1373–1391. [CrossRef]
8. Sheng, N.; Zhu, C.Y.; Sakai, H.; Akiyama, T.; Nomura, T. Synthesis of Al-25 wt% Si@Al$_2$O$_3$@Cu microcapsules as phase change materials for high temperature thermal energy storage. *Sol. Energy Mater. Sol. Cells* **2019**, *191*, 141–147. [CrossRef]
9. Jamekhorshid, A.; Sadrameli, S.M.; Farid, M. A review of microencapsulation methods of phase change materials (PCMs) as a thermal energy storage (TES) medium. *Renew. Sustain. Energy Rev.* **2014**, *31*, 531–542. [CrossRef]
10. Su, W.G.; Darkwa, J.; Kokogiannakis, G. Review of solid-liquid phase change materials and their encapsulation technologies. *Renew. Sustain. Energy Rev.* **2015**, *48*, 373–391. [CrossRef]
11. Alva, G.; Lin, Y.X.; Liu, L.K.; Fang, G.Y. Synthesis, characterization and applications of microencapsulated phase change materials in thermal energy storage: A review. *Energy Build.* **2017**, *144*, 276–294. [CrossRef]
12. Huang, X.; Zhu, C.Q.; Lin, Y.X.; Fang, G.Y. Thermal properties and applications of microencapsulated PCM for thermal energy storage: A review. *Appl. Therm. Eng.* **2019**, *147*, 841–855. [CrossRef]
13. Peng, H.; Wang, J.H.; Zhang, X.W.; Ma, J.; Shen, T.T.; Li, S.L.; Dong, B.B. A review on synthesis, characterization and application of nanoencapsulated phase change materials for thermal energy storage systems. *Appl. Therm. Eng.* **2021**, *185*, 677–683. [CrossRef]
14. Konuklu, Y.; Ostry, M.; Paksoy, H.O.; Charvat, P. Review on using microencapsulated phase change materials (PCM) in building applications. *Energy Build.* **2015**, *106*, 134–155. [CrossRef]
15. Jurkowska, M.; Szczygiel, I. Review on properties of microencapsulated phase change materials slurries (mPCMS). *Appl. Therm. Eng.* **2016**, *98*, 365–373. [CrossRef]
16. Aklujkar, P.S.; Kandasubramanian, B. A review of microencapsulated thermochromic coatings for sustainable building applications. *J. Coat. Technol. Res.* **2021**, *18*, 19–37. [CrossRef]

17. Cheng, C.; Gong, F.Y.; Fu, Y.R.; Liu, J.; Qiao, J.G. Effect of polyethylene glycol/polyacrylamide graft copolymerizaton phase change materials on the performance of asphalt mixture for road engineering. *J. Mater. Res. Technol.* **2021**, *15*, 1970–1983. [CrossRef]
18. Marani, A.; Madhkhan, M. Thermal performance of concrete sandwich panels incorporating phase change materials: An experimental study. *J. Mater. Res. Technol.* **2021**, *12*, 760–775. [CrossRef]
19. Liu, W.; Zhang, X.L.; Ji, J.; Wu, Y.F.; Liu, L. A Review on Thermal Properties Improvement of Phase Change Materials and Its Combi-nation with Solar Thermal Energy Storage. *Energy Technol.* **2021**, *9*, 2100169. [CrossRef]
20. Alehosseini, E.; Jafari, S.M. Micro/nano-encapsulated phase change materials (PCMs) as emerging materials for the food industry. *Trends Food Sci. Technol.* **2019**, *91*, 116–128. [CrossRef]
21. Petrulis, D.; Petrulyte, S. Potential use of microcapsules in manufacture of fibrous products: A review. *J. Appl. Polym. Sci.* **2019**, *136*, 47066. [CrossRef]
22. Iqbal, K.; Khan, A.; Sun, D.M.; Ashraf, M.; Rehman, A.; Safdar, F. Phase change materials, their synthesis and application in tex-tiles—A review. *J. Text. Inst.* **2019**, *110*, 625–638. [CrossRef]
23. Wang, W.; Fang, S.J.; Zhang, L.P.; Mao, Z.P. Infrared stealth property study of mesoporous carbon-aluminum doped zinc oxide coated cotton fabrics. *Text. Res. J.* **2015**, *85*, 1065–1075. [CrossRef]
24. Kim, T.; Bae, J.Y.; Lee, N.; Cho, H.H. Metamaterials: Hierarchical Metamaterials for Multispectral Camouflage of Infrared and Micro-waves. *Adv. Funct. Mater.* **2019**, *29*, 1807319. [CrossRef]
25. Zhou, X.; Xin, B.; Liu, Y. Research progress on infrared stealth fabric. *J. Phys. Conf. Serie* **2021**, *1790*, 012058. [CrossRef]
26. Hao, L.C.; Xiao, H.; Liu, W.; Xu, J.; Li, S.X. Research development of thermal infrared camouflage textiles. *J. Text. Res.* **2014**, *35*, 158–164.
27. Fang, S.J.; Wang, W.; Yu, X.L.; Xu, H.; Zhong, Y.; Sui, X.F. Preparation of ZnO:(Al, La)/polyacrylonitrile (PAN) nonwovens with low infrared emissivity via electrospinning. *Mater. Lett.* **2015**, *143*, 120–123. [CrossRef]
28. Jeong, S.M.; Ahn, J.; Choi, Y.K.; Lim, T.; Seo, K.; Hong, T. Development of a wearable infrared shield based on a polyure-thane-antimony tin oxide composite fiber. *NPG Asia Mater.* **2020**, *12*, 32. [CrossRef]
29. Kang, J.; Chen, Y.J.; Cui, Y.; Zhu, B.; Yu, Y.; Wei, H.Y. Preparation of Bionic Porous Zirconia Fiber by Microemulsion Electrospinning and Its Infrared Stealth Property. *Russ. J. Inorg. Chem.* **2021**, *66*, 510–515. [CrossRef]
30. Wang, H.; Ma, Y.Y.; Qiu, J.; Wang, J.; Zhang, H.; Li, Y.X. Multifunctional PAN/Al-ZnO/Ag Nanofibers for Infrared Stealth, Self-Cleaning, and Antibacterial Applications. *ACS Appl. Nano Mater.* **2022**, *5*, 782–790. [CrossRef]
31. Lim, T.; Jeong, S.M.; Seo, K.; Pak, J.H.; Choi, Y.K.; Ju, S. Development of fiber-based active thermal infrared camouflage textile. *Appl. Mater. Today* **2020**, *20*, 100624. [CrossRef]
32. Chu, H.T.; Zhang, Z.C.; Liu, Y.J.; Leng, J.S. Silver particles modified carbon nanotube paper/glassfiber reinforced polymer composite material for high temperature infrared stealth camouflage. *Carbon* **2016**, *98*, 557–566. [CrossRef]
33. Liu, Z.D.; Heng, Z.G.; Zhang, H.R.; Zhou, J.; Chen, Y.; Liang, M. Synergistic Action of Polyethylene Glycol/Expanded Graphite/Cellulose Nanofibers with Superior Infrared Stealth Performance. *J. Macromol. Sci. Part B—Phys.* **2021**, *60*, 485–499. [CrossRef]
34. Wang, W.; Zhang, L.P.; Fang, S.J.; Xu, H.; Zhong, Y.; Mao, Z.P. Low-Emitting Property of Lanthanum Aluminate and Its Application in Infrared Stealth. *Sci. Adv. Mater.* **2015**, *7*, 1649–1656. [CrossRef]
35. Mao, Z.P.; Yu, X.L.; Zhang, L.P.; Zhong, Y.; Xu, H. Novel infrared stealth property of cotton fabrics coated with nano ZnO: (Al, La) particles. *Vacuum* **2014**, *104*, 111–115. [CrossRef]
36. Mao, Z.P.; Wang, W.; Liu, Y.; Zhang, L.P.; Xu, H.; Zhong, Y. Infrared stealth property based on semiconductor (M)-to-metallic (R) phase transition characteristics of W-doped VO_2 thin films coated on cotton fabrics. *Thin Solid Films* **2014**, *558*, 208–214. [CrossRef]
37. Xu, R.; Wang, W.; Yu, D. Preparation of silver-plated Hollow Glass Microspheres and its application in infrared stealth coating fabrics. *Prog. Org. Coat.* **2019**, *131*, 1–10. [CrossRef]
38. Xu, R.; Wang, W.; Yu, D. A novel multilayer sandwich fabric-based composite material for infrared stealth and super thermal insulation protection. *Compos. Struct.* **2019**, *212*, 58–65. [CrossRef]
39. Li, L.F.; Xu, W.L.; Wu, X.; Liu, X.; Li, W.B. Fabrication and characterization of infrared-insulating cotton fabrics by ALD. *Cellulose* **2017**, *24*, 3981–3990. [CrossRef]
40. Gu, J.; Wang, W.; Yu, D. Temperature control and low infrared emissivity double-shell phase change microcapsules and their ap-plication in infrared stealth fabric. *Prog. Org. Coat.* **2021**, *159*, 106439. [CrossRef]
41. Geng, X.Y.; Gao, Y.; Wang, N.; Han, N.; Zhang, X.X.; Li, W. Intelligent adjustment of light-to-thermal energy conversion efficiency of thermo-regulated fabric containing reversible thermochromic MicroPCMs. *Chem. Eng. J.* **2021**, *408*, 127276. [CrossRef]
42. Li, J.; Zhu, X.Y.; Wang, H.C.; Lin, P.C.; Jia, L.S.; Li, L.J. Synthesis and properties of multifunctional microencapsulated phase change material for intelligent textiles. *J. Mater. Sci.* **2021**, *56*, 2176–2191. [CrossRef]
43. Hassabo, A.G. New approaches to improving thermal regulating property of cellulosic fabric. *Carbohydr. Polym.* **2014**, *101*, 912–919. [CrossRef]
44. Zuravliova, S.V.; Stygiene, L.; Krauledas, S.; Minkuviene, G.; Sankauskaite, A.; Abraitiene, A. The Dependance of Effectiveness of In-corporated Microencapsulated Phase Change Materials on Different Structures of Knitted Fabrics. *Fiber Polym.* **2015**, *16*, 1125–1133. [CrossRef]
45. Xu, R.; Xia, X.M.; Wang, W.; Yu, D. Infrared camouflage fabric prepared by paraffin phase change microcapsule with Good thermal insulating properties. *Colloid. Surf. A Physicochem. Eng. Asp.* **2020**, *591*, 124519. [CrossRef]

46. Huang, F.L.; Xiao, Y.; Liu, W.T.; Ning, J.X.; Lu, Z.N. Infrared Stealth Fabric with Multilayer Composite Structure and Preparation Method Thereof. Chinese Patent CN106758163A, 31 May 2017.
47. Chen, X.B.; Jiang, C.S.; Jiang, L.H.; Lin, J.M.; Huang, Z.D.; Zheng, R.S. Research Process of Phase Change Microcapsule Materials for Textile. *Contemp. Chem. Ind.* **2021**, *50*, 162–165.
48. Liu, G.J.; Shi, F.; Zhang, G.Q.; Zhou, L. Preparation of phase change wax@polyvinyl alcohol thermo-regulated finishing agents and its applications on cotton fabrics. *J. Mater. Eng.* **2020**, *48*, 97–102.
49. Ma, H.J.; Tian, B.H.; He, Y. Research progress of preparation and application of MCPCM. *New Chem. Mater.* **2020**, *48*, 20–23.
50. Mondal, S. Phase change materials for smart textiles—An overview. *Appl. Therm. Eng.* **2008**, *28*, 1536–1550. [CrossRef]
51. Yang, A.D. *Preparation of Paraffin Phase Change Micro-Encapsulates and Its Application in Thermal Stealth Coatings*; Beijing University of Technology: Beijing, China, 2009.
52. Zhu, P. *Functional Fiber and Functional Textile*; China Textile & Apparel Press: Beijing, China, 2006.
53. Wang, L.; Liu, C.Y.; Xu, G.Y.; Xiang, S.S.; Shi, M.Y.; Zhang, Y.J. Influences of morphology and floating rate of CeO_2 fillers on controlling infrared emissivity of the epoxy-silicone resin based coatings. *Mater. Chem. Phys.* **2019**, *229*, 380–386. [CrossRef]
54. Chen, X. *Research on Preparation and Application of Infrared Low Emissivity Coatings*; Xiamen University: Xiamen, China, 2019.
55. *ISO 34-1*; Rubber, Vulcanized or Thermoplastic-Determination of Tear Strength-Part 1: Trouser, Angle and Crescent Test Pieces. International Organization for Standardization: Geneva, Switzerland, 2022.
56. *ISO 4604*; Reinforcement Fabrics—Determination of Conventional Flexural Stiffness—Fixed-Angle Flexometer Method. International Organization for Standardization: Geneva, Switzerland, 2011.
57. *ISO 12947-3*; Textiles—Determination of the Abrasion Resistance of Fabrics by the Martindale Method—Part 3: Determination of Mass Loss. International Organization for Standardization: Geneva, Switzerland, 1998.

Disclaimer/Publisher's Note: The statements, opinions and data contained in all publications are solely those of the individual author(s) and contributor(s) and not of MDPI and/or the editor(s). MDPI and/or the editor(s) disclaim responsibility for any injury to people or property resulting from any ideas, methods, instructions or products referred to in the content.

Article

Wearable Pressure Sensor Using Porous Natural Polymer Hydrogel Elastomers with High Sensitivity over a Wide Sensing Range

Fan Xiao [1,†], Shunyu Jin [2,†], Wan Zhang [1], Yingxin Zhang [1], Hang Zhou [3] and Yuan Huang [1,*]

1. School of Microelectronics Science and Technology, Sun Yat-Sen University, Guangzhou 510275, China
2. Hefei National Research Center for Physical Sciences at the Microscale, University of Science and Technology of China, Hefei 230026, China
3. School of Electronic and Computer Engineering, Peking University Shenzhen Graduate School, Shenzhen 518055, China
* Correspondence: huangy723@mail.sysu.edu.cn
† These authors contributed equally to this work.

Citation: Xiao, F.; Jin, S.; Zhang, W.; Zhang, Y.; Zhou, H.; Huang, Y. Wearable Pressure Sensor Using Porous Natural Polymer Hydrogel Elastomers with High Sensitivity over a Wide Sensing Range. *Polymers* 2023, 15, 2736. https://doi.org/10.3390/polym15122736

Academic Editor: Tsuyoshi Michinobu

Received: 16 May 2023
Revised: 14 June 2023
Accepted: 16 June 2023
Published: 19 June 2023

Copyright: © 2023 by the authors. Licensee MDPI, Basel, Switzerland. This article is an open access article distributed under the terms and conditions of the Creative Commons Attribution (CC BY) license (https://creativecommons.org/licenses/by/4.0/).

Abstract: Wearable pressure sensors capable of quantifying full-range human dynamic motion are pivotal in wearable electronics and human activity monitoring. Since wearable pressure sensors directly or indirectly contact skin, selecting flexible soft and skin-friendly materials is important. Wearable pressure sensors with natural polymer-based hydrogels are extensively explored to enable safe contact with skin. Despite recent advances, most natural polymer-based hydrogel sensors suffer from low sensitivity at high-pressure ranges. Here, by using commercially available rosin particles as sacrificial templates, a cost-effective wide-range porous locust bean gum-based hydrogel pressure sensor is constructed. Due to the three-dimensional macroporous structure of the hydrogel, the constructed sensor exhibits high sensitivities (12.7, 5.0, and 3.2 kPa^{-1} under 0.1–20, 20–50, and 50–100 kPa) under a wide range of pressure. The sensor also offers a fast response time (263 ms) and good durability over 500 loading/unloading cycles. In addition, the sensor is successfully applied for monitoring human dynamic motion. This work provides a low-cost and easy fabrication strategy for fabricating high-performance natural polymer-based hydrogel piezoresistive sensors with a wide response range and high sensitivity.

Keywords: pressure sensor; porous hydrogels; natural polymer

1. Introduction

Wearable pressure sensors capable of detecting human body movements are crucial in muscle motion analysis, speech recognition, disease diagnosis, and health monitoring [1–6]. To meet diverse demands in different human body movements monitoring applications, such as subtle joint-bending detection and heavy foot tapping monitoring, pressure sensors are desired to possess high sensitivity over a wide sensing range. Wearable pressure sensors have been developed based on several major sensing mechanisms, including piezoresistive [7,8], capacitive [9], and piezoelectric [10] mechanisms. Among these, piezoresistive pressure sensors based on applied pressure transduction in an electrical conductance of sensing materials have been widely studied due to their easy signal acquisition, simple process, and simple circuit integration [11].

Since wearable pressure sensors directly or indirectly contact skin, selecting flexible soft and skin-friendly materials is important. Hydrogels, a kind of soft material consisting of cross-linked networks of hydrophilic polymers in water, have drawn particular interest for their liquid-like transport and solid-like mechanical properties [12]. There are two categories of hydrogels: natural polymer-based hydrogels and synthetic polymer-hydrogels [12]. Natural polymers were widely used for biomedicine and wearable electronics due to their appealing merits of multi-functionality, ease of accessibility, and biocompatibility. Natural polymer-based hydrogels have high biodegradability and biocompatibility, offering a breakthrough in wearable electronics [12–15]. Several studies reported wearable pressure sensors with natural polymer-based hydrogels, such as the locust bean gum (LBG)-based [16], cellulosic [17], and chitosan-based [18] hydrogels. For example, Shen et al. used a unique kind of cellulose-rich material with high molecular weight which was isolated directly and simply from wood through rapid dissolution in ionic liquid at temperatures above the glass transition of lignin to prepare a macroporous, compressible lignocellulosic hydrogel [17]. The assembled hydrogel sensors possess a wide responsive stress range (above 0.35 MPa) [17]. Yang et al. reported a wearable pressure sensor based on a resilient, anti-fatigue and freezing-tolerant chitosan-poly (hydroxyethyl acrylamide) double-network hydrogel [18]. The wearable sensor exhibited a sensitive and large-range detection capacity, together with long-term stability and wide operating temperature range [18]. We also reported an elastic and biocompatible hybrid network hydrogel by cross-linking LBG, polyvinyl alcohol (PVA), and carbon nanotubes (CNTs) [16]. A pressure sensor with a double-rough surface of LBG-based hydrogel exhibits a high sensitivity (20.5 kPa^{-1}) at the low-pressure range (0.1–1 kPa). However, as the cross-linked network structure inside the hydrogels improves their robustness, the pressure sensitivity significantly decreases in a higher-pressure range (2.28 kPa^{-1}, 1–10 kPa; 0.24 kPa^{-1}, 10–100 kPa). Thus, developing wearable pressure sensors based on elastic natural polymer-based hydrogels capable of detecting a wide pressure range while maintaining high sensitivity remains challenging.

Many reports have shown that porous structure inside elastomers can effectively distribute the applied pressure, thus increasing the sensing range of wearable pressure sensors and improving their sensitivity at high-pressure ranges [6,19–22]. Porous elastomers are prepared by the sacrificial template method by mixing elastomer solution with templates, such as salt, sugar, and citric acid monohydrate, and dissolving the mixture in water after solidifying [6,19,22–25]. For example, Sang et al. reported a porous composite foam via a simple heat molding of conductive fillers and elastomer together with commercially available popcorn salts followed by water-assisted salt removal [6]. Zhao et al. reported a multilayered and graded-porosity polydimethylsiloxane/silver nanoparticle sponge, which can be fabricated using the sacrificial template method of mixing polydimethylsiloxane solution with citric acid monohydrate templates [19]. Lo et al. reported an elastomeric sponge-based sensor based on a porous polydimethylsiloxane sponge, which was fabricated from a sugar cube sacrificial template [24]. However, salt, sugar, and citric acid monohydrate are soluble in hydrogels; thus, they are unsuitable for porous natural polymer-based hydrogels. Consequently, developing porous natural polymer-based hydrogels using a suitable and economic sacrificial template is highly desired for wearable pressure sensors.

In this study, a highly porous LBG-based hydrogel is synthesized from commercially available rosin particles as the sacrificial template. These particles are insoluble in water but can be leached in ethanol. For an optimized porosity of the LBG-based hydrogel, the wearable pressure sensor based on porous LBG-based hydrogels sandwiched between two carbon cloth electrodes can monitor a wide range of pressure with high sensitivities (12.7, 5.0, and 3.2 kPa^{-1} under 0.1–20, 20–50, and 50–100 kPa). The proposed sensor's sensitivities at the high-pressure range are superior to those of the previous hydrogel sensors. The sensor's wide pressure range with high sensitivities is ascribed to a three-dimensional macroporous structure with dense large-sized pores (~100–150 μm). The sensor also offers a fast response time (263 ms) and good durability over 500 loading/unloading cycles. Using the sensor for various human movement applications is demonstrated as a proof-of-concept. A low-cost and facile craft route to fabricate highly porous natural polymer-based hydrogels is provided, which can be useful for high-performance wearable pressure sensor applications.

2. Materials and Methods

2.1. Preparation of Porous LBG-Based Hydrogels and Original LBG-Based Hydrogels

Porous LBG-based hydrogels were prepared by mixing 4 g PVA (1750 ± 50, Shanghai Yuanye Bio-Technology Co., Ltd., Shanghai, China), 300 mg CNTs (XFNANO Materials

Tech Co., Ltd., Nanjing, China), and 40 mL deionized water. The mixture was heated to 100 °C under vigorous stirring until fully dissolved PVA. Then 1 g LBG (Aladdin, Shanghai, China) was added to the mixture. The rosin particles and the mixture solution were stirred at specific ratios until reaching homogeneous dissolution. The mixture was transferred to an environmental cabinet at −20 °C for 12 h to form LBG/PVA/CNTs/rosin hydrogels. The hydrogels were immersed in ethanol for 24 h to dissolve rosin particles.

Original LBG-based hydrogels were prepared by mixing 4 g PVA (1750 ± 50, Shanghai Yuanye Bio-Technology Co., Ltd.), 300 mg CNTs (XFNANO Materials Tech Co., Ltd.), and 40 mL deionized water. The mixture was heated to 100 °C under vigorous stirring until fully dissolved PVA. Then 1 g LBG (Aladdin) was added to the mixture. The mixture was transferred to an environmental cabinet at −20 °C for 12 h to form LBG/PVA/CNTs hydrogels.

2.2. Materials Characterization

The hydrogels' microstructure was examined by a scanning electron microscope (SEM, Zeiss Crossbeam 350 Germany). The porosities of the porous hydrogels was examined by an X-ray microscope (XRM, Zeiss, Xradia 520 Versa, Germany). In brief, 3D CT images of the hydrogels were obtained by XRM and the porosities of the porous hydrogels can be identified in the 3D CT images by different contrast automatically by Dragonfly Pro software. The compressive mechanical properties of porous hydrogels were tested by ZQ-990LB (China) equipped with a force gauge (maximum force: 100 N). The compressive tests were performed on a cylindrical sample (diameter: 1 cm, height: 4.13 mm) at 10 mm min^{-1}. The weight retention test was conducted at 25 °C and 15% humidity.

2.3. Electrical Characterization

Hydrogels were tailored into 1 cm × 1 cm square pieces. The sensors with hydrogels sandwiched between two carbon cloth electrodes are fabricated. To evaluate the pressure sensing performance, a testing machine ZQ-990LB equipped with a force gauge (maximum force: 100 N) was used to apply pressure to the device, and the corresponding current signals were collected by a digital source meter (Keithley 2635B, America) at the working voltage of 1.5 V.

3. Results and Discussion

3.1. Fabrication and Characterization of Porous and Original Hydrogels

The fabrication process of porous hydrogels is schematically illustrated in Figure 1. The processes start with PVA and CNTs dissolved in deionized water, followed by the addition of LBG. This mixture solution and rosin particles were mixed under continuous stirring until reaching homogeneous dispersion. The hydrogels were formed by freezing the mixture. Since the rosin particles were barely soluble in ethanol, their amount in the hydrogels was well preserved. Removing the rosin particles with ethanol resulted in a highly porous structure with pore sizes similar to the grain size of rosin particles. The procedure involved a simple hand mixing of LBG, CNTs, and PVA with low-cost rosin in non-toxic solvents. Based on different grain sizes of rosin particles, porous structures with various geometries can be achieved, which may influence the mechanical and piezoresistive properties of the hydrogels. Here, common commercial rosin particles with sizes of 100–150 μm (Figure S1, Supporting Information) were used as an example to demonstrate the method's feasibility. For comparison, original hydrogels were also prepared by a similar method without introducing rosin particles. The major interactions in the hydrogels are illustrated in our previous work [16]. The high mechanical strength and elasticity of PVA made it a useful primary polymer network. The LBG and PVA molecules containing large numbers of hydrogen bonds directly formed intra- and inter-hydrogen bonds. CNTs were essential for constructing a conductive 3D network.

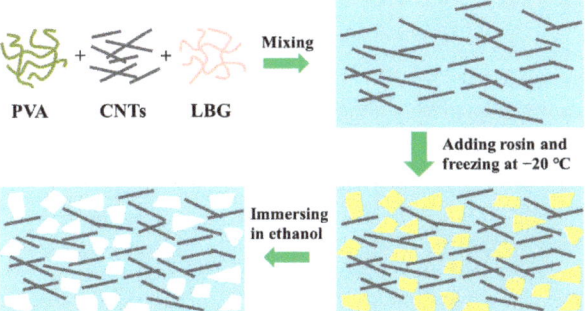

Figure 1. Schematic illustrations of the porous hydrogels' fabrication process.

The microstructure of the hydrogels is observed by SEM (Figure 2a,b). A few cracks and small pores (sub 10 μm level, red circle) are visible in the cross-sectional images of the original hydrogels (Figure 2a). In contrast, the porous hydrogels show a different porous morphology with large pore sizes (100–150 μm) agree with the grain size of rosin particles (Figure 2b). XRM evaluates smaller micropore sizes in porous hydrogels. The tomography of $100 \times 100 \times 100$ μm^3 highlights three-dimensional morphological features of porous hydrogels and the sub 10 μm level pores (Figure 2c). The SEM and XRM results suggest that both sub 10 μm level pores and 100–150 μm level pores are inside the porous hydrogel. Due to three-dimensional macroporous structures, the porous hydrogels demonstrated promising compressive elasticity that can be compressed and recovered to the original shape after releasing the compressive force (Figure 2d).

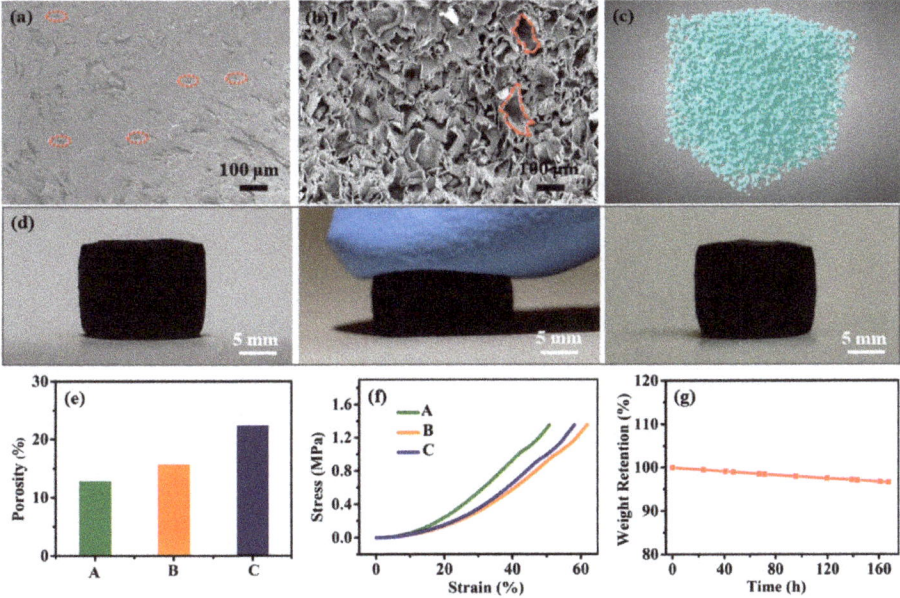

Figure 2. Cross-sectional SEM images of (**a**) original and (**b**) porous hydrogels. Small pores (sub 10 μm level) are marked with red circle in (**a**). Large pore (100–150 μm) are marked with red circle in (**b**). (**c**) X-ray microtomography of porous hydrogels. (**d**) The compressive recovery behavior of porous hydrogels. (**e**) The porosity of hydrogel A, B, and C. (**f**) Compressive stress–strain curves of all three hydrogel samples. (**g**) Weight variation of hydrogel B at 25 °C and 15% humidity.

The weight fraction of rosin particles in the mixture (LBG, PVA, CNTs, deionized water, and rosin particles) essentially controls the porosities of the porous hydrogel, where rosin particles are dissolved after soaking in ethanol solution. XRM is also employed to evaluate the porous hydrogels' porosity. Figure 2e shows a concomitant increase in porosity with the weight fraction of rosin particles: 12.8% for the hydrogel with 0.5% weight fraction of rosin particles (denoted as hydrogel A), 15.7% for the hydrogel with 1.5% weight fraction of rosin particles (denoted as hydrogel B), and 22.4% for the hydrogel with 3.0% weight fraction of rosin particles (denoted as hydrogel C).

To investigate the influence of porosity on mechanical softness, typical stress–strain curves of porous hydrogels from compression are depicted in Figure 2f. The higher porosity suggests that more pores are closed under pressure, and stiffness decreases. Thus, when the porosity increases from 12.8% to 15.7%, the compressive strain at a compressive stress of 1.35 MPa increases from 50.7% for hydrogel A to 61.8% for hydrogel B. However, when the porosity increases to 22.4%, the compressive strain of hydrogel C decreases to 58.1%, attributed to the residual rosin particles in hydrogel C (Figure S2, Supporting Information), which are not completely removed by ethanol.

Moreover, the long-term stability of hydrogels is still a challenge to be addressed. The weight retention of hydrogel B was measured with the weight ratio of W_t (weight of hydrogel B at time t) and W_0 (the initial weight of hydrogel B) at 25 °C. As shown, hydrogel B maintains approximately 96.6% of its initial weight after 7 days of storage, indicating the desirable durability of hydrogel for long-term use (Figure 2g).

3.2. Sensing Performance of the Sensors

The three-dimensional macroporous structure of porous hydrogels using rosin particles as the sacrificial template demonstrates exceptional compressive elasticity and a high-compressive strain, enabling their potential application as piezoresistive pressure sensors. The piezoresistive sensors with original or porous hydrogels sandwiched between two carbon cloth electrodes were fabricated and investigated under compressive pressure stimuli. The sensitivity (S) is defined as $S = \delta((I - I_0)/I_0)/\delta P$, where I_0 is the baseline current under no pressure, I is the responsive current under applied pressure, and P is the applied pressure [16].

The response curves of sensors are plotted in Figure 3a. Comparing the sensors with original hydrogels and porous hydrogels reveals that the sensor with original hydrogels exhibits a relatively low current change in response to pressure. The improved sensing performance of porous hydrogel sensors (hydrogels A, B, C) suggests that the macroporous hydrogel structure endows the sensor to respond to pressure more effectively. Moreover, the sensor's sensitivity and measurement range depend on the hydrogels' porosity. Compared with the sensor with hydrogel A (12.8% porosity), the sensor with hydrogel B (15.7% porosity) exhibits a greater current change in response to pressure. The higher porosity could further increase the current by forming more current paths with increased pressure. However, the monotonous increase in the porosity of a porous structure cannot automatically improve the sensitivity. The lower sensitivity of the sensor with hydrogel C (22.4% porosity) is observed compared with that of the sensor with hydrogel B (15.7% porosity). This is because residual rosin particles increase gradually with increased rosin content (Figure S2, Supporting Information). Due to the residual rosin particles, the stiffness of hydrogel C increases (Figure 2f), generating fewer current paths under the same pressure as hydrogel B. Thus, the complete dissolution of rosin particles and the formation of an effective porous structure is formed from 1.5% to 3.0% rosin in the mixture (LBG, PVA, CNTs, deionized water, and rosin particles); however, the further addition of rosin particles lowers the sensors' performance.

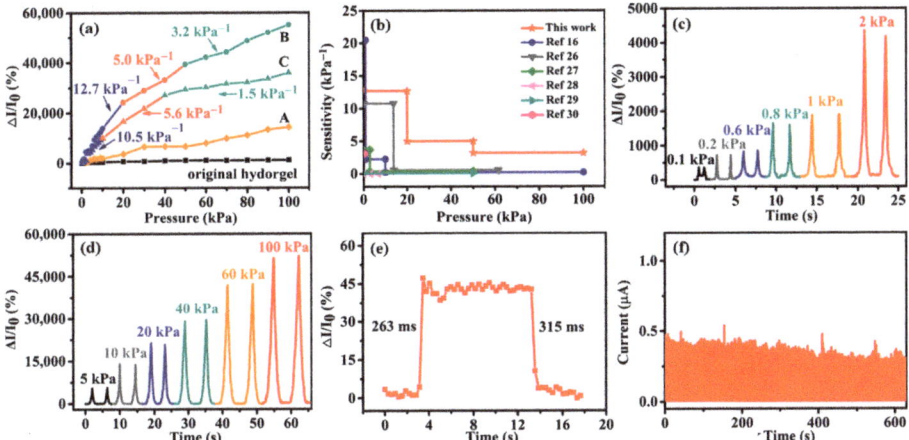

Figure 3. (a) The relative current changes for the sensors with original hydrogels and hydrogels A, B, and C; the slope of the curve indicates sensitivity. (b) Detection range and sensitivity of our sensor with hydrogel B in comparison with other piezoresistive sensors based on hydrogels as reported in references [16,26–30]. (c) The relative current changes for the sensor with hydrogel B under low serial pressures (0.1–2 kPa). (d) The relative current changes for the sensor with hydrogel B under high serial pressures (5–100 kPa). (e) The sensor's response time and recovery time with hydrogel B. (f) Stability of the sensor with hydrogel B at 3 kPa for 500 cycles.

Based on the lowest stiffness of hydrogel B, the sensor monitors pressure more sensitively over the wide pressure range. In the pressure range below 20 kPa, the sensor shows a high sensitivity of 12.7 kPa^{-1}. It shows a slightly lower sensitivity of 5.0 kPa^{-1} in the 20–50 kPa pressure range and maintains a 3.2 kPa^{-1} in the 50–100 kPa. Compared to our previously reported sensors (double-rough surface LBG/PVA/CNTs hydrogel) [16], the hydrogel B-based sensor exhibits better sensitivity over a wide linear detection range (Figure 3b). For example, the sensitivity of this sensor at low pressure (12.7 kPa^{-1}, 1–20 kPa) is higher than that of a double-rough surface LBG/PVA/CNTs hydrogel-based sensor (2.28 kPa^{-1}, 1–10 kPa). The sensitivity detection range of the sensor based on hydrogel B (1–20 kPa) is much wider than that of the double-rough surface hydrogel sensor (1–10 kPa). Furthermore, the sensitivities of this sensor at a high-pressure range (50–100 kPa) are 13.3-fold higher than that of the sensor with double-rough surface LBG/PVA/CNTs hydrogel. These results indicate that porous structure can improve sensitivity and the detection range simultaneously, and it is crucial in the sensor's response to high pressure. Compared to other hydrogel-based sensors [16,26–30], our sensor's sensitivities at the high-pressure range are superior to those of the previous hydrogel sensors (Figure 3b).

Due to its high sensitivity and wide sensing range, the hydrogel B-based sensor was selected for subsequent experiments. To verify the reliability of the sensor, different dynamic pressures were applied. Figure 3c,d shows that the increased applied pressure simultaneously raises the current response. Hence, the sensor can accurately distinguish between different levels of force in the pressure range of 0.1–2 kPa (Figure 3c) and 5–100 kPa (Figure 3d). Additionally, it shows a detection limit of 100 Pa. The response/relaxation time of the device was also analyzed. The proposed sensor immediately responds when a key is placed on it. The response and recovery time is 263 and 315 ms, respectively (Figure 3e). Durability is a critical aspect for wearable piezoresistive sensors. In order to test the durability of hydrogel B-based sensor, 500 loading/unloading cycles at 3 kPa and 20 kPa were performed. The results are shown Figures 3f and S3 (Supporting Information), which indicates the long-term durability and stability of hydrogel B-based sensor. We also compared the response time, detection limit, and durability of our sensor and those of the previous hydrogel sensors [16,26–30] (Table S1, Supporting Information). Further studies are needed to improve the response time, detection limit, and durability of our sensor.

3.3. The Sensing Mechanism of the Sensor

The above results indicate that the sensor with porous hydrogels effectively responds to pressure more than original hydrogels. Small pores (sub 10 μm level) are present in original and porous hydrogels. However, there is a visible difference between the two hydrogels at high pressure (Figure 4a,b). Original hydrogels with small pores are highly compact at low pressure, and consequently, the changes in conductive pathways are minimal under compression (Figure 4a). In contrast, porous hydrogels with both small and large pores (100–150 μm) allow more compaction and maintain high piezoresistivity due to forming more conductive pathways under increased pressure (Figure 4b). The structure of porous hydrogels with dense large-sized pores facilitates the closure of large pores at a high-pressure range, allowing the measurement of high pressures.

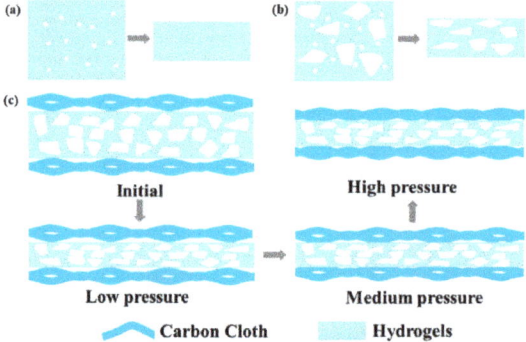

Figure 4. Schematic illustration of the piezoresistive mechanism of (**a**) original and (**b**) porous hydrogels, and (**c**) the sensor structure with porous hydrogels and its analysis mechanism.

The schematic pressure-sensing models of porous hydrogels sandwiched between two carbon cloth electrodes are illustrated in Figure 4c. The deformation of hydrogel B and the carbon cloth and the changing processes of the contact points under external force are also presented in Figure 4c. When low pressure (0.1–20 kPa) is applied to the sensor, the pores are squeezed to a high degree, forming many conductive paths in hydrogel B. Therefore, the output signal changes significantly, contributing to a high sensitivity of 12.7 kPa^{-1} in the low-pressure range. When pressure increases to a medium range, the deformation of the pores in hydrogels reaches saturation, while the contact area between the carbon cloth and hydrogels increases. At this stage, the decreasing rate in the total resistance of the sensor slows down, and the sensitivity decreases to 5.0 kPa^{-1} (20–50 kPa). A further increase in pressure compresses the carbon fiber in carbon cloth, thereby reducing the resistance. At this stage, the sensor only has a lower sensitivity of 3.2 kPa^{-1} in the pressure range of 50–100 kPa.

3.4. Applications of the Sensor

The good performances of the proposed sensor have potential prospects in many application fields, including wearable electronic products and human activity monitoring systems. We further demonstrate that the sensor's response based on hydrogel B is potentially useful in practical applications. The sensor is attached to each joint to monitor the finger (Figure 5a), wrist (Figure 5b), and elbow bending (Figure 5c). The signal waveform remains consistent under the same bending angle, while a significant change in current is observed with different joint-bending angles. The pressure produced by joint bending increases with the bending angle; therefore, when the bending angle increases, the sensor generates stronger signals, thus increasing the variation. The sensors are attached to the fingers to simulate touch sensation. Figure 5d shows that the sensor is attached to the index finger. The changes in pressure signals are observed when standard weights of 1, 2, 5, 10, and 20 g are held by hand in sequence. The sensor easily measures the force applied by the finger, suggesting that the sensor can play an important role in the tactile perception of the manipulator. Furthermore, the sensor can also monitor human microexpression attached near the eyes (Figure 5e), and the eye-opening/closing states are identified with good sensitivity. Interestingly, the sensors could distinguish different handwriting samples (Figure 5e). Due to the difference in writing power and direction, the handwriting samples of 1, 2, 3, 4, and 5 differ in number and peak shape, presenting an opportunity to realize handwriting anti-counterfeiting applications. Foot-stepping pressure is important information in biomechanics, healthcare, recovery, and diagnosis, especially for athletes who exercise extensively, teenagers in development, and patients suffering from Parkinson's disease or diabetic foot ulcers [19]. Figure S4 (Supporting Information) shows the foot-stepping pressures detected by using the sensor. Because of the broad measuring range, the sensor can detect different stepping intensities, such as the weak step and the heavy step. The result demonstrates the sensor has potential applications in footwear electronics.

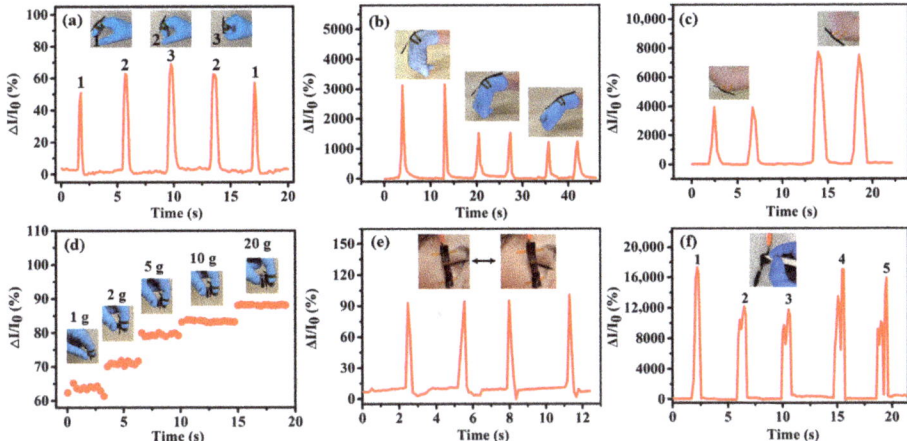

Figure 5. The signal responses arise from (**a**) finger bending, (**b**) wrist bending, and (**c**) elbow bending. (**d**) The signal responses of the sensors attached to the finger-holding standard weights of 1, 2, 5, 10, and 20 g. The signal responses from (**e**) eye blinking and (**f**) different handwriting.

An individual pressure sensor only provides limited information due to its low coverage area and integrated properties. Individual hydrogel B was integrated into a 3 × 3 pixel sensor array on a polyethylene terephthalate (PET) substrate to perceive the spatial distributions of pressure by employing silver glue to connect with the copper foil as the electrode (Figure 6). The 3 × 3 pixel sensor array was connected to a signal management circuit (Figure S5, Supporting Information), which included a micro-controller unit, an analog-to-digital converter (ADC), channel selection, and communication interface. Using the circuit diagram shown in Figure S5, the real-time signal can be transmitted to a computer via a communication interface. In this process, by applying pressure on the sensor array, the electrical resistance of the on-site sensor changes and the responsive current under applied pressure is recorded, generating the color contrast mapped with local pressure distribution. When two fingers are pressed on the sensor array (Figure 6a,b), the position and pressure of the finger are accurately determined through the responsive current changes (Figure 6c,d), consistent with the finger position. These results suggest the potential of the sensor arrays in E-skin devices for next-generation wearable electronics.

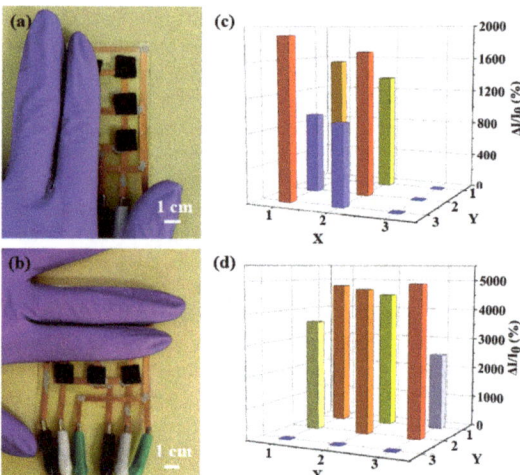

Figure 6. (**a**,**b**) Photograph of a sensor array (3 × 3 pixel) with different devices pressed by two fingers, and (**c**,**d**) and their corresponding signals.

4. Conclusions

A simple, environment-friendly, low-cost, high-performance wearable piezoresistive pressure sensor has been developed using a highly porous LBG-based hydrogel. By tuning the porosity of porous LBG-based hydrogel (12.8%, 15.7%, 22.4%), the hydrogel with a porosity of 15.7% demonstrates excellent sensing performance. The porous LBG-based hydrogel offers a wide pressure detection range with high sensitivities (12.7, 5.0, and 3.2 kPa^{-1} under 0.1–20, 20–50, and 50–100 kPa), a fast response time (263 ms), and good durability over 500 loading/unloading cycles. These excellent properties are ascribed to the three-dimensional macroporous structure formed with dense large-sized pores (100–150 μm), facilitating the compression of large pores over a wide pressure range. Additionally, the sensor has promising practical applications in monitoring and identifying human activities. Their sensitivity and wide-range properties endow the porous LBG-based hydrogel with great potential for application in various wearable sensors. This work provides a good strategy for fabricating high-performance wearable pressure sensors in human motion monitoring with high sensitivity over a wide response range.

Supplementary Materials: The following supporting information can be downloaded at: https://www.mdpi.com/article/10.3390/polym15122736/s1, Figure S1: SEM image of the commercial rosin particles; Figure S2: Cross-sectional SEM image of hydrogel C; Figure S3: Stability of the sensor with hydrogel B at 20 kPa for 500 cycles; Figure S4: Pressure signals when a foot steps on and steps off; Figure S5: A schematic diagram of a sensor array with a signal management circuit system; Table S1: Response time, detection limit, and durability of our sensor in comparison with other piezoresistive sensors based on hydrogels as reported in references.

Author Contributions: Conceptualization, F.X. and S.J.; methodology, F.X. and S.J.; software, F.X.; validation, F.X. and W.Z.; formal analysis, Y.Z.; investigation, Y.Z.; resources, Y.H.; data curation, F.X.; writing—original draft preparation, Y.H.; writing—review and editing, H.Z.; visualization, Y.H.; supervision, Y.H.; project administration, Y.H.; funding acquisition, Y.H. All authors have read and agreed to the published version of the manuscript.

Funding: This work is supported by the National Natural Science Foundation of China (No. 62101605). This work was partially carried out at the USTC Center for Micro and Nanoscale Research and Fabrication. Yuan Huang would like to acknowledge support from Zinergy Shenzhen Ltd.

Institutional Review Board Statement: This study did not require ethical approval.

Data Availability Statement: The data are available upon reasonable request.

Conflicts of Interest: The authors declare no conflict of interest.

References

1. Yue, Y.; Liu, N.; Su, T.; Cheng, Y.; Liu, W.; Lei, D.; Cheng, F.; Ge, B.; Gao, Y. Self-Powered Nanofluidic Pressure Sensor with a Linear Transfer Mechanism. *Adv. Funct. Mater.* **2023**, *33*, 2211613. [CrossRef]
2. Meng, K.; Xiao, X.; Wei, W.; Chen, G.; Nashalian, A.; Shen, S.; Xiao, X.; Chen, J. Wearable Pressure Sensors for Pulse Wave Monitoring. *Adv. Mater.* **2022**, *34*, 2109357. [CrossRef] [PubMed]
3. Wang, X.-M.; Tao, L.-Q.; Yuan, M.; Wang, Z.-P.; Yu, J.; Xie, D.; Luo, F.; Chen, X.; Wong, C. Sea Urchin-like Microstructure Pressure Sensors with an Ultra-broad Range and High Sensitivity. *Nat. Commun.* **2021**, *12*, 1776. [CrossRef]
4. Huang, C.-Y.; Yang, G.; Huang, P.; Hu, J.-M.; Tang, Z.-H.; Li, Y.-Q.; Fu, S.-Y. Flexible Pressure Sensor with an Excellent Linear Response in a Broad Detection Range for Human Motion Monitoring. *ACS Appl. Mater. Interfaces* **2023**, *15*, 3476–3485. [CrossRef] [PubMed]
5. Cao, W.; Luo, Y.; Dai, Y.; Wang, X.; Wu, K.; Lin, H.; Rui, K.; Zhu, J. Piezoresistive Pressure Sensor Based on a Conductive 3D Sponge Network for Motion Sensing and Human–Machine Interface. *ACS Appl. Mater. Interfaces* **2023**, *15*, 3131–3140. [CrossRef]
6. Sang, Z.; Ke, K.; Manas-Zloczower, I. Design Strategy for Porous Composites Aimed at Pressure Sensor Application. *Small* **2019**, *15*, 1903487. [CrossRef] [PubMed]
7. Wang, S.; Deng, W.; Yang, T.; Ao, Y.; Zhang, H.; Tian, G.; Deng, L.; Huang, H.; Huang, J.; Lan, B.; et al. Bioinspired MXene-Based Piezoresistive Sensor with Two-stage Enhancement for Motion Capture. *Adv. Funct. Mater.* **2023**, *33*, 2214503. [CrossRef]
8. Ma, Y.; Zhao, K.; Han, J.; Han, B.; Wang, M.; Tong, Z.; Suhr, J.; Xiao, L.; Jia, S.; Chen, X. Pressure Sensor Based on a Lumpily Pyramidal Vertical Graphene Film with a Broad Sensing Range and High Sensitivity. *ACS Appl. Mater. Interfaces* **2023**, *15*, 13813–13821.
9. Ha, K.-H.; Zhang, W.; Jang, H.; Kang, S.; Wang, L.; Tan, P.; Hwang, H.; Lu, N. Highly Sensitive Capacitive Pressure Sensors over a Wide Pressure Range Enabled by the Hybrid Responses of a Highly Porous Nanocomposite. *Adv. Mater.* **2021**, *33*, 2103320. [CrossRef]
10. Lu, J.; Hu, S.; Li, W.; Wang, X.; Mo, X.; Gong, X.; Liu, H.; Luo, W.; Dong, W.; Sima, C.; et al. A Biodegradable and Recyclable Piezoelectric Sensor based on a Molecular Ferroelectric Embedded in a Bacterial Cellulose Hydrogel. *ACS Nano* **2022**, *16*, 3744–3755.
11. Liu, Y.; Wu, B.; Zhang, Q.; Li, Y.; Gong, P.; Yang, J.; Park, C.B.; Li, G. Micro/nano-structure Skeleton Assembled with Graphene for Highly Sensitive and Flexible Wearable Sensor. *Compos. Part A* **2023**, *165*, 107357. [CrossRef]
12. Tong, R.; Chen, G.; Tian, J.; He, M. Highly Stretchable, Strain-Sensitive, and Ionic-Conductive Cellulose-Based Hydrogels for Wearable Sensors. *Polymers* **2019**, *11*, 2067. [CrossRef] [PubMed]
13. Wang, J.; Huang, Y.; Liu, B.; Li, Z.; Zhang, J.; Yang, G.; Hiralal, P.; Jin, S.; Zhou, H. Flexible and Anti-freezing Zinc-ion Batteries Using a Guar-gum/sodium-alginate/ethylene-glycol Hydrogel Electrolyte. *Energy Storage Mater.* **2021**, *41*, 599–605. [CrossRef]
14. Cao, L.; Zhao, Z.; Wang, X.; Huang, X.; Li, J.; Wei, Y. Tough, Antifreezing, and Conductive Hydrogel Based on Gelatin and Oxidized Dextran. *Adv. Mater. Technol.* **2022**, *7*, 2101382. [CrossRef]
15. Wu, M.; Wang, X.; Xia, Y.; Zhu, Y.; Zhu, S.; Jia, C.; Guo, W.; Li, Q.; Yan, Z. Stretchable Freezing-tolerant Triboelectric Nanogenerator and Strain Sensor Based on Transparent, Long-term Stable, and Highly Conductive Gelatin-based Organohydrogel. *Nano Energy* **2022**, *95*, 106967. [CrossRef]
16. Huang, Y.; Liu, B.; Zhang, W.; Qu, G.; Jin, S.; Li, X.; Nie, Z.; Zhou, H. Highly Sensitive Active-powering Pressure Sensor Enabled by Integration of Double-rough Surface Hydrogel and Flexible Batteries. *Npj Flex. Electron.* **2022**, *6*, 92. [CrossRef]
17. Shen, X.; Zheng, L.; Tang, R.; Nie, X.; Wang, Z.; Jin, C.; Sun, Q. Double-network Hierarchical-porous Piezoresistive Nanocomposite Hydrogel Sensors Based on Compressive Cellulosic Hydrogels Deposited with Silver Nanoparticles. *ACS Sustain. Chem. Eng.* **2020**, *8*, 7480–7488. [CrossRef]
18. Yang, Y.; Yang, Y.; Cao, Y.; Wang, X.; Chen, Y.; Liu, H.; Gao, Y.; Wang, J.; Liu, C.; Wang, W.; et al. Anti-freezing, Resilient and Tough Hydrogels for Sensitive and Large Range Strain and Pressure Sensors. *Chem. Eng. J.* **2021**, *403*, 126431.
19. Zhao, S.; Zhu, R. High Sensitivity and Broad Range Flexible Pressure Sensor Using Multilayered Porous PDMS/AgNP Sponge. *Adv. Mater. Technol.* **2019**, *4*, 1900414. [CrossRef]
20. Bae, K.; Kim, M.; Kang, Y.; Sim, S.; Kim, W.; Pyo, S.; Kim, J. Dual-Scale Porous Composite for Tactile Sensor with High Sensitivity over an Ultrawide Sensing Range. *Small* **2022**, *18*, 2203193. [CrossRef] [PubMed]
21. Song, Y.; Chen, H.; Su, Z.; Chen, X.; Miao, L.; Zhang, J.; Cheng, X.; Zhang, H. Highly Compressible Integrated Supercapacitor–Piezoresistance-Sensor System with CNT–PDMS Sponge for Health Monitoring. *Small* **2017**, *13*, 1702091. [CrossRef] [PubMed]
22. Zhu, D.; Wang, S.H.; Zhou, X. Recent Progress in Fabrication and Application of Polydimethylsiloxane Sponges. *J. Mater. Chem. A* **2017**, *5*, 16467–16497. [CrossRef]
23. Yu, C.; Yu, C.; Cui, L.; Song, Z.; Zhao, X.; Ma, Y.; Jiang, L. Facile Preparation of the Porous PDMS Oil-Absorbent for Oil/Water Separation. *Adv. Mater. Interfaces* **2017**, *4*, 1600862. [CrossRef]
24. Lo, L.-W.; Zhao, J.; Wan, H.; Wang, Y.; Chakrabartty, S.; Wang, C. A Soft Sponge Sensor for Multimodal Sensing and Distinguishing of Pressure, Strain, and Temperature. *ACS Appl. Mater. Interfaces* **2022**, *14*, 9570–9578. [CrossRef]

25. Wu, K.; Li, X. Wearable Pressure Sensor for Athletes' Full-range Motion Signal Monitoring. *Mater. Res. Express* **2020**, *7*, 105003.
26. Lu, Y.; Qu, X.; Zhao, W.; Ren, Y.; Si, W.; Wang, W.; Wang, Q.; Huang, W.; Dong, X. Highly Stretchable, Elastic, and Sensitive MXene-based Hydrogel for Flexible Strain and Pressure Sensors. *Research* **2020**, *1*, 2038560. [CrossRef] [PubMed]
27. Yin, M.; Zhang, Y.; Yin, Z.; Zheng, Q.; Zhang, A.P. Micropatterned Elastic Gold-nanowire/polyacrylamide Composite Hydrogels for Wearable Pressure Sensors. *Adv. Mater. Technol.* **2018**, *3*, 1800051. [CrossRef]
28. Ge, G.; Zhang, Y.; Shao, J.; Wang, W.; Si, W.; Huang, W.; Dong, X. Stretchable, Transparent, and Self-patterned Hydrogel-based Pressure Sensor for Human Motions Detection. *Adv. Funct. Mater.* **2018**, *28*, 1802576. [CrossRef]
29. Qin, Z.; Sun, X.; Yu, Q.; Zhang, H.; Wu, X.; Yao, M.; Liu, W.; Yao, F.; Li, J. Carbon Nanotubes/Hydrophobically Associated Hydrogels as Ultrastretchable, Highly Sensitive, Stable Strain, and Pressure Sensors. *ACS Appl. Mater. Interfaces* **2020**, *12*, 4944–4953. [CrossRef]
30. Gao, Y.; Liu, D.; Xie, Y.; Song, Y.; Zhu, E.; Shi, Z.; Yang, Q.; Xiong, C. Flexible and Sensitive Piezoresistive Electronic Skin Based on TOCN/PPy Hydrogel Films. *J. Appl. Polym. Sci.* **2021**, *138*, 51367. [CrossRef]

Disclaimer/Publisher's Note: The statements, opinions and data contained in all publications are solely those of the individual author(s) and contributor(s) and not of MDPI and/or the editor(s). MDPI and/or the editor(s) disclaim responsibility for any injury to people or property resulting from any ideas, methods, instructions or products referred to in the content.

Article

The Mechanical Properties of Silicone Rubber Composites with Shear Thickening Fluid Microcapsules

Chun Wei, Xiaofei Hao, Chaoying Mao, Fachun Zhong and Zhongping Liu *

Institute of Chemical Materials, China Academy of Engineering Physics (CAEP), Mianyang 621900, China; weichun20@gscaep.ac.cn (C.W.); haoxiaofei163@163.com (X.H.); chaoyingmao@caep.cn (C.M.); zhongfachun@caep.cn (F.Z.)
* Correspondence: liuzp@caep.cn

Abstract: In this study, Sylgard 184 silicone rubber (SylSR) matrix composites with shear thickening fluid (STF) microcapsules (SylSR/STF) were fabricated. Their mechanical behaviors were characterized by dynamic thermo-mechanical analysis (DMA) and quasi-static compression. Their damping properties increased with the addition of STF into the SR in DMA tests and the SylSR/STF composites presented decreased stiffness and an obvious positive strain rate effect in the quasi-static compression test. Moreover, the impact resistance behavior of the SylSR/STF composites was tested by the drop hammer impact test. The addition of STF enhanced the impact protective performance of silicone rubber, and the impact resistance increased with the increase of STF content, which should be ascribed to the shear thickening and energy absorption of STF microcapsules in the composites. Meanwhile, in another matrix, hot vulcanized silicone rubber (HTVSR) with a mechanical strength higher than Sylgard 184, the impact resistance capacity of its composite with STF (HTVSR/STF) was also examined by the drop hammer impact test. It is interesting to note that the strength of the SR matrix obviously influenced the enhancement effect of STF on the impact resistance of SR. The stronger the strength of SR, the better the effect of STF on improving the impact protective performance of SR. This study not only provides a new method for packaging STF and improving the impact resistance behavior of SR, but is also beneficial for the design of STF-related protective functional materials and structures.

Keywords: shear thickening fluid; silicone rubber; microcapsule; impact resistance behavior

Citation: Wei, C.; Hao, X.; Mao, C.; Zhong, F.; Liu, Z. The Mechanical Properties of Silicone Rubber Composites with Shear Thickening Fluid Microcapsules. Polymers 2023, 15, 2704. https://doi.org/10.3390/polym15122704

Academic Editors: Jiangtao Xu and Sihang Zhang

Received: 14 May 2023
Revised: 1 June 2023
Accepted: 5 June 2023
Published: 16 June 2023

Copyright: © 2023 by the authors. Licensee MDPI, Basel, Switzerland. This article is an open access article distributed under the terms and conditions of the Creative Commons Attribution (CC BY) license (https://creativecommons.org/licenses/by/4.0/).

1. Introduction

Shear thickening fluid (STF), as a lightweight, intelligent, and efficient impact protective material, has attracted much research interest [1,2]. It is a solid–liquid suspension system composed of a high concentration of particles and liquid oligomers. In the steady state, STF appears as a viscous liquid with fluidity, but its viscosity will increase sharply to a solid-like state after being impacted. This liquid–solid conversion is rapid and reversible, accompanied by a large amount of impact energy dissipation. The various mechanisms of the shear thickening phenomenon, such as order-disorder transition theory, hydro-cluster theory, jamming theory, and friction contact theory, have been proposed by researchers [3,4]. Moreover, the reasons for energy dissipation of STFs under impact loading have been considered to be viscous damping [5], fraction between clusters [6], cracking of the jammed network [7,8], and extrusion deformation, cracking, and crushing of particles under high impact pressure [9,10]. Due to their flexible and excellent energy absorption properties, STF-treated high-performance fabrics such as Kevlar and UHMWPE have attracted much attention in applications of body protection and exhibit enhanced bullet-proof and stab-resistant properties [11–13].

Since STFs are liquids without a fixed three-dimensional shape under normal conditions and are sensitive to the environment due to their easy moisture absorption, they need to be properly encapsulated in practical applications. To encapsulate STFs, Zhang et al. added a small amount of polyethylene imine into STF and dripped it into an MDI isocyanate prepolymer, and then prepared polyurea-walled macroscopic STF capsules [14]. Zhang Xin et al. further developed three methods to prepare STF microcapsules and realized the reinforcement of the polyurea shells [15]. Liu et al. added a sodium alginate solution and Span 20 to the STF paraffin solution containing Span 80, then the mixture was dropped into anhydrous calcium chloride aqueous solution using a syringe to prepare STF capsules [16]. Kaczorowski et al. adopted polypropylene glycol diacrylate as a monomer to prepare a slightly crosslinked shear thickening liquid organic gel and then embedded it into the uncured polyurethane mixture to form STF/polyurethane elastomer composites [17]. Soutrenon et al. used the vacuum resin infusion method to infiltrate STFs into foam and then covered the outer layer with silicone rubber to achieve the encapsulation of STF/foam composites [18].

Silicone rubber (SR) is a potential material for packaging and storing STFs due to its good barrier performance against water and air and chemical inertness to STFs [19]. Meanwhile, because of its excellent mechanical properties such as high elasticity, low impedance, viscoelasticity, low-temperature resistance, aging resistance, and flexibility, silicone rubber is often used as an impact energy absorption material in the weapon, shipping, electronics, and machinery industries [20]. Developing methods to improve the impact energy absorption performance of silicone rubber materials also has important practical significance in promoting the application of silicone rubber for impact protection.

In our previous work, the impact protective property of the SR matrix composites with shear thickening fluid microcapsules was studied under high-strain-rate loadings. It was found that this composite was a promising flexible material for impact protection due to its flexibility at a low strain rate (10^{-3} s^{-1}) but higher stiffness at a high impact loading rate (3500 s^{-1}) [21]. In this work, their mechanical properties were characterized by dynamic thermo-mechanical analysis (DMA) measurements, quasi-static compression tests, and drop hammer impact tests. It was found that the addition of STFs can improve the impact protection performance of silicone rubber and the impact resistance increased with the increase of STF content. The composites presented decreased stiffness and an obvious positive strain rate effect in the quasi-static compression test. Notably, the strength of the silicone rubber matrix influenced the enhancement effect of STFs on the impact resistance of silicone rubber.

2. Materials and Methods

2.1. Materials

Tetraethoxysilane (TEOS, 99.9%) and ammonia water ($NH_3 \cdot H_2O$, 25–28%) were purchased from Chengdu Kelong Chemical Reagent Co., LTD. Ethanol (EtOH, 96%) and polyethylene glycol (PEG, Mw = 200) were both purchased from China Sinopharm Chemical Reagent Co., LTD. Sylgard 184 silicone elastomer kit, consisting of a base agent (part A) and a curing agent (part B), is a kind of hydro-silylated liquid silicone rubber purchased from Dow Corning Co., LTD. Methyl vinyl silicone rubber (MVMQ, 110-2, Mw = 6.4×10^5), which contains 0.17 mol% vinyl groups on the backbone chain, was commercially obtained from Dongjue Fine Chemicals (Nanjing). Hydroxyl silicone oil (GY-209-3) was provided by the Chenguang Research Institute of Chemical Industry, China. Dicumyl peroxide (DCP), a vulcanized agentof hot vulcanized silicone rubber, was purchased from Aladdin. The reinforcing filler fumed silica (AS200, hydrophilic) was obtained from Evonik Degussa, Germany.

2.2. SiO_2 Preparation

SiO_2 was synthesized by a modified Stöber method as follows. Firstly, 16.25 mL ethanol, 9.0 mL ammonia, and 24.75 mL water were introduced into a beaker under magnetic stirring at 800 rpm at room temperature. Then, a mixture of 6 mL TEOS and 44 mL ethanol was quickly added to the breaker. After 5 min, the stirring speed was changed to 400 rpm, and the reaction was maintained for 2 h. Finally, the ethanol in the solution was removed under vacuum and the SiO_2 was obtained.

2.3. STF Preparation

The STF was prepared by dispersing SiO_2 into a solution of PEG/ethanol under sonication mixing for 4 h, and the amounts were set to be 68:32:600 (weight by weight, w/w) of SiO_2/PEG/ethanol. After homogeneous mixing, the ethanol was removed by rotary evaporation, and the viscous STF composed of PEG and SiO_2 was finally obtained.

2.4. SR/STF Composites Preparation

The SR/STF composites were fabricated by emulsifying the STF in silicone rubber and then vulcanizing the mixture. The schematic diagram for preparing the SR/STF composites is depicted in Figure 1.

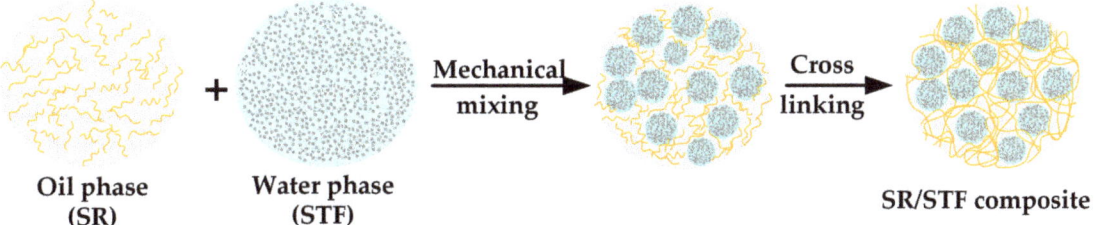

Figure 1. Schematic diagram for preparing the SR/STF composites.

In detail, when Sylgard 184 silicone rubber was used as the matrix, a certain amount of the STF was added into Sylgard 184 silicone rubber ($m_A:m_B$ = 10:1) and mechanically mixed at 300 rpm for 5 min, and then defoamed by a vacuum oven. After that, the mixture was poured into a mold and cured in an air-dry oven at 80 °C for 2 h, then SylSR/STF composites were obtained. The mass fractions of the STF in the SylSR/STF composites were 0%, 10%, 20%, 30%, and 40%, respectively. For brevity, they were abbreviated as SylSR, SylSR/STF-10, SylSR/STF-20, SylSR/STF-30, and SylSR/STF-40, respectively.

In addition, the hot vulcanized silicone rubber was also adopted as another silicone rubber matrix. The hot vulcanized silicone rubber/STF composite was prepared by homogeneously mixing the hot vulcanized rubber component (methyl vinyl silicone rubber, fumed silica, hydroxyl silicone oil, and DCP with a weight ratio of 100:40:4:2) with STF using a double-roller mixing machine. Then, they were vulcanized in a flat vulcanizing machine at 165 °C for 12 min and the hot vulcanized silicone rubber matrix composites with shear thickening fluid microcapsules were prepared, marked as HTVSR/STF. The vulcanized silicone rubber materials without the STF were also prepared by the same method, and it was labeled as HTVSR.

2.5. Characterization

The microstructural characteristics of SiO_2 were analyzed by field-emission scanning electron microscopy (FE-SEM, Ultra55, Carl Zeiss Ltd., Oberkochen, Germany) at a 10 KV acceleration voltage and attenuated total reflection-Fourier transform infrared spectroscopy (ATR–FTIR, Nicolet 800, Thermo Fisher Scientific, Waltham, MA, USA) at a resolution of 4 cm^{-1} for a total of 32 scans with a scan wave between 400 and 4000 cm^{-1}. The particle-size distribution was analyzed by Image J. The rheological properties of the STF and PEG were tested using a Kinexus Pro rotary rheometer (Malvern, UK). The diameter of the lamina was 40 mm with a cone angle of 1°. The spacing was 0.03 mm, and the temperature was 25 °C. The microstructures of the composites were characterized using a Axio Lab.A1 (Zeiss, Jena, Germany) optical microscope (OM) with a CCD Camera and FE-SEM. The dynamic thermo-mechanical properties were analyzed using a DMA Q800 (TA, New Castle, DE, USA) in compression mode. It was a cylindrical sample with a 13 mm diameter and 3 mm height. In temperature scan tests, the temperature range was −50~150 °C, with a heating rate of 5 °C/min, frequency of 1 Hz, and amplitude of 5 μm. In frequency scan tests, the frequency range was 0.1–200 Hz, with a temperature of 25 °C and an amplitude of 5 μm. The quasi-static uniaxial compression tests were carried out on Instron-5582 electronic universal testing machine at room temperature. The samples were in cylindrical sizes with a diameter of 29 mm and a height of 12.5 mm. The loading rates of the test were set to be 0.5, 5, 50, and 200 mm/min, respectively, which corresponded to the quasi-static engineering strain rates of 0.00067 s^{-1}, 0.0067 s^{-1}, 0.067 s^{-1}, and 0.267 s^{-1}, respectively. The maximum engineering strain of the compressed specimen was about 0.6. The drop weight impact tests were used to evaluate the impact resistance behavior of materials and conducted on the drop hammer impact tester conforming to the EN1621-1-2012 standard [22]. The samples were in rectangular blocks with sizes of 150 mm × 100 mm × 4.80 mm (length × width × thickness). The drop hammer weight was 4.977 kg, the fall height was 42.6 cm, and the impact energy was 20 J.

3. Results and Discussion

3.1. Microstructural Characteristics of SiO_2

The microstructure of the SiO_2 particles and their size distribution was investigated by SEM. As shown in Figure 2a, the SiO_2 particles were nearly spherical monodisperse with an average size of about 300 nm. The surface state of SiO_2 particles was characterized via FTIR, and its typical absorbance spectra curve is shown in Figure 2b. The peaks at 794 cm^{-1} and 1058 cm^{-1} were assigned to the symmetric and asymmetric stretching vibrations of Si-O-Si bridges, respectively. The broad peak between 3700 cm^{-1} and 2800 cm^{-1} was the stretching vibration of Si-OH groups [23]. This result indicates that the SiO_2 prepared in this work is hydrophilic.

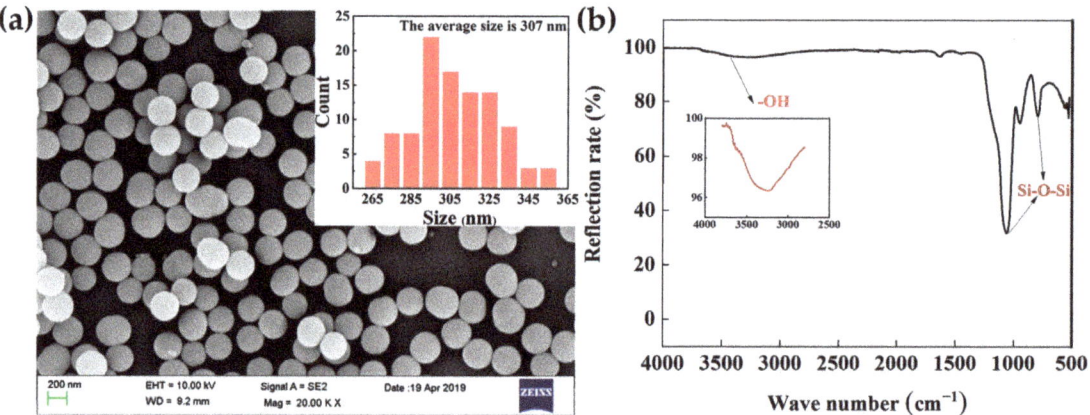

Figure 2. (**a**) SEM micrograph of SiO_2 particles and their size distribution graph, and (**b**) FTIR curve of SiO_2.

3.2. Rheological Behavior of STF

Figure 3 presents the rheological behavior of the as-prepared STF and PEG. As a Newtonian fluid, PEG exhibited the same low viscosity value (0.5 Pa·s) at different shear rates. In contrast, the STF showed a shear thinning at low shear rates with the viscosity value decreasing from the initial 318 Pa·s to 6 Pa·s, while when the shear rate increased beyond the "threshold" of 13 s^{-1}, the viscosity increased steeply into 381 Pa·s and the STF became extremely viscous. Note that this was not the highest viscosity value for STF; the experiment was stopped automatically at higher strain rates and the viscosity values were not recorded due to the self-protection of the rheometer from high torque forces. This is the typical curve of a discontinuous shear thickening fluid according to previous reports [2,5]. Hence, it should be concluded that the as-prepared STF will undergo a liquid–solid conversion and then dissipate impact energy under the impact loading.

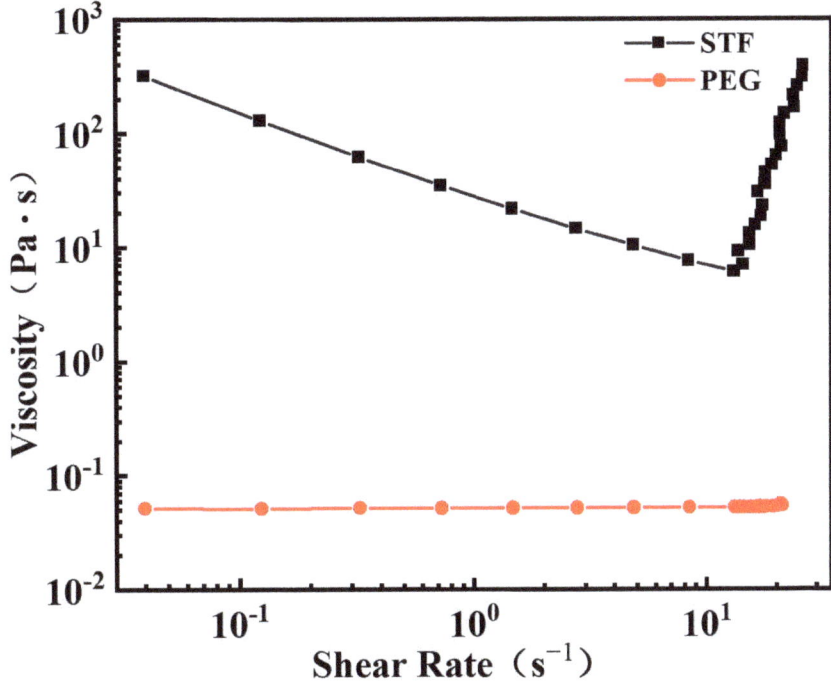

Figure 3. The rheological behaviors of STF and PEG.

3.3. Microstructure of SylSR/STF Composites

The microstructure of the SylSR/STF composites was first characterized using an optical microscope. As can be seen from the optical microscopy photos in Figure 4, STF was dispersed in the silicone rubber matrix in the form of spherical microcapsules for all the SylSR/STF composites. The microcapsules in all SylSR/STF composites had a wide diameter distribution from 5 to 25 μm. In particular, the mean diameters of the STF microcapsules for SylSR/STF-10, SylSR/STF-20, SylSR/STF-30, and SylSR/STF-40 were 13.7 ± 5.0, 13.2 ± 4.2, 13.4 ± 4.5, and 10.2 ± 2.1 μm, respectively. To further investigate the STF microcapsules, the SylSR/STF composites were made brittle and broken using liquid nitrogen, and then their fracture surface properties were investigated by FE-SEM. As shown in Figure 5, silica microspheres in the matrix of silicone rubber exhibited a state of aggregation rather than the average distribution state. In particular, it was observed that the silica microspheres were in STF microcapsules (Figure 5b). Therefore, the results indicate the composites prepared in this work were indeed composed of STF microcapsules and silicone rubber matrix, rather than a simple homogeneous mixture of SiO_2, PEG, and silicone rubber. Since SiO_2 microspheres and PEG, the components of the STF, are hydrophilic, the STF itself is a hydrophilic liquid. Considering the hydrophobicity of silicone rubber, a water-in-oil emulsion would be formed when the STF and silicone rubber are mixed under mechanical stirring [24]. Then, the water-in-oil configuration could be fixed with the help of the silicone rubber curing process, which resulted in the microstructure formation of STF microcapsules in the silicone rubber matrix.

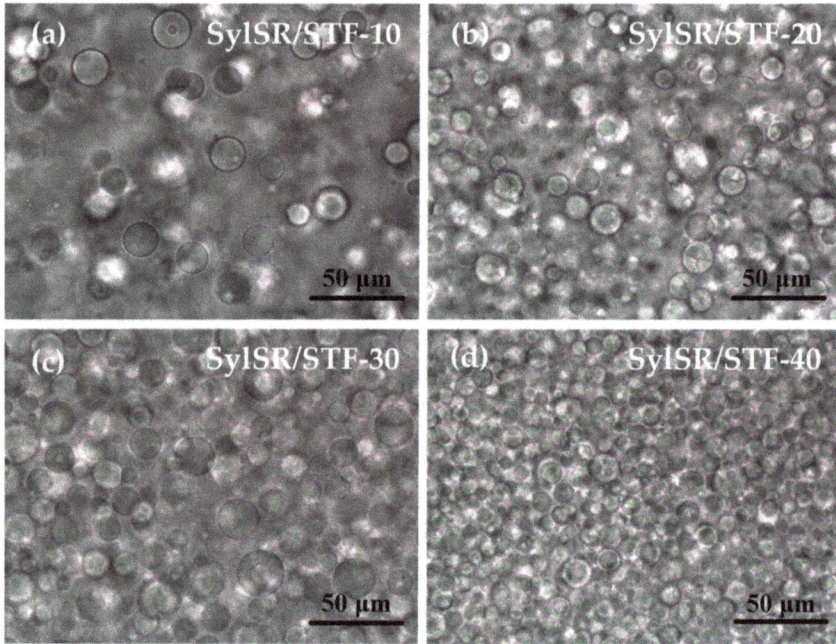

Figure 4. Optical micrographs of SylSR/STF composites with different mass fractions of STF. (**a**) 10%, (**b**) 20%, (**c**) 30%, and (**d**) 40%.

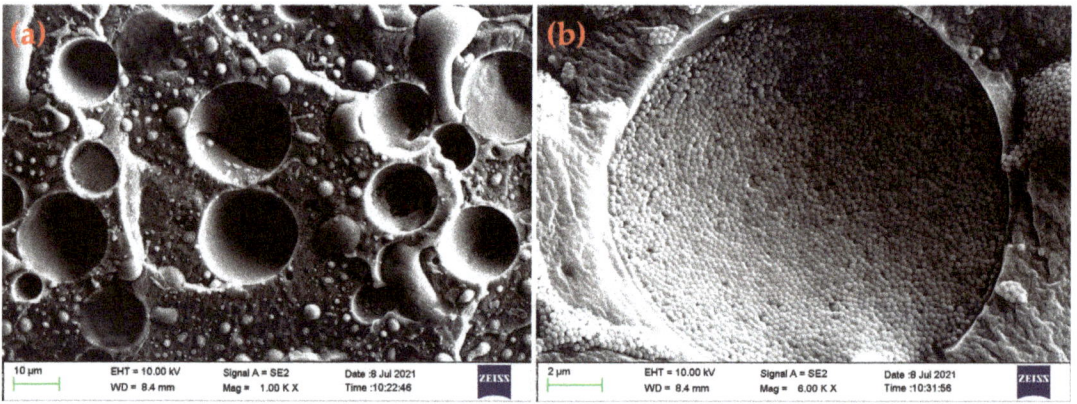

Figure 5. SEM images of (**a**) the fracture surface of a typical SylSR/STF composite and (**b**) the enlarged image of a STF microcapsule.

3.4. Mechanical Properties and Impact Resistance Behavior of SylSR/STF Composites

Dynamic thermo-mechanical analysis (DMA) measurements were used to investigate the damping properties by calculating the damping factor Tan δ of the SylSR/STF composites at different temperatures and frequencies, and the results are shown in Figure 6. It was found that the values of Tan δ for SylSR/STF-30 and SylSR/STF-40 were higher than that of silicone rubber in the temperature range of −50~150 °C. In addition, at the tested frequency range, 0.1–200 Hz, when the weight fractions of the STF were 10% and 20%, the values of Tan δ of the SylSR/STF composites were slightly lower than silicone rubber. When the mass fractions of the STF were increased to 30% and 40%, the Tan δ values of the SylSR/STF composites were greatly increased. Based on these results, it was concluded that STFs could markedly improve the damping properties of silicone rubber, which means that SylSR/STF composites have a better protection ability than silicone rubber itself when they are applied in changeable temperature environments and complex impact frequency domains.

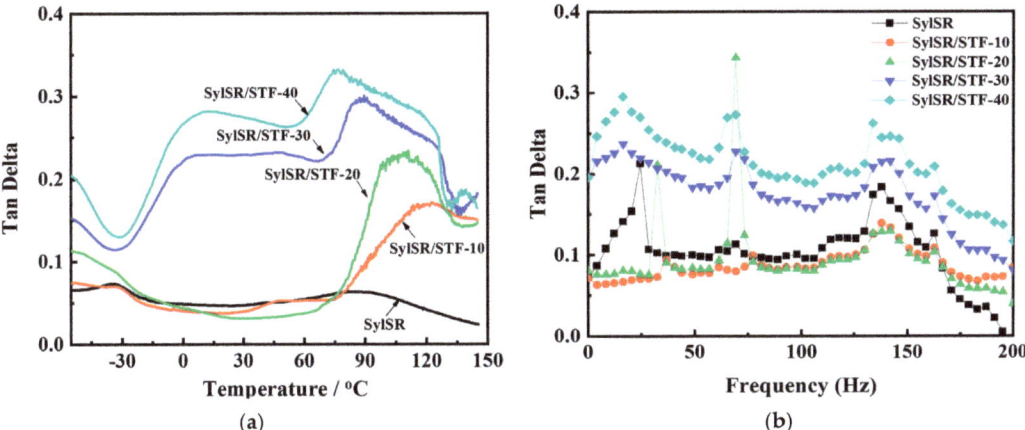

Figure 6. (a) Tan δ versus temperature curves and (b) Tan δ versus frequency curves of SylSR/STF composites.

The mechanical behavior of the SylSR/STF composites under low strain rates was studied using compression tests on a universal testing machine. The engineering stress–strain relationships of the SylSR/STF composites under loading rates ranging from 0.5 mm/min to 200 mm/min are shown in Figures 7 and 8. The results indicate that all the five kinds of composites show nonlinear elasticity for strains smaller than 60%. In addition, at the same strain, the stiffness increased with the increase of strain rate, as shown in Figure 7, indicating an obvious positive strain-rate sensitivity of the SylSR/STF composites. As shown in Figure 8, the stiffness of the composite decreased with the increasing mass fraction of the STF microcapsules under low strain rates. According to the mechanical behavior of the STF, below the critical strain rates of the STF it acts as liquid and exhibits slight shear thinning behavior, leading to the weakening effect when its content increases. The lower stiffness of the composite indicates the higher flexibility of the material with the increasing mass fraction of the STF microcapsules, which is strongly required for the design of soft impact protective structures [25].

The dynamic impact experiments conducted on drop hammer impact tester were used to evaluate the impact resistance behavior of SylSR/STF composites under low-speed impacts. In these experiments, two different kinds of silicone rubber, i.e., hydro-silylated liquid silicone rubber (Sylgard 184) and hot vulcanized silicone rubber (HTVSR), were used as the matrix of the SylSR/STF composites. The experiments were conducted with the same 48 kN initial impact force, and the contact forces tested behind the specimens were adopted to evaluate their impact resistance behaviors. Every specimen was tested at least three times. Figure 9a shows the results of the average value of contact force for the different tested specimens.

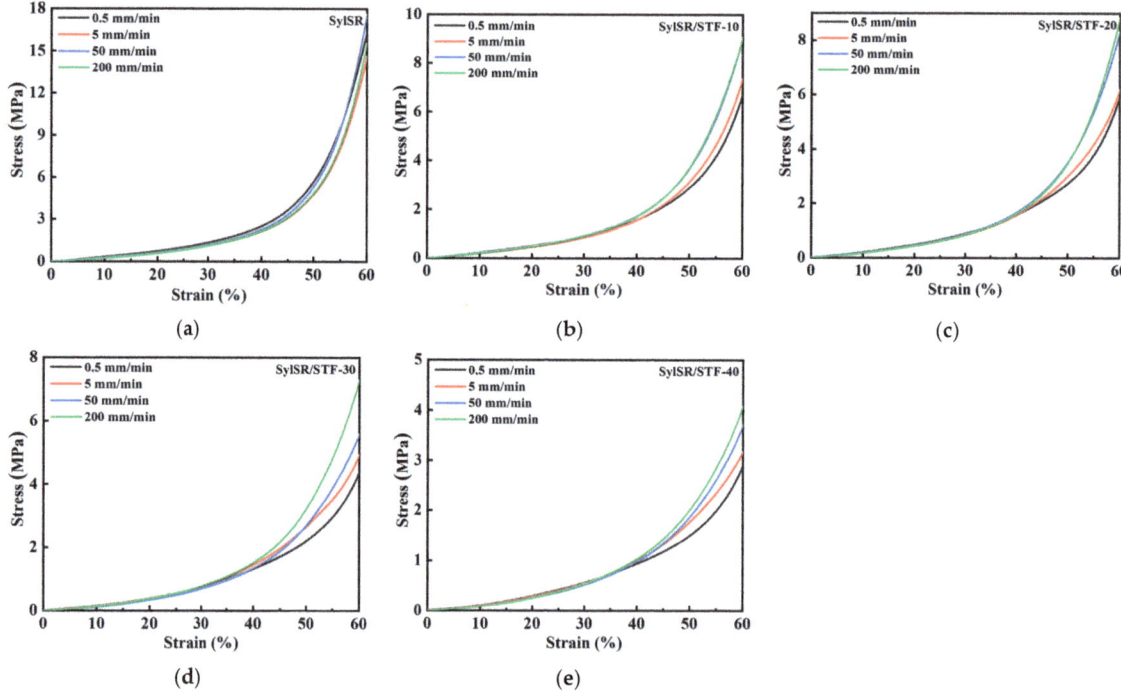

Figure 7. Quasi-static compressive stress–strain curves of all specimens at various strain rates. The specimens were (**a**) SylSR, (**b**) SylSR/STF-10, (**c**) SylSR/STF-20, (**d**) SylSR/STF-30, and (**e**) SylSR/STF-40.

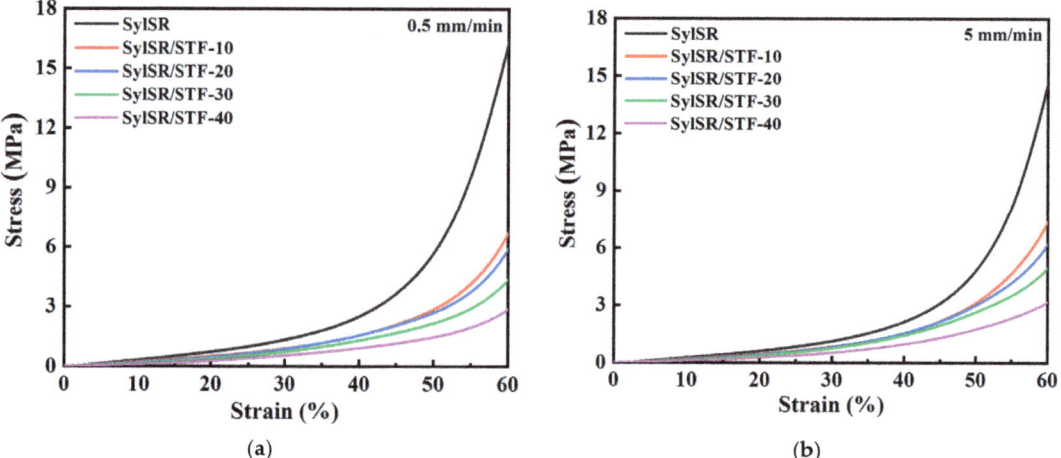

Figure 8. *Cont.*

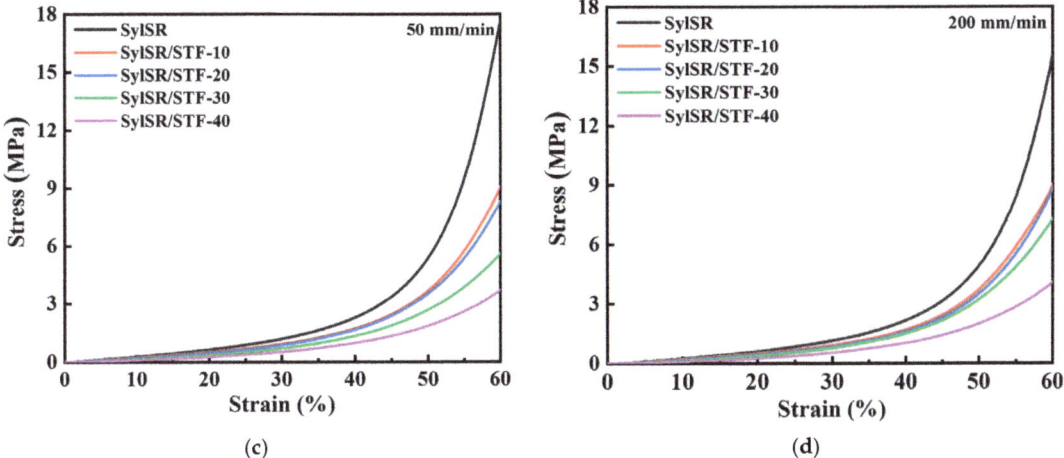

Figure 8. Quasi-static compressive stress–strain curves of all specimens at various strain rates. The loading rates were (**a**) 0.5 mm/min, (**b**) 5 mm/min, (**c**) 50 mm/min, and (**d**) 200 mm/min.

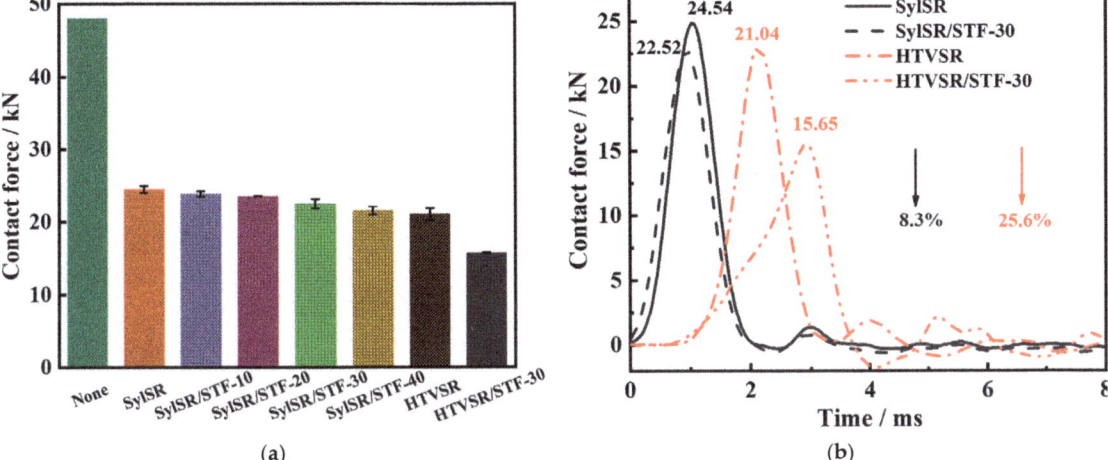

Figure 9. (**a**) Contact force for all specimens tested in drop hammer impact tests, and (**b**) typical contact force–time curves of SylSR, SylSR/STF-30, HTVSR, and HTVSR/STF-30.

For the SylSR/STF composites, with the increase in the STF mass fraction, the contact force was reduced. When the fraction of the STF increased to 30%, the contact force was reduced by 8.3%, from 24.54 kN to 22.52 kN; the typical curves are presented in Figure 9b. This result indicated that the STF can enhance the impact resistance properties of silicone rubber, and the higher the mass fraction of the STF in SylSR/STF composites resulted in a better impact protection effect. Since it was reported that the shear thickening effect of STFs can dissipate impact energy, with an increase in the STF fraction, more energy would be absorbed [5,9,13]. Therefore, SylSR/STF-30 presented a lower value of contact force compared with SylSR itself.

To further evaluate the effect of the mechanical strength of the silicone rubber matrix on the impact resistance of composites with STF microcapsules, another silicone rubber, HTVSR, was selected as a matrix. The mechanical strength of the HTVSR matrix is higher than that of the Sylgard 184 silicone rubber matrix in the SylSR/STF composites [26]. The

HTVSR/STF composites presented a significant decrease in the value of the contact force. Their data are also shown in Figure 9a and their typical curves are shown in Figure 9b. The value of the contact force for HTVSR/STF-30 was reduced by 25.6% in comparison with HTVSR, from 21.04 kN to 15.65 kN, and the contact force reduction degree for HTVSR/STF-30 in comparison with HTVSR was almost four times of that of SylSR/STF-30 in comparison with SylSR. Hence, it can be concluded that the strength of the silicone rubber matrix has an obvious influence on the effect of improving the impact protective performance of silicone rubber with the addition ofSTF. The stronger the strength of the silicone rubber, the better the effect of STF on improving the impact-protective performance of the composites. The enhancement effect of STF on the impact resistance of silicone rubber composites should be ascribed to the shear thickening and the energy absorption of STF microcapsulesunder external force impact, which are influenced by the strength of the silicone rubber.

4. Conclusions

In this work, SR/STF composites were successfully prepared by a simple mechanical mixing and curing process. STF was dispersed in a silicone rubber matrix in the form of microcapsules, and the mechanical test results showed that the STF improves the impact resistance behavior of silicone rubber. The specific conclusions are as follows:

(1) The damping properties of silicone rubber can be enhanced by the addition of STF and the increase of the fraction of the STF increases the value of Tan δ.
(2) The SR/STF composites presented decreased stiffness and an obvious strain rate effect at low strain rates (from 0.5 mm/min to 200 mm/min) in quasi-static compression experiments.
(3) The addition of STF can improve the impact protection performance of silicone rubber and the impact resistance increased with the increase in the STF mass fraction.
(4) The enhancement effect of STF on the impact resistance of silicone rubber is influenced by the strength of the silicone rubber matrix. The stronger the strength of silicone rubber, the better the effect of STF on improving the impact protective performance of the silicone rubber.

Author Contributions: Conceptualization, F.Z. and Z.L.; methodology, X.H. and Z.L.; investigation, C.W. and X.H.; resources, X.H.; data curation, C.W.; writing—original draft preparation, C.W.; writing—review and editing, F.Z. and Z.L.; supervision, C.M. and Z.L.; project administration, F.Z.; funding acquisition, Z.L. All authors have read and agreed to the published version of the manuscript.

Funding: This paper was funded by National Natural Science Foundation of China (No. 51803197 and No. 52003260).

Institutional Review Board Statement: Not applicable.

Informed Consent Statement: Not applicable.

Data Availability Statement: All data from this study are presented in the paper.

Acknowledgments: The authors wish to express their appreciation to Tao Liu at the Institute of Chemical Materials, China Academy of Engineering Physics (CAEP), for providing the HTV silicone rubber material.

Conflicts of Interest: The authors declare no conflict of interest.

Abbreviations

SylSR	Sylgard 184 silicone rubber
HTVSR	hot vulcanized silicone rubber
STF	shear thickening fluid
MVMQ	methyl vinyl silicone rubber
DCP	dicumyl peroxide
OM	optical microscope
FE-SEM	field-emission scanning electron microscopy
DMA	dynamic thermo-mechanical analysis

References

1. Peters, I.R.; Majumdar, S.; Jaeger, H.M. Direct observation of dynamic shear jamming in dense suspensions. *Nature* **2016**, *532*, 214–217. [CrossRef] [PubMed]
2. Liu, H.; Fu, K.; Cui, X.; Zhu, H.; Yang, B. Shear Thickening Fluid and Its Application in Impact Protection: A Review. *Polymers* **2023**, *15*, 2238. [CrossRef] [PubMed]
3. Wei, M.; Lin, K.; Sun, L. Shear thickening fluids and their applications. *Mater. Des.* **2022**, *216*, 110570. [CrossRef]
4. Zarei, M.; Aalaie, J. Application of shear thickening fluids in material development. *J. Mater. Res. Technol.* **2020**, *9*, 10411–10433. [CrossRef]
5. Zhang, X.Z.; Li, W.; Gong, X. The rheology of shear thickening fluid (STF) and the dynamic performance of an STF-filled damper. *Smart Mater. Struct.* **2008**, *17*, 035027. [CrossRef]
6. Wang, L.; Du, Z.; Fu, W.; Wang, P. Study of mechanical property of shear thickening fluid (STF) for soft body-armor. *Mater. Res. Express* **2021**, *8*, 045021. [CrossRef]
7. Roché, M.; Myftiu, E.; Johnston, M.C.; Kim, P.; Stone, H.A. Dynamic Fracture of Nonglassy Suspensions. *Phys. Rev. Lett.* **2013**, *110*, 148304. [CrossRef]
8. Moon, J.Y.; Dai, S.; Chang, L.; Lee, J.S.; Tanner, R.I. The effect of sphere roughness on the rheology of concentrated suspensions. *J. Non-Newton. Fluid Mech.* **2015**, *223*, 233–239. [CrossRef]
9. Wu, X.; Zhong, F.; Yin, Q.; Huang, C. Dynamic response of shear thickening fluid under laser induced shock. *Appl. Phys. Lett.* **2015**, *106*, 071903. [CrossRef]
10. Wu, X.; Yin, Q.; Huang, C. Experimental study on pressure, stress state, and temperature-dependent dynamic behavior of shear thickening fluid subjected to laser induced shock. *J. Appl. Phys.* **2015**, *118*, 173102. [CrossRef]
11. Li, W.; Xiong, D.; Zhao, X.; Sun, L.; Liu, J. Dynamic stab resistance of ultra-high molecular weight polyethylene fabric impregnated with shear thickening fluid. *Mater. Des.* **2016**, *102*, 162–167. [CrossRef]
12. Gürgen, S.; Kuşhan, M.C. The stab resistance of fabrics impregnated with shear thickening fluids including various particle size of additives. *Compos. Part A Appl. Sci. Manuf.* **2017**, *94*, 50–60. [CrossRef]
13. Mawkhlieng, U.; Majumdar, A. Deconstructing the role of shear thickening fluid in enhancing the impact resistance of high-performance fabrics. *Compos. Part B Eng.* **2019**, *175*, 107167. [CrossRef]
14. Zhang, H.; Zhang, X.; Chen, Q.; Li, X.; Wang, P.; Yang, E.H.; Duan, F.; Gong, X.; Zhang, Z.; Yang, J. Encapsulation of shear thickening fluid as an easy-to-apply impact-resistant material. *J. Mater. Chem. A* **2017**, *5*, 22472–22479. [CrossRef]
15. Zhang, X.; Zhang, H.; Wang, P.; Chen, Q.; Li, X.; Zhou, Y.; Gong, X.; Zhang, Z.; Yang, E.H.; Yang, J. Optimization of shear thickening fluid encapsulation technique and dynamic response of encapsulated capsules and polymeric composite. *Compos. Sci. Technol.* **2019**, *170*, 165–173. [CrossRef]
16. Liu, X.; Huo, J.L.; Li, T.T.; Peng, H.K.; Lin, J.H.; Lou, C.W. Investigation of the Shear Thickening Fluid Encapsulation in an Orifice Coagulation Bath. *Polymers* **2019**, *11*, 519. [CrossRef]
17. Kaczorowski, M.; Ronowicz, M.; Rokicki, G. Organogels containing immobilized shear thickening fluid and their composites with polyurethane elastomer. *Smart Mater. Struct.* **2019**, *28*, 035034. [CrossRef]
18. Soutrenon, M.; Michaud, V. Impact properties of shear thickening fluid impregnated foams. *Smart Mater. Struct.* **2014**, *23*, 035022. [CrossRef]
19. Soutrenon, M.; Michaud, V.; Manson, J.A. Influence of Processing and Storage on the Shear Thickening Properties of Highly Concentrated Monodisperse Silica Particles in Polyethylene Glycol. *Appl. Rheol.* **2013**, *23*, 54865.
20. Shit, S.C.; Shah, P. A Review on Silicone Rubber. *Natl. Acad. Sci. Lett.* **2013**, *36*, 355–365. [CrossRef]
21. Zhu, J.Q.; Gu, Z.P.; Liu, Z.P.; Zhong, F.C.; Wu, X.Q.; Huang, C.G. Silicone rubber matrix composites with shear thickening fluid microcapsules realizing intelligent adaptation to impact loadings. *Compos. Part B Eng.* **2022**, *247*, 110312. [CrossRef]
22. Kaewpradit, P.; Kongchoo, A.; Chonlathan, P.; Lehman, N.; Kalkornsurapranee, E. Impact absorbing kneepad prepared from natural rubber. *IOP Conf. Ser. Mater. Sci. Eng.* **2020**, *773*, 012063. [CrossRef]
23. Li, G.; Zhao, T.; Zhu, P.; He, Y.; Sun, R.; Lu, D.; Wong, C.-P. Structure-property relationships between microscopic filler surface chemistry and macroscopic rheological, thermo-mechanical, and adhesive performance of SiO_2 filled nanocomposite underfills. *Compos. Part A Appl. Sci. Manuf.* **2019**, *118*, 223–234. [CrossRef]
24. Zhu, M.; Yang, C.; Han, T.; Hu, C.; Wu, Y.; Si, T.; Liu, J. An encapsulation–reduction–catalysis confined all-in-one microcapsule for lithium–sulfur batteries displaying a high capacity and stable temperature tolerance. *Mater. Chem. Front.* **2021**, *5*, 4565–4570. [CrossRef]
25. Wang, Q.; Wang, S.; Chen, M.; Wei, L.; Dong, J.; Sun, R. Anti-puncture, frigostable, and flexible hydrogel-based composites for soft armor. *J. Mater. Res. Technol.* **2022**, *21*, 2915–2925. [CrossRef]
26. Santiago-Alvarado, A.; Cruz-Felix, A.; González-García, J.; Sánchez-López, O.; Mendoza, A.; Hernández-Castillo, I. Polynomial fitting techniques applied to opto-mechanical properties of PDMS Sylgard 184 for given curing parameters. *Mater. Res. Express* **2020**, *7*, 045301. [CrossRef]

Disclaimer/Publisher's Note: The statements, opinions and data contained in all publications are solely those of the individual author(s) and contributor(s) and not of MDPI and/or the editor(s). MDPI and/or the editor(s) disclaim responsibility for any injury to people or property resulting from any ideas, methods, instructions or products referred to in the content.

Article

Interaction Mechanism of Composite Propellant Components under Heating Conditions

Jiahao Liang [1], Jianxin Nie [1,*], Haijun Zhang [2], Xueyong Guo [1], Shi Yan [1] and Ming Han [3]

[1] State Key Laboratory of Explosion Science and Technology, Beijing Institute of Technology, Beijing 100081, China
[2] Xi'an Modern Control Technology Research Institute, Xi'an 710065, China
[3] The Eighth Military Representative Office of Air Force Equipment Ministry, Beijing 100843, China
* Correspondence: niejx@bit.edu.cn

Citation: Liang, J.; Nie, J.; Zhang, H.; Guo, X.; Yan, S.; Han, M. Interaction Mechanism of Composite Propellant Components under Heating Conditions. *Polymers* 2023, *15*, 2485. https://doi.org/10.3390/polym15112485

Academic Editors: Jiangtao Xu and Sihang Zhang

Received: 3 May 2023
Revised: 22 May 2023
Accepted: 26 May 2023
Published: 28 May 2023

Copyright: © 2023 by the authors. Licensee MDPI, Basel, Switzerland. This article is an open access article distributed under the terms and conditions of the Creative Commons Attribution (CC BY) license (https://creativecommons.org/licenses/by/4.0/).

Abstract: To examine the interactions between two binder systems—hydroxyl-terminated polybutadiene (HTPB) and hydroxyl-terminated block copolyether prepolymer (HTPE)—as well as between these binders and ammonium perchlorate (AP) at various temperatures for their susceptibility to varying degrees of thermal damage treatment, the thermal characteristics and combustion interactions of the HTPB and HTPE binder systems, HTPB/AP and HTPE/AP mixtures, and HTPB/AP/Al and HTPE/AP/Al propellants were studied. The results showed that the first and second weight loss decomposition peak temperatures of the HTPB binder were, respectively, 85.34 and 55.74 °C higher than the HTPE binder. The HTPE binder decomposed more easily than the HTPB binder. The microstructure showed that the HTPB binder became brittle and cracked when heated, while the HTPE binder liquefied when heated. The combustion characteristic index, S, and the difference between calculated and experimental mass damage, ΔW, indicated that the components interacted. The original S index of the HTPB/AP mixture was 3.34×10^{-8}; S first decreased and then increased to 4.24×10^{-8} with the sampling temperature. Its combustion was initially mild, then intensified. The original S index of the HTPE/AP mixture was 3.78×10^{-8}; S increased and then decreased to 2.78×10^{-8} with the increasing sampling temperature. Its combustion was initially rapid, then slowed. Under high-temperature conditions, the HTPB/AP/Al propellants combusted more intensely than the HTPE/AP/Al propellants, and its components interacted more strongly. A heated HTPE/AP mixture acted as a barrier, reducing the responsiveness of solid propellants.

Keywords: composite propellant; thermal damage treatment; thermal weight loss; interaction; combustion characteristics

1. Introduction

Hydroxyl-terminated polybutadiene (HTPB) propellant is currently the most important type of composite propellant and has been applied in various rocket motor models in China and abroad. HTPB not only improves the specific impulse of the propellant but also has a widely adjustable range of burning rates, good mechanical properties, a simple manufacturing process, and abundant raw materials [1–6]. Thus, it is one of the mainstream composite propellants in use. However, a new type of insensitive solid propellant has been developed that uses hydroxyl-terminated block copolyether prepolymer (HTPE) as a binder. It has good desensitisation performance, energy characteristics, and application performance [7]. Owing to its significant insensitivity under hazardous conditions such as slow heating, HTPE propellant is intended to replace HTPB propellant [8–12]. Based on their performance, HTPB and HTPE propellants can be applied in a wide range of environments. However, with the increasing complexity of the service environment, the possibility of accidental reaction and the degree of harm of the motor in the process of use is also increasing. Therefore, it is an important development direction to improve the survivability of solid rocket motors in a complex environment in the future.

A composite solid propellant, which has certain mechanical properties, is manufactured by mixing and curing oxidants (such as ammonium perchlorate (AP) and ammonium nitrate), combustion agents (such as Al powder), and polymer binders. Unexpected thermal decomposition and the energy release of propellants can occur in stimulating environments subjected to heating conditions. A typical example is an external fire causing a warehouse to warm up slowly; the heating process can lead to the thermal decomposition of the propellant. The thermal decomposition behaviour of propellants, especially the interaction between components, significantly affects their combustion characteristics after ignition. The thermal decomposition process is defined as the initial stage of ignition and combustion [13], and the decomposition characteristics of propellants have a profound effect on their combustion characteristics [14]. Therefore, the thermal analysis and cocombustion of energetic materials, such as solid propellants, is crucial not only to understand the thermal decomposition behaviour of propellants but also to conduct an in-depth evaluation of the effect of propellant exothermic decomposition on potential hazards during heating processes [15,16].

The thermal decomposition and combustion characteristics of propellants are closely related to the interactions between other components of the propellant [17]. Extensive research has been conducted on the thermal decomposition properties of AP [18,19], HTPB [20–23], HTPE [24], HTPB/AP propellant [25,26], and HTPE/AP propellant [16,27,28], laying an important foundation for understanding the thermal decomposition, ignition, and combustion characteristics of HTPB and HTPE propellants. However, similar research has not been conducted on the interaction between the propellant components during heating. During the heating process, various degrees of thermal decomposition of the propellant components occur, and their interactions are variable. Therefore, understanding the interactions between the components during the heating process is crucial for understanding the thermal decomposition behaviour of propellants and the effect of exothermic decomposition on potential hazards.

In this study, the interactions and cocombustion between two binder systems (HTPB and HTPE) at various temperatures were studied. The interactions and cocombustion between the incremental components of the propellants were evaluated by using thermogravimetric–Fourier transform infrared–mass spectrometry (TG–FTIR–MS) and other testing methods. In addition, the interactions between the binders and AP were analysed, which can provide a theoretical basis for further understanding and research on the thermal safety of propellants under heating conditions.

2. Materials and Methods

2.1. Materials

To study the interaction and cocombustion between propellants and their components via thermal analysis characteristics, the following incremental component formulations were designed: HTPB binder, HTPB/AP mixture, HTPB/AP/Al propellant, HTPE binder, HTPE/AP mixture, and HTPE/AP/Al propellant. The mass ratio of the binder to AP particles in the mixture and propellant was 18:82, and the composition and content of their respective formulations are listed in Table 1. The samples used in this experiment were developed and prepared by the Beijing Institute of Technology.

2.2. Equipment and Conditions Methods

To analyse the interaction and cocombustion between the components of the propellant, an experimental device for slow-heating propellant tables [29] was used to heat the samples (Table 1) at a rate of 0.2 °C/min. One sample was removed when the temperature listed in Table 2 was reached. The heated samples were analysed by using a simultaneous thermal analyser infrared–mass spectrometer (Netzsch—STA449F3, FTIR Nicolet iS20, Netzsch—QMS 403, Beijing, China, accessed from www.eceshi.com, accessed on 14 February 2023.). Approximately 3.0 mg of the sample was heated from an initial temperature of 45 °C to 800 °C at a rate of 10 K/min. The purge gas was high-purity argon with a gas flow rate of 240 mL/min. The testing range of the mass spectrometer was 0–300 m/z. Scanning electron microscopy (SEM; Hitachi, S-4800, Beijing, China, accessed from www.eceshi.com, accessed on 15 May 2023.) was used to visually analyse the micromorphology of the components of the propellant.

Table 1. Composition of the sample formula (wt.%).

System	Sample	HTPB	DOA [a]	TDI [b]	AP	Al	
HTPB system	HTPB binder	42.00	54.67	3.33			
	AP				100.00		
	HTPB/AP mixture	7.56	9.84	0.60	82.00		
	HTPB/AP/Al propellant	6.30	8.20	0.50	68.00	17.00	
		HTPE	A3 [c]	IPDI [d]	Butanetriol	AP	Al
HTPE system	HTPE binder	37.33	54.67	7.33	0.67		
	HTPE/AP mixture	6.72	9.84	1.32	0.12	82.00	
	HTPE propellant	5.60	8.20	1.10	0.10	68.00	17.00

[a] dioctyl adipate; [b] toluene diisocyanate; [c] bis (2,2-dinitropropyl) formal and bis (2,2-dinitropropyl) formal acetal mixture; [d] isophorone diisocyanate.

Table 2. Propellant sampling temperatures for various degrees of thermal damage treatment.

System	Sample	Experiment Number	Sampling Temperature
	AP	1#	original
		2#	160 °C
		3#	180 °C
		4#	220 °C
HTPB system	HTPB binder	5#	original
		6#	160 °C
		7#	180 °C
		8#	220 °C
	HTPB/AP mixture	9#	original
		10#	160 °C
		11#	180 °C
		12#	220 °C
	HTPB/AP/Al propellant	13#	original
		14#	160 °C
		15#	180 °C
		16#	220 °C
HTPE system	HTPE binder	17#	original
		18#	160 °C
	HTPE/AP mixture	19#	original
		20#	160 °C
		21#	180 °C
		22#	220 °C
	HTPE/AP/Al propellant	23#	original
		24#	160 °C
		25#	180 °C
		26#	220 °C

3. Results and Discussion

3.1. Thermal Decomposition of a Single-Component Propellant

3.1.1. Thermal Decomposition of the HTPB Binder at Various Sampling Temperatures

Figure 1 shows the TG–MS–FTIR curves of the HTPB binder films heated to various temperatures (original samples and samples with sampling temperatures of 160 °C, 180 °C, and 220 °C, respectively). The TG/DTG curve (differential thermogravimetry (DTG), a curve that differentiates each point on a TG curve with respect to temperature coordinates to the first degree) in Figure 1a shows that the thermal weight loss process of the film was completed in two stages.

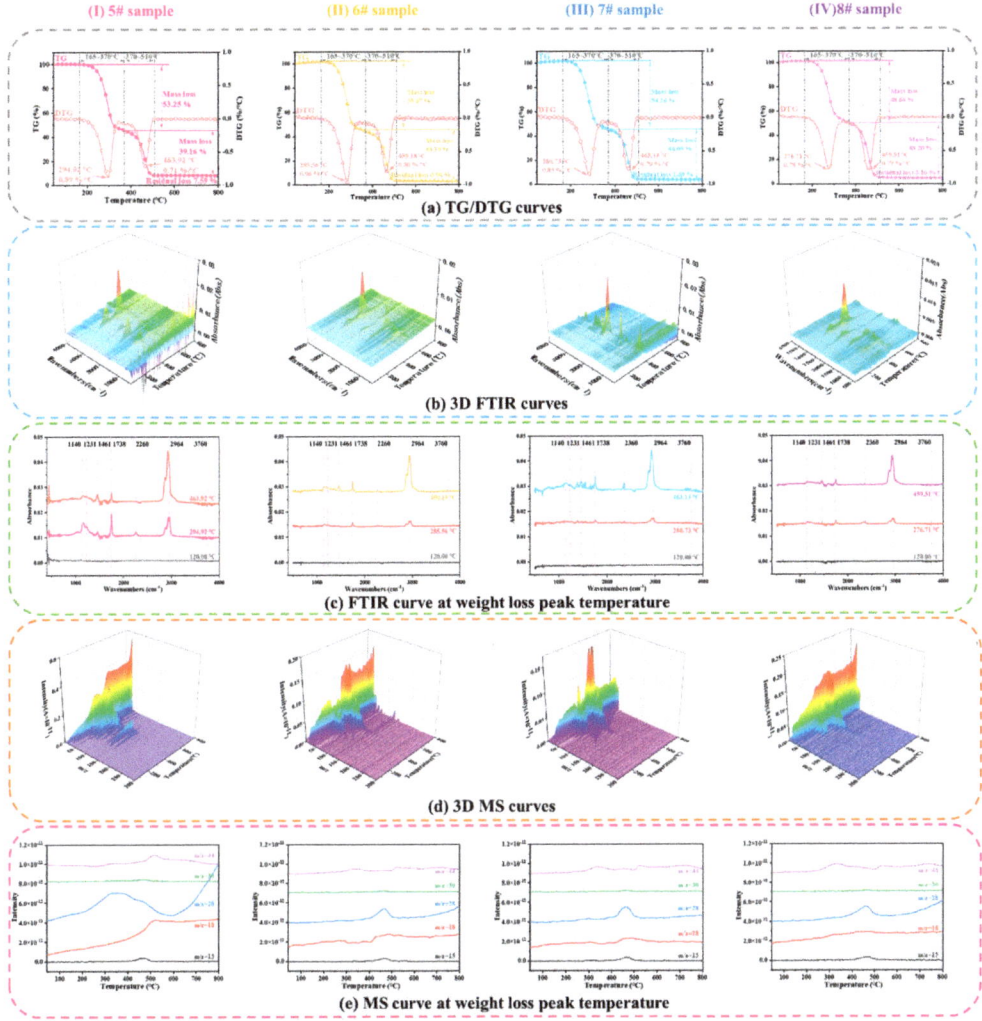

Figure 1. TG−FTIR−MS curves of the HTPB binder films heated to various temperatures.

For sample 5# HTPB binder, the first stage of weight loss occurs between 165 and 370 °C, with a maximum peak temperature at 294.92 °C for weight loss decomposition and a maximum weight loss rate of 0.89%/°C. The second stage occurs between 370 and 510 °C, with a maximum peak temperature at 463.92 °C for weight loss decomposition and a maximum weight loss rate of 0.71%/°C. The first stage occurs mainly because of the decomposition and volatilisation of DOA and TDI in the film when heated, and the second stage occurs mainly because of the chain-breaking decomposition and volatilisation of the HTPB polymer [30].

For sample 6#, the first stage of weight loss occurs between 165 and 370 °C, with a weight loss of 55.67%. The maximum peak temperature for weight loss decomposition is 285.56 °C, and the maximum weight loss rate is 0.96%/°C. The second stage occurs between 370 and 510 °C. During this stage, the weight loss is 43.39%, the maximum peak temperature of weight loss decomposition is 459.18 °C, and the maximum weight loss rate is 0.80%/°C.

For sample 7#, the first stage of weight loss occurs between 165 and 370 °C, with a weight loss of 54.26%. The maximum peak temperature for weight loss decomposition is 280.73 °C, and the maximum weight loss rate is 0.85%/°C. The second stage occurs between 370 and 510 °C, with a weight loss of 44.05%, maximum peak temperature of weight loss decomposition of 463.13 °C, and maximum weight loss rate of 0.79%/°C.

For sample 8#, the first stage of weight loss occurs between 165 and 370 °C, with a weight loss of 48.64%. The maximum peak temperature for weight loss decomposition is 276.71 °C, and the maximum weight loss rate is 0.78%/°C. The second stage occurs between 370 and 510 °C. During this stage, the weight loss is 48.20%, the maximum peak temperature of weight loss decomposition is 459.51 °C, and the maximum weight loss rate is 0.79%/°C. These results indicate that as the sampling temperature increases, the peak temperature of the first stage of weight loss of the HTPB films and their weight loss rate slightly decrease, but almost no effect is observed in the second stage of weight loss.

The weight loss peak temperatures of the HTPB binder in the FTIR curve reveal the corresponding groups of each characteristic absorption peak as follows: 2874–2964 cm^{-1} for C-H (2964 cm^{-1} for the asymmetric stretching vibration peak of C-H on CH_3 and 2874 cm^{-1} for the symmetric stretching vibration peak of C-H on CH_3); 2260 cm^{-1} for N_2O; 1738 cm^{-1} for the stretching vibration peak of C=O; 1461 cm^{-1} for the in-plane bending vibration peak of C-H on CH_2; 1231 cm^{-1} for the amide band III (this peak also represents the stretching vibration peak of C-N, which is a strong characteristic of polyurethane when it exists together with band II, the stretching vibration peak of C=O, and the stretching vibration peak of C-O described below), 1140–1178 cm^{-1} for the C-O stretching vibration peak; 966 cm^{-1} for (transform 1,4) -CH=CH- on the C-H out-of-plane bending vibration peak; and 910 cm^{-1} for (1,2-)—CH=CH_2 on the C-H out-of-plane bending vibration peak. Among them, the absorption peak at 1738 cm^{-1} was formed by the superposition of C=O absorption in polyurethane and DOA. The absorption peaks at 1178 and 1140 cm^{-1} were formed by the superposition of C-O absorption peaks in polyurethane and DOA. The absorption peaks at 1535 and 1231 cm^{-1} represented the characteristic peaks of polyurethane hard segment urethane bonds, and the absorption peaks at 966 and 910 cm^{-1} represented the characteristic peaks of HTPB polymer. Using mass spectrometry, the gaseous products in the first weight loss stage of the HTPB films were determined to be CH_3-containing gases, CO, CO_2, N_2O, and NO. During the second weight loss stage, the concentration of gas-containing CH_3 increased significantly, whereas those of CO_2 and CO decreased significantly.

Based on the decomposition peak temperatures of samples 6#, 7#, and 8# in the FTIR curves, the characteristic absorption peaks in the first stage were observed to have decreased; for example, the amide III band at 1231 cm^{-1} and the stretching vibration peak of C-O between 1140 and 1178 cm^{-1}. As the sampling temperature increased, the gaseous products in the HTPB binder film contained CH_3 gas, and the volatilisation of CO, CO_2, N_2O, and NO started. Furthermore, as the temperature increased, the DOA and TDI in the HTPB binder film were thermally decomposed and volatilised, leaving only those substances that were difficult to volatilise. When reheated, the substances that had not been completely volatilised in the first stage continued to volatilise; thus, the peak temperature in the first stage of weight loss decreased slightly. In contrast, reheating had almost no effect on the second weight loss stage.

SEM was used to analyse the apparent morphology of HTPB binder films at different sampling temperatures, as shown in Figure 2. It can be seen that the morphology of sample 5# is smooth and rich in viscoelasticity. Sample 6# begins to undergo changes, and as the components in the binder undergo thermal decomposition and volatilisation, the binder becomes brittle and cracks on the surface. As the sampling temperature reaches 180 °C, there are more cracks on the surface of sample 7#. When the sampling temperature is 220 °C, the 8# sample becomes more brittle, forming a bumpy surface.

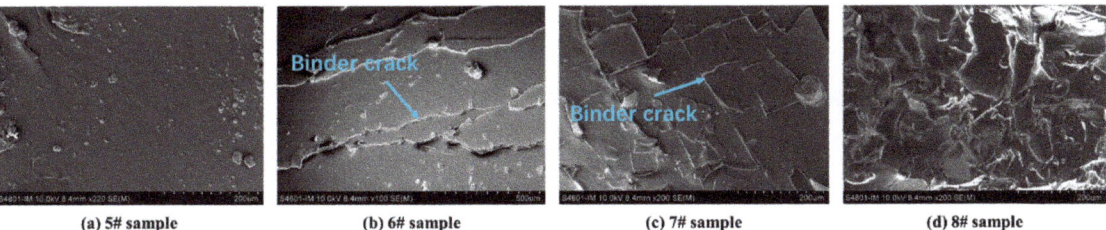

(a) 5# sample   (b) 6# sample   (c) 7# sample   (d) 8# sample

Figure 2. SEM morphology of the HTPB binder films heated to various temperatures.

3.1.2. Thermal Decomposition of the HTPE Binder at Various Sampling Temperatures

Figure 3 shows the TG–FTIR–MS curves of the HTPE binder films heated to various temperatures. In Figure 3a, (I) and (II) show the TG/DTG curves of samples 17# and 18#, respectively. In Figure 3a, (III) shows the residual HTPE binder film samples after combustion, and (IV) shows the TG/DTG curves of component A3. The thermal weight loss process of the HTPE binder film can be observed to be completed in two stages. In the first stage, when the sampling temperature increases, the peak temperature of the weight loss of the HTPE binder film increases slightly, and the weight loss rate decreases slightly. In contrast, little effect is observed in the second weight loss stage. As the HTPE binder ignited before the temperature reached 180 °C, HTPE samples were not collected at temperatures exceeding this. Furthermore, the energetic plasticiser A3 was added to the HTPE binder film, which lowered the reaction temperature of the HTPE binder owing to the volatilisation and decomposition heat release of A3; A3 is observed to be completely volatilised between 150 °C and 269 °C.

Based on the positions of the main absorption peaks, their corresponding groups in the FTIR spectrum can be determined to be as follows: 2874–2964 cm^{-1} for C-H (where the peaks at 2964 and 2874 cm^{-1} correspond to the asymmetric and symmetric stretching vibration peaks, respectively, of C-H in CH_3); 2260 cm^{-1} corresponding to N_2O; 1738 cm^{-1} to the stretching vibration peak of C=O; and 1140–1178 cm^{-1} to the stretching vibration peak of C-O. Based on the intensity of the infrared absorption peak of the decomposition product, the decomposition product can be determined to be mainly composed of small molecular ethers, alkanes, and a small amount of aldehydes. Based on MS, the first stage of weight loss of HTPE can be determined to comprise mainly the pyrolysis of the A3 plasticiser, and the second stage of weight loss can be determined to comprise the pyrolysis of the HTPE polymer colloid.

Figures 1 and 3 indicate that although the thermal decomposition process of both binder films is completed in two stages, the first weight loss decomposition peak temperature of the HTPE binder is 209.58 °C, whereas that of the HTPB binder is 294.92 °C, which is 85.34 °C higher than that of the HTPE binder. The second weight loss decomposition peak temperature of the HTPE binder is 408.18 °C, whereas that of the HTPB binder is 463.92 °C, 55.74 °C higher than that of the HTPE binder. Therefore, compared to HTPB binders, HTPE binders decompose more easily. Such different decomposition peak temperatures are bound to impact the thermal decomposition and cocombustion interactions of propellants.

SEM was used to analyse the apparent morphology of HTPE binder films at different sampling temperatures, as shown in Figure 4. It can be seen that the morphology of sample 17# is wrinkled and elastic. As the components in the binder undergo thermal decomposition and volatilisation, sample 18# begins to liquefy, making the binder more viscous and smoothing the surface wrinkles. It can be seen that as the sampling temperature increases, the HTPB binder begins to become brittle, while the HTPE binder becomes sticky, which may cause the HTPE binder to adhere to the surface of AP particles and affect the interaction between the two components.

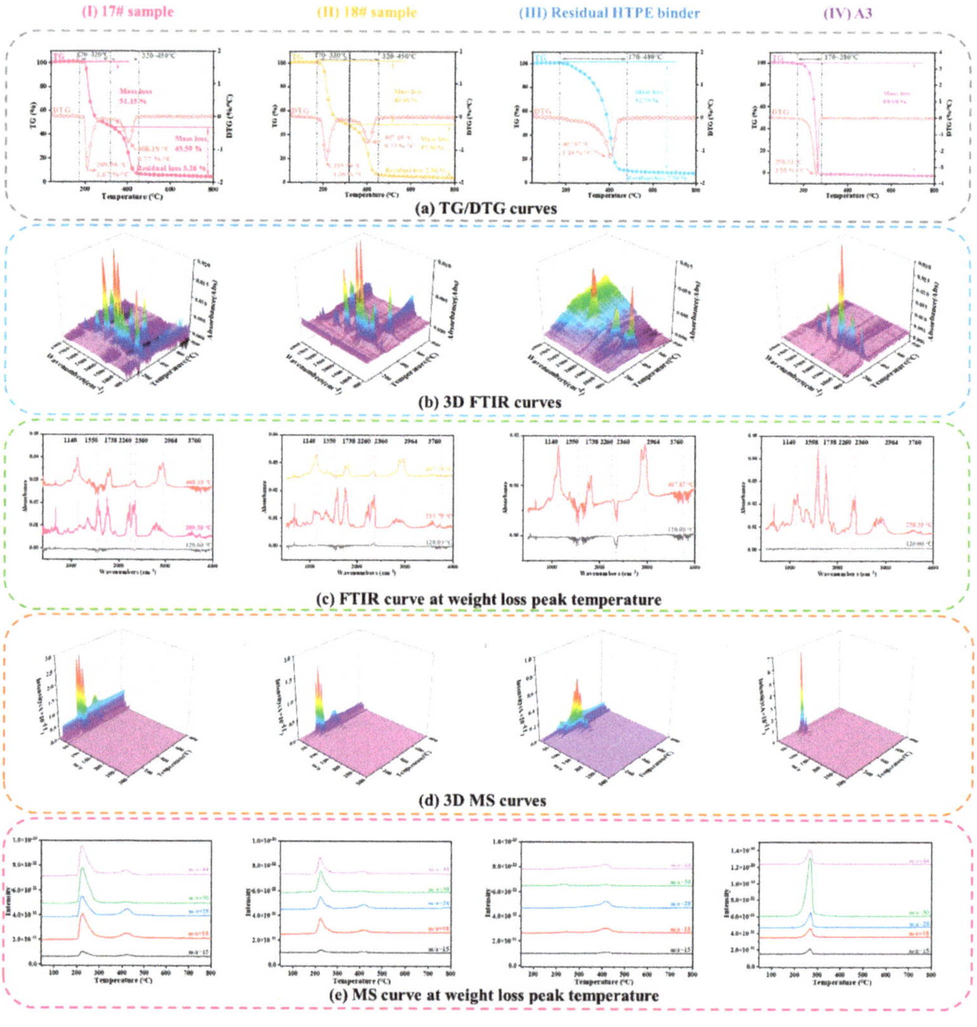

Figure 3. TG–FTIR–MS curves of the HTPE binder films heated to various temperatures.

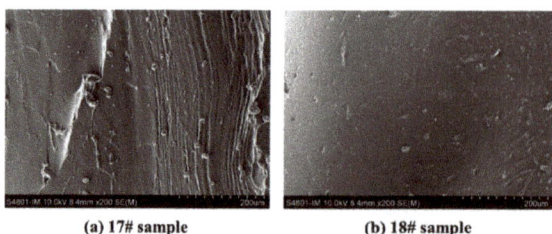

Figure 4. SEM morphology of the HTPE binder films heated to various temperatures.

3.1.3. Thermal Decomposition of AP Particles at Various Sampling Temperatures

Figure 5 shows the TG–FTIR-MS curves of the AP particles heated to various temperatures. Figure 5a shows that two stages exist in the thermal weight loss process of AP particles: the low- and high-temperature weight loss stages. By increasing sampling temperature, the DTG low- and high-temperature decomposition peaks of AP advance slightly. Compared to sample 1#, the starts of the low-temperature decomposition of samples 2# and 3# are delayed because the sampling temperature consumes a portion of the defective AP nuclei during the period when the temperature is between 160 and 180 °C; therefore, the start of low-temperature decomposition is delayed during reheating. However, the low-temperature decomposition peak of sample 4# is observed 20 °C earlier than those of samples 2# and 3# because the AP particles generate pores, and the specific surface area increases when the sampling temperature is 220 °C, resulting in AP dissociation.

Based on the FTIR curves of the gaseous products decomposed from AP at the weight loss peak temperatures in Figure 5b,c, the wave numbers of N_2O (2238 and 2201 cm^{-1}), NO_2 (1630 and 1598 cm^{-1}), H_2O (3500–4000 cm^{-1}), and HCl (2700–3012 cm^{-1}) can be determined by combining the data from (d) and (e). This analysis reveals that the main gaseous products of AP thermal decomposition are N_2O and NO_2.

It can be seen from Figure 5c that during the low-temperature weight loss stage, the NO_2 absorption intensity of sample 1# is 0.0215, and the N_2O absorption intensity is 0.0405. The NO_2 absorption intensity of sample 2# is 0.0215, and the N_2O absorption intensity is 0.0359. The NO_2 absorption intensity of sample 3# is 0.0201, and the N_2O absorption intensity is 0.0246. The NO_2 absorption intensity of sample 4# is 0.0193, and the N_2O absorption intensity is 0.0226. Their ratios are 1.88, 1.67, 1.22, and 1.17, respectively. As the sampling temperature increases, the absorption intensity ratio of NO_2 and N_2O gradually decreases during the low-temperature weight loss stage. During the high-temperature weight loss stage, the NO_2 absorption intensity of sample 1# is 0.0594, and the N_2O absorption intensity is 0.0514. The NO_2 absorption intensity of sample 2# is 0.0608, and the N_2O absorption intensity is 0.0560. The NO_2 absorption intensity of sample 3# is 0.0604, and the N_2O absorption intensity is 0.0530. The NO_2 absorption intensity of sample 4# is 0.0640, and the N_2O absorption intensity is 0.0511. Their ratios are thus 0.87, 0.92, 0.87, and 0.80, respectively. As the sampling temperature increases, the absorption intensity ratio of NO_2 and N_2O does not change significantly during the high-temperature weight loss stage.

SEM was used to analyse the apparent morphology of AP particles at different sampling temperatures, as shown in Figure 6. It can be seen that the surface of sample 1# is smooth and free from pores. Sample 2# began to undergo changes and the surface became uneven, but there were no pores. As the sampling temperature reaches 180 °C, there is a trend of increasing pores in sample 3#. When the sampling temperature is 220 °C, as the degree of decomposition deepens, sample 4# forms a porous AP structure.

Moreover, the changes in the intensity ratio of N_2O to NO_2 during the two weight loss stages indicate that a competitive relationship exists between the formation reaction of N_2O and NO_2 during the thermal decomposition of AP. This is consistent with the observations in reference [19]. The absorbance of N_2O and NO_2 during the low-temperature weight loss stage is greater than that of NO_2, indicating that the products of N_2O play a dominant role in the low-temperature weight loss process. However, during the high-temperature weight loss stage, the absorption intensity of NO_2 is greater than that of N_2O, indicating that the products of NO_2 dominate high-temperature weight loss.

The first step of the thermal decomposition of AP is the dissociation of NH_4ClO_4 through proton transfer to form adsorbed NH_3 and $HClO_4$. Low-temperature thermal decomposition mainly occurs as a reaction between NH_3 and $HClO_4$ adsorbed onto the surface of the particles. At low temperatures, the decomposition products of $HClO_4$ cannot oxidise NH_3, and the remaining adsorbed NH_3 covers the AP surface. When the particle surfaces are completely covered by NH_3, low-temperature thermal decomposition causes a weight loss of approximately 30%.

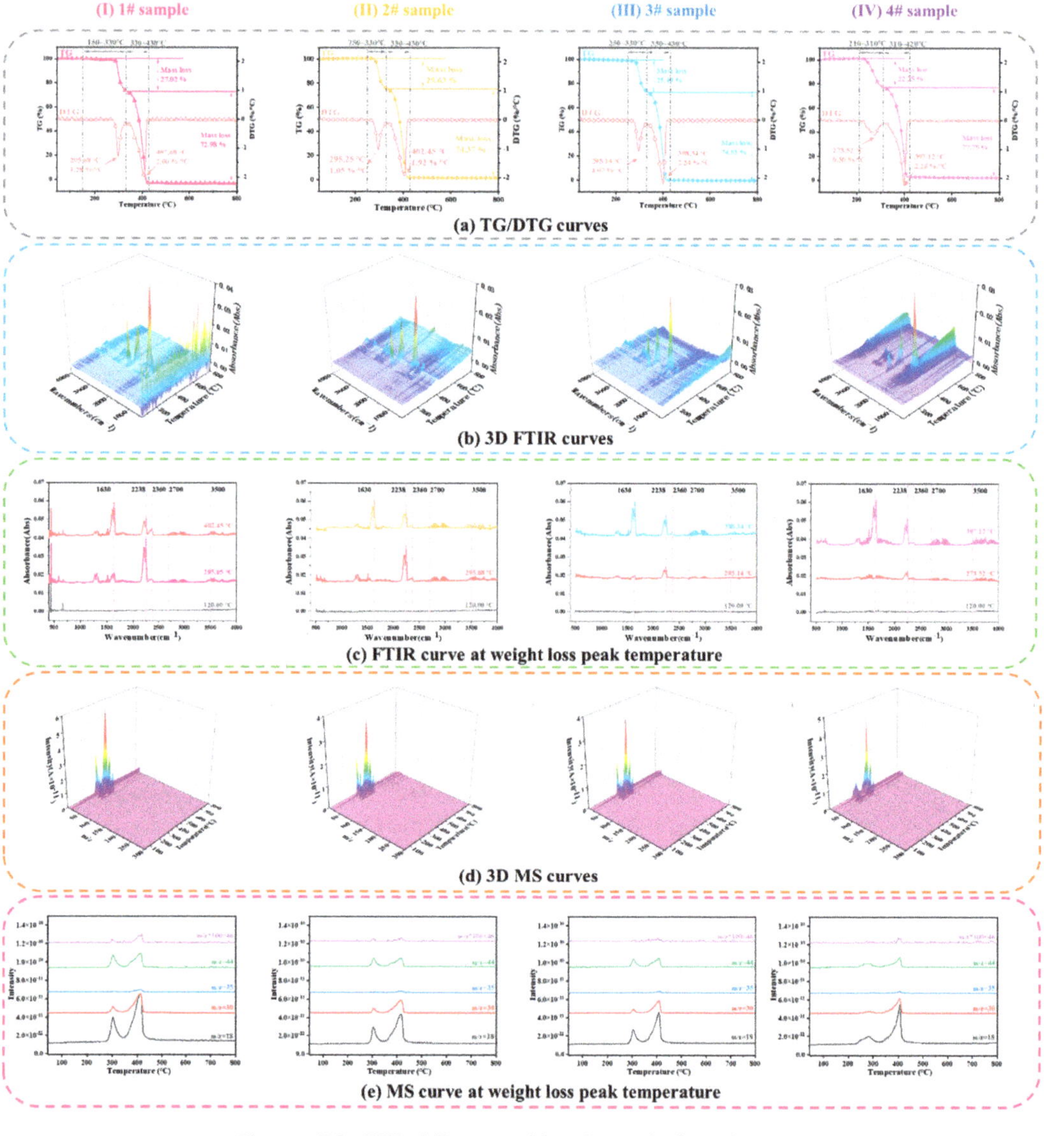

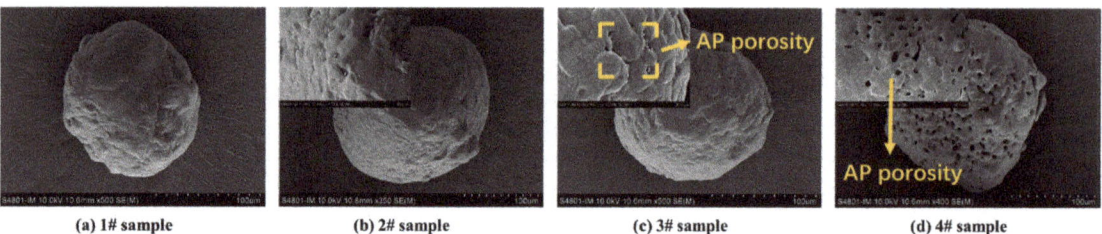

Figure 5. TG−FTIR−MS curves of the AP particles heated to various temperatures.

Figure 6. SEM morphology of the AP particles films heated to various temperatures.

The high-temperature thermal decomposition process of AP is mainly a gaseous-phase reaction: the adsorbed NH_3 and $HClO_4$ are pyrolysed and absorbed into the gaseous phase. In the gaseous phase, $HClO_4$ further decomposes to generate oxidation products, whereas NH_3 is oxidised by the oxidation products decomposed by $HClO_4$ to generate the final products. As the sampling temperature increases, the severity of the partial decomposition of AP increases. When reheated, the specific surface area increases due to the presence of pores in the particles, and NH_4ClO_4 dissociation is more likely to occur, resulting in a decrease in the initial reaction temperature [31].

Although only 30% of AP decomposes at low temperatures, with the solid residue after decomposition is still AP, its physical properties change significantly and form a relatively stable porous material. However, when the temperature rises to 350–400 °C, AP undergoes high-temperature decomposition, releasing a large amount of energy. The thermal weight loss data of the HTPB and HTPE binders indicate that the first weight loss temperature of both HTPB and HTPE are low (less than 350 °C), whereas the second weight loss temperature exceeds 350 °C. Therefore, the two binders may interact at both the low- and high-temperature decomposition of AP.

3.2. Study on Cocombustion of Propellant Component

After analysing and understanding the thermal decomposition characteristics of individual components of propellant, the cocombustion and interaction are analysed and researched.

3.2.1. Cocombustion of the HTPB Binder and AP Particles

Figure 7 shows the TG and DTG curves of the HTPB/AP mixture and HTPB/AP/Al propellant, respectively. The thermal weight loss process of the original HTPB/AP mixture sample can be observed to be divided into three stages. The first stage is in the temperature range of 145–273 °C, with a maximum weight loss peak temperature at 201.65 °C and a gentle peak shape. The maximum weight loss rate is 0.31%/°C, and the weight loss is approximately 15.75%. This stage mainly results from the breakage and decomposition of the binder chain in HTPB. The second stage is in the temperature range of 273–330 °C, with a maximum peak temperature at 294.65 °C and a sharp peak shape. The weight loss is approximately 23.15%, and the maximum weight loss decomposition rate is 0.96%/°C. This stage consists mainly of the low-temperature decomposition of AP. The third stage is in the temperature range of 330–414 °C, with a peak temperature at 401.65 °C and a sharp peak shape. The maximum weight loss decomposition rate is 1.88%/°C, and the weight loss is approximately 61.10%. This stage consists mainly of the high-temperature decomposition of AP.

The thermal weight loss process of the HTPB/AP mixture samples heated to 160 °C can also be divided into three stages. The first stage is in the temperature range of 146–270 °C, with a maximum weight loss peak temperature at 225.16 °C and a gentle peak shape. The maximum weight loss rate is 0.21%/°C, and the weight loss is approximately 10.62%. This stage results mainly from the breaking and decomposition of the binder chain in HTPB. The second stage is in the temperature range of 270–328 °C, with a peak temperature at 295.36 °C and a sharp peak shape. The weight loss is approximately 25.89%, and the maximum weight loss decomposition rate is 0.96 /°C. This stage consists mainly of the low-temperature decomposition of AP. The third stage is in the temperature range of 328–408 °C, with a peak temperature at 387.16 °C and a sharp peak shape. The maximum weight loss decomposition rate is 1.55%/°C, and the weight loss is approximately 63.49%. This stage consists mainly of the high-temperature decomposition of AP.

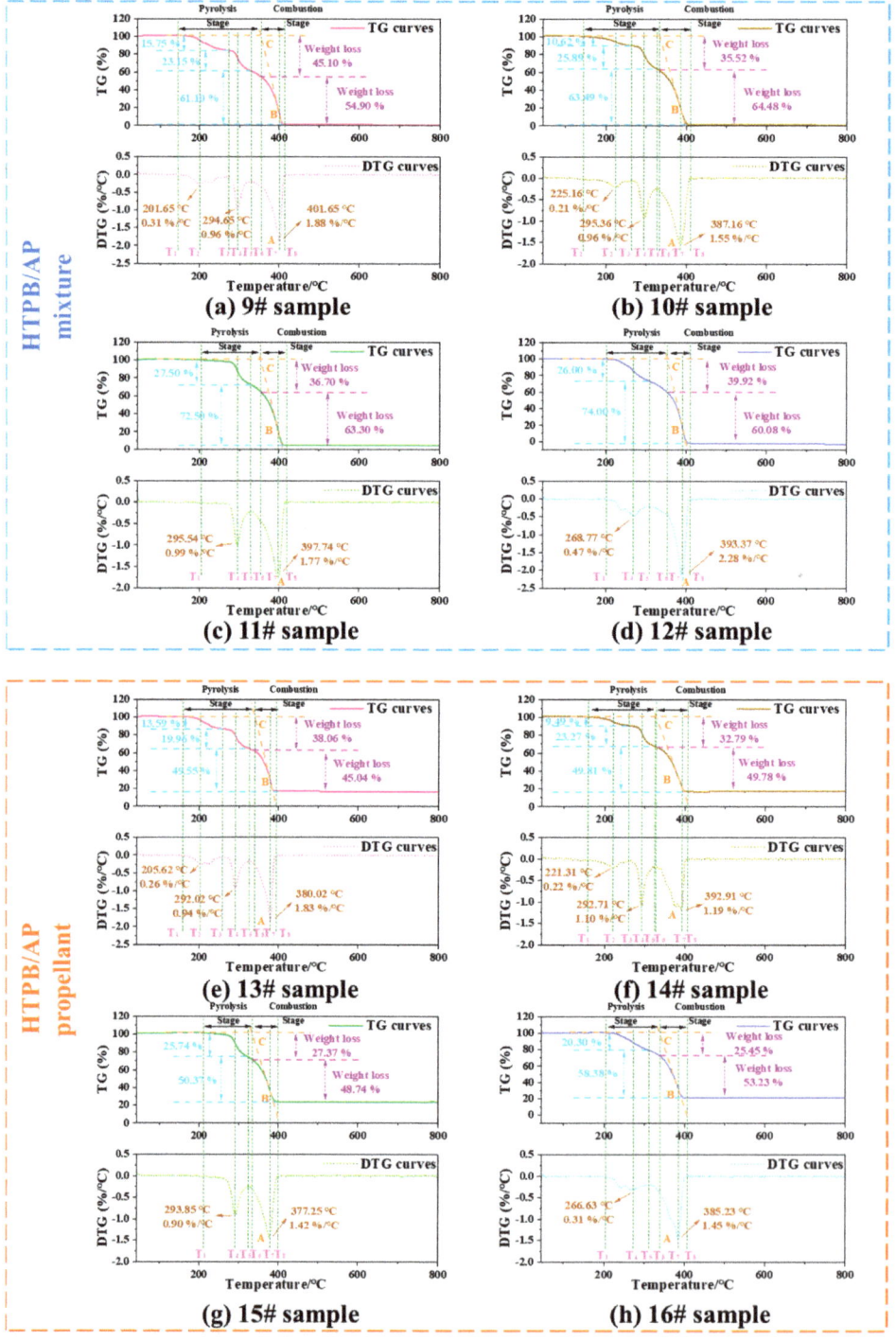

Figure 7. TG and DTG curves of the HTPB/AP mixtures and HTPB/AP/Al propellants at various sampling temperatures.

The thermal weight loss process of the HTPB/AP mixture samples heated to 180 °C can be divided into two stages: the first stage is in the temperature range of 202–328 °C, with a peak temperature at 295.54 °C and a sharp peak shape. The maximum rate of weight loss is 0.99%/°C, and the weight loss is approximately 27.50%. This stage mainly results from the continued decomposition of the binder chain that is not completely decomposed in HTPB and the low-temperature decomposition of AP. The second stage is in the temperature range of 328–415 °C, with a peak temperature at 397.74 °C and a sharp peak shape. The maximum weight loss decomposition rate is 1.77%/°C, and the weight loss is approximately 72.50%. This stage consists mainly of the high-temperature decomposition of AP.

The thermal weight loss process of the HTPB/AP mixture sample heated to 220 °C can also be divided into two stages. The first stage is in the temperature range of 210–306 °C, with a maximum weight loss temperature at 268.77 °C and a gentle peak shape. The maximum weight loss rate is 0.47%/°C, and the weight loss is approximately 26.00%. This stage results mainly from the continued decomposition of the binder chain that is not fully decomposed in HTPB and the low-temperature decomposition of AP. The second stage is in the temperature range of 306–406 °C, with a peak temperature at 393.37 °C and a sharp peak shape. The maximum weight loss decomposition rate is 2.28%/°C, and the weight loss is approximately 74.00%. This stage consists mainly of the high-temperature decomposition of AP.

The thermal weight loss process of the original HTPB/AP/Al propellant sample can be divided into three stages. The first stage is in the temperature range of 149–265 °C, with a maximum weight loss peak temperature at 205.62 °C and a gentle peak shape. The maximum weight loss rate is 0.26%/°C, and the weight loss is approximately 13.59%. This stage results mainly from the breaking of the binder chain in HTPB. The second stage is in the temperature range of 265–326 °C, with a peak temperature at 292.02 °C and a sharp peak shape. The weight loss is approximately 19.96%, and the maximum weight loss decomposition rate is 0.94%/°C. This stage consists mainly of the low-temperature decomposition of AP. The third stage is in the temperature range of 326–391 °C, with a peak temperature at 380.02 °C and a sharp peak shape. The maximum weight loss decomposition rate is 1.83%/°C, and the weight loss is approximately 49.55%. This stage consists mainly of the high-temperature decomposition of AP.

The thermal weight loss process of the HTPB/AP/Al propellant samples heated to 160 °C can also be divided into three stages. The first stage is in the temperature range of 158–264 °C, with a maximum weight loss peak temperature at 221.31 °C and a gentle peak shape. The maximum weight loss rate is 0.22%/°C, and the weight loss is approximately 9.49%. This stage results mainly from the fracture and decomposition of the binder chain in HTPB. The second stage is in the temperature range of 264–325 °C, with a peak temperature at 292.71 °C and a sharp peak shape. The weight loss is approximately 23.27%, and the maximum weight loss decomposition rate is 1.10%/°C. This stage consists mainly of the low-temperature decomposition of AP. The third stage is in the temperature range of 325–401 °C, with a peak temperature at 392.91 °C and a sharp peak shape. The maximum weight loss decomposition rate is 1.19%/°C, and the weight loss is approximately 49.81%. This stage consists mainly of the high-temperature decomposition of AP.

The thermal weight loss process of the HTPB/AP/Al propellant samples heated to 180 °C can be divided into two stages. The first stage is in the temperature range of 197–324 °C, with a peak temperature at 293.85 °C and a sharp peak shape. The maximum rate of weight loss is 0.90%/°C, and the weight loss is approximately 25.74%. This stage results mainly from the continued decomposition of the binder chain that is not fully decomposed in HTPB and the low-temperature decomposition of AP. The second stage is in the temperature range of 324–399 °C, with a peak temperature at 377.25 °C and a sharp peak shape. The maximum weight loss decomposition rate is 1.42%/°C, and the weight loss is approximately 0.37%. This stage consists mainly of the high-temperature decomposition of AP.

The thermal weight loss process of the HTPB/AP/Al propellant samples heated to 220 °C can also be divided into two stages. The first stage is in the temperature range of 206–310 °C, with a maximum peak temperature at 266.63 °C and a gentle peak shape. The maximum rate of weight loss is 0.31%/°C, and the weight loss is approximately 20.30%. This stage results mainly from the continued decomposition of the binder chain that is not completely decomposed in HTPB and the low-temperature decomposition of AP. The second stage is in the temperature range of 310–400 °C, with a peak temperature at 385.23 °C and a sharp peak shape. The maximum weight loss decomposition rate is 1.45%/°C, and the weight loss is approximately 58.38%. This stage consists mainly of the high-temperature decomposition of AP.

When the sampling temperature is 180 °C and 220 °C, the thermal decomposition process can be divided into two stages. However, it can be divided into three stages when unheated and when the sampling temperature is 160 °C. This is because the binder in the mixture decomposes when the temperature exceeds 180 °C, meaning that a decomposition stage of the binder is lacking. This is because during the preparation of the sample, the first stage of binder decomposition has been completed. Therefore, it made the first-stage and second-stage original decomposition temperature higher.

SEM was used to analyse the apparent morphology of HTPB/AP mixtures and HTPB/AP/Al propellants at different sampling temperatures, as shown in Figure 8. It can be seen that as the sampling temperature increases, cracks begin to appear in the adhesive of sample 11# at 180 °C, while cracks have already appeared in sample 14# at 160 °C. It indicates that the adhesive has decomposed at this time, and in sample 15#, it can be seen that the adhesive is filled with pores while AP particles have no detailed changes. This also indicates that the adhesive decomposes first during the heating process. In samples 12# and 16#, pores are observed in AP particles. Both the decomposition of the binder and the pore structure of AP particles will have an impact on the combustion of the propellant.

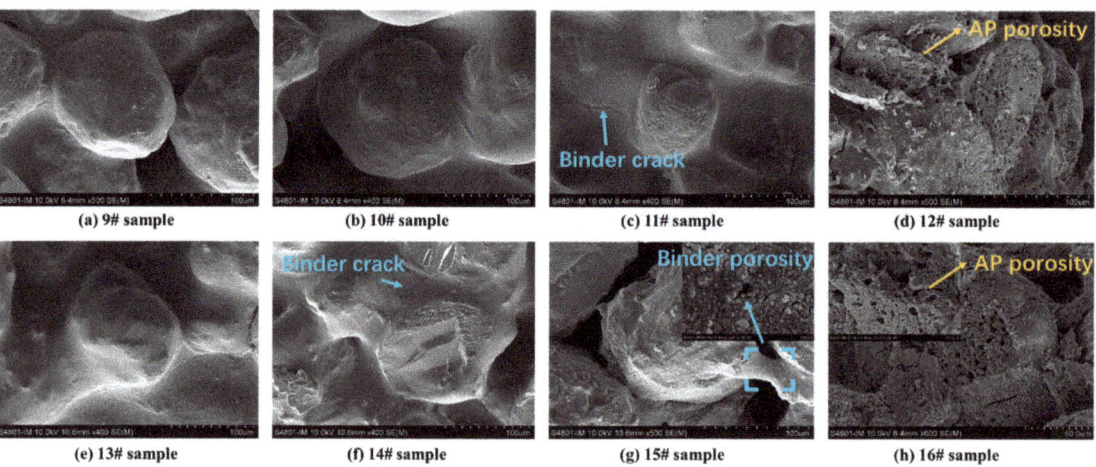

Figure 8. SEM morphology of the HTPB/AP mixtures and HTPB/AP/Al propellants heated to various temperatures.

The characteristic temperature is an important characteristic parameter in the heating process of propellants. As shown in Figure 7, T_1 is the temperature at which the propellant begins to decompose, T_2 is the temperature corresponding to the first peak of the weight loss rate, T_3 is the end temperature of the first stage of weight loss, T_4 is the temperature corresponding to the second peak of the weight loss rate, T_5 is the end temperature of the second stage of weight loss, T_6 is the temperature at which the propellant begins to burn (ignition temperature), T_7 is the temperature corresponding to the third peak of the weight loss rate, and T_8 is the temperature at which all combustible elements in the propellant are burned out. The ignition temperature is defined as the temperature corresponding to the intersection point C of the TG baseline and the tangent line of the TG descent point B corresponding to the peak point A on the DTG curve [32–34]. The heating process of the propellant can be mainly divided into two stages: the first stage is thermal decomposition and the second stage is combustion after ignition. Understanding the cocombustion behaviour of propellant components is important for investigating the interactions between the propellant components.

Figure 7 shows that the heating weight loss process of the propellant can be divided into two stages: the thermal decomposition before ignition of the propellant and the combustion stage after ignition. The first weight loss stage of the HTPB/AP mixture is characterised by a slow weight loss, which is 35% to 45% higher than the binder content in the propellant. This indicates that the first stage results not only from the thermal decomposition of the binder but also from the low-temperature decomposition of AP particles. The second stage is characterised by a rapid weight loss of 55% to 65%, mainly owing to the combustion of AP oxidants in the propellant. The weight loss of the HTPB/AP/Al propellant in the first stage is 32% to 40%, which is also higher than the binder content in the propellant. The second stage includes the thermal decomposition of the binder and the low-temperature decomposition of the AP particles. The weight loss in the second stage is 45% to 53%. The material remaining after the second stage consists of Al powder and a reaction residue.

To analyse the combustion characteristics of the propellants comprehensively, the flammability index, S, is defined as follows [35].

The combustion at lower heating rates can be determined by chemical reaction kinetics. According to Arrhenius' law,

$$\frac{dW}{dt} = A exp\left(-\frac{E}{RT}\right), \quad (1)$$

where dW/dt is the combustion rate (%/°C), A is the pre-exponential factor (min^{-1}), E is the activation energy (kJ/mol), and T is the temperature (K).

From Equation (1), the following derivation can be obtained:

$$\frac{R}{E}\frac{d}{dT}\left(\frac{dW}{dt}\right) = \frac{dW}{dt}\frac{1}{T^2}. \quad (2)$$

At the ignition temperature, the aforementioned formula becomes

$$\frac{R}{E}\frac{d}{dT}\left(\frac{dW}{dt}\right)_{T=T_i} = \left(\frac{dW}{dt}\right)_{T=T_i}\frac{1}{T_i^2}. \quad (3)$$

Equation (3) can be converted as follows:

$$\frac{R}{E}\frac{d}{dT}\left(\frac{dW}{dt}\right)_{T=T_i}\frac{\left(\frac{dW}{dt}\right)_{max}\left(\frac{dW}{dt}\right)_{mean}}{\left(\frac{dW}{dt}\right)_{T=T_i} T_h} = \frac{\left(\frac{dW}{dt}\right)_{max}\left(\frac{dW}{dt}\right)_{mean}}{T_i^2 T_h}, \quad (4)$$

where $(dW/dt)_{max}$ is the maximum combustion rate (%/°C), $(dW/dt)_{mean}$ is the average combustion rate (%/°C), $(dW/dt)_{T=T_i}$ is the combustion rate at the ignition temperature (%/°C), T_i is the ignition temperature (°C), and T_h is the burnout temperature (°C). R/E represents the reactivity of the propellant: the greater the value, the faster the reaction speed. At the ignition temperature, $d/dT\ (dW/dt)_{T=T_i}$ is the percentage of combustion rate conversion: the greater the value, the more rapid the ignition. Moreover, at the ignition temperature, $(dW/dt)_{max}/(dW/dt)_{T=T_i}$ is the ratio of the maximum combustion rate to the combustion rate. Furthermore, $(dW/dt)_{mean}/T_H$ represents the ratio of the average combustion rate to the burnout temperature: the greater the value, the faster the propellant burns. The product of the aforementioned terms reflects the combustion characteristics of the propellant, and its flammability index S is defined as

$$S = \frac{\left(\frac{dW}{dt}\right)_{max}\left(\frac{dW}{dt}\right)_{mean}}{T_i^2 T_h}. \tag{5}$$

Here, T_i and T_h are T_6 and T_8, respectively. The calculated flammability indices of the propellant samples are listed in Table 3.

Table 3. Flammability indices of the HTPB/AP mixture and HTPB/AP/Al propellant at various sampling temperatures.

Propellant Samples	Experiment Number	$\left(\frac{dW}{dt}\right)_{max}$ (%/°C)	$\left(\frac{dW}{dt}\right)_{mean}$ (%/°C)	T_i (°C)	T_h (°C)	$S \times 10^{-8}$
HTPB/AP mixture	9#	1.88	0.92	353.65	413.45	3.34
	10#	1.55	0.81	333.16	407.96	2.77
	11#	1.77	0.96	352.34	415.34	3.30
	12#	2.28	1.17	353.77	406.57	4.24
HTPB/AP/Al propellant	13#	1.83	0.84	339.22	391.82	3.41
	14#	1.19	0.66	326.11	401.91	1.84
	15#	1.42	0.71	330.85	399.05	2.31
	16#	1.45	0.66	332.83	400.23	2.42

Table 3 shows that the S index of sample 9# is 3.34×10^{-8}, and S decreases to 2.77×10^{-8} as the sampling temperature increases to 160 °C. This occurs because the HTPB binder in the propellant is thermally decomposed, thus weakening the interaction between the binder and AP particles. As the sampling temperature continues to increase, the S index gradually increases to 4.24×10^{-8}; although the HTPB binder has decomposed and the interaction is weak, pores are generated inside the sample at that time, which, in return, increases the specific surface area. To put it another way, even the interaction is weak, while the increase in the specific surface area could also enlarge the interaction effect. Therefore, the combustion characteristics are first mild and then become intense. The HTPB/AP/Al propellants exhibit the same trend as the mixtures, indicating that the addition of the Al powder has no significant effect on the interaction between the binder and AP other than performing a catalytic role.

3.2.2. Cocombustion of the HTPE Binder and AP Particles

Figure 9 shows the TG and DTG curves of the HTPE/AP mixture and HTPE/AP/Al propellant, respectively. The thermal weight loss process of the original HTPE/AP mixture sample and samples 20#, 21#, and 22# can all be observed to be divided into three stages.

For the original HTPE/AP mixture sample, the first stage is in the temperature range of 174–274 °C, with a maximum weight loss peak temperature at 252.95 °C and a gentle peak shape. The maximum weight loss rate is 0.24%/°C, and the weight loss is approximately 11.16%. This stage results mainly from the breaking and decomposition of the binder chain in HTPE. The second stage is in the temperature range of 274–327 °C, with a maximum peak temperature at 294.75 °C and a sharp peak shape. The weight loss is approximately 21.88%, and the maximum weight loss decomposition rate is 0.93%/°C. This stage consists mainly of the low-temperature decomposition of AP. The third stage is in the temperature range of 327–411 °C, with a peak temperature at 395.75 °C and a sharp peak shape. The maximum weight loss decomposition rate is 2.00%/°C, and the weight loss is approximately 66.96%. This stage consists mainly of the high-temperature decomposition of AP.

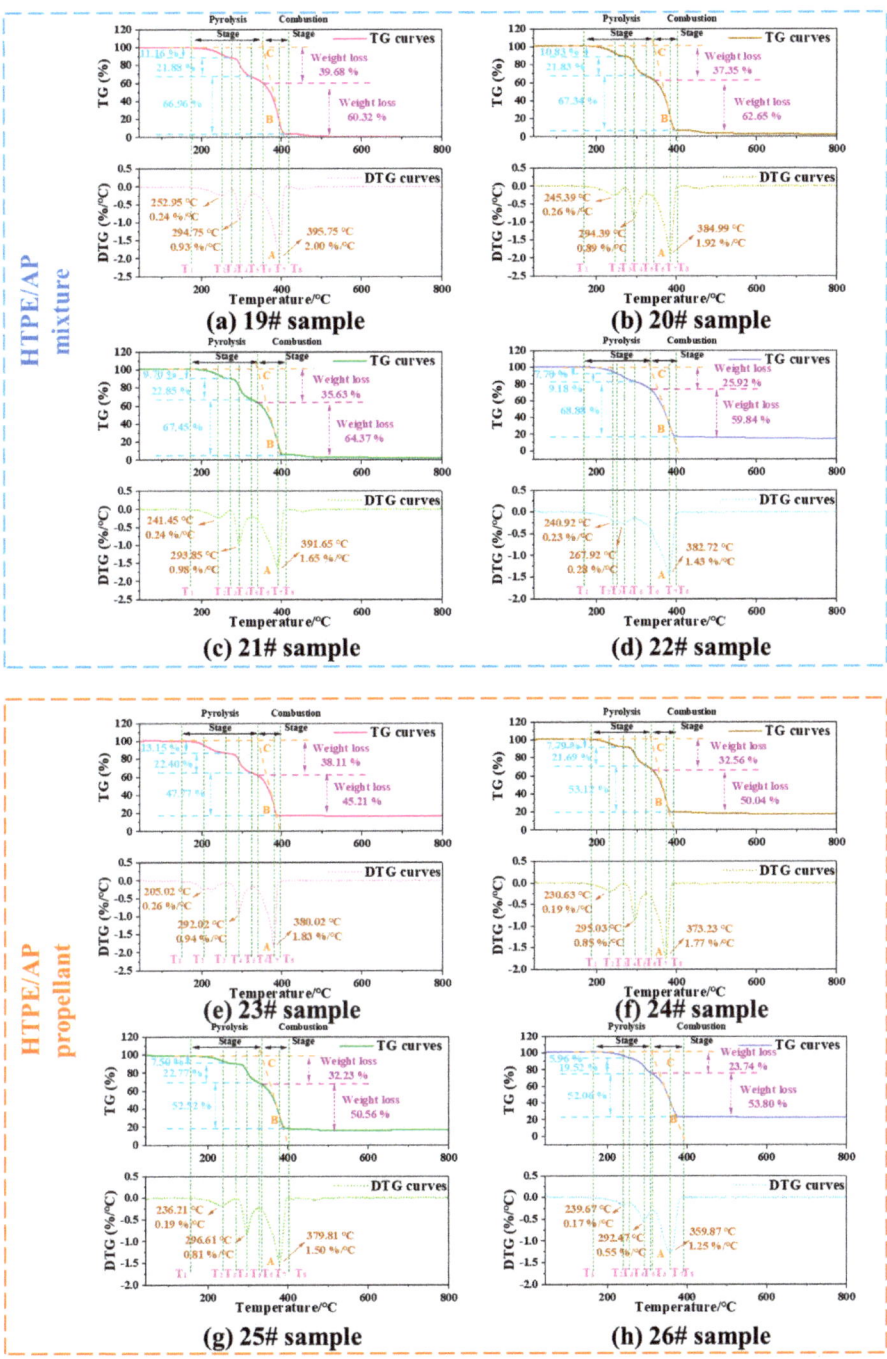

Figure 9. TG and DTG curves of the HTPE/AP mixtures and HTPE/AP/Al propellants at various sampling temperatures.

For sample 20#, the first stage is in the temperature range of 173–270 °C, with a maximum weight loss peak temperature at 245.39 °C and a gentle peak shape. The maximum weight loss rate is 0.26%/°C, and the weight loss is approximately 10.83%. The second stage is in the temperature range of 270–324 °C, with a maximum peak temperature at 294.39 °C and a sharp peak shape. The weight loss is approximately 21.83%, and the maximum weight loss decomposition rate is 0.89%/°C. Finally, the third stage is in the temperature range of 324–399 °C, with a peak temperature at 384.99 °C and a sharp peak shape. The maximum weight loss decomposition rate is 1.92%/°C, and the weight loss is approximately 67.34%

For sample 21#, the first stage is in the temperature range of 173–270 °C, with a maximum weight loss peak temperature at 241.45 °C and a gentle peak shape. The maximum weight loss rate is 0.24%/°C, and the weight loss is approximately 9.70%. The second stage is in the temperature range of 270–324 °C, with a peak temperature at 293.85 °C and a sharp peak shape. The weight loss is approximately 22.85%, and the maximum weight loss decomposition rate is 0.98%/°C. The third stage is in the temperature range of 324–408 °C, with a peak temperature at 391.65 °C and a sharp peak shape. The maximum weight loss decomposition rate is 1.65%/°C, and the weight loss is approximately 67.45%.

For sample 22#, the first stage is in the temperature range of 171–251 °C, with a maximum weight loss peak temperature at 240.92 °C, maximum weight loss rate of 0.23%/°C, and weight loss of approximately 7.70%. The second stage is connected to the first stage, with a weight loss of approximately 9.18% in the temperature range of 251–293 °C. The third stage is in the temperature range of 293–405 °C, with a peak temperature at 382.72 °C and a sharp peak shape. The maximum weight loss decomposition rate is 1.43%/°C, and the weight loss is approximately 68.88%.

On account of the defects on the AP crystal surface, a small number of AP molecules at the defect sites readily dissociate into NH_3 and $HClO_4$ via proton transfer at lower temperatures. Furthermore, as a strong acid, $HClO_4$ readily reacts with the oxygen atoms of the ether bond in the HTPE molecular chain to form a salt, which makes the thermal stability of the ether bond decrease. Therefore, the earlier initial decomposition temperature of the first decomposition stage of the HTPE/AP mixture may have been caused by the small number of AP decomposition products promoting the decomposition of the HTPE binder.

The thermal weight loss process of the original HTPE/AP/Al propellant sample and samples 24#, 25#, and 26# can all be divided into three stages. For the original HTPE/AP/Al propellant sample, the first stage is in the temperature range of 136–259 °C, with a maximum weight loss peak temperature at 205.02 °C and a gentle peak shape. The maximum weight loss rate is 0.26%/°C, and the weight loss is approximately 13.15%. This stage results mainly from the breaking of the binder chains in HTPE. The second stage is in the temperature range of 259–326 °C, with a peak temperature at 292.02 °C and a sharp peak shape. The weight loss is approximately 22.40%, and the maximum weight loss decomposition rate is 0.94%/°C. This stage consists mainly of the low-temperature decomposition of AP. The third stage is in the temperature range of 326–392 °C, with a peak temperature at 380.02 °C and a sharp peak shape. The maximum weight loss decomposition rate is 1.83%/°C, and the weight loss is approximately 47.77%. This stage consists mainly of the high-temperature decomposition of AP.

For sample 24#, the first stage is in the temperature range of 170–270 °C, with a maximum weight loss peak temperature at 230.03 °C and a gentle peak shape. The maximum weight loss rate is 0.19%/°C, and the weight loss is approximately 7.79%. The second stage is in the temperature range of 270–324 °C, with a peak temperature at 295.03 °C and a sharp peak shape. The weight loss is approximately 21.69%, and the maximum weight loss decomposition rate is 0.85%/°C. The third stage is in the temperature range of 324–391 °C, with a peak temperature at 373.23 °C and a sharp peak shape. The maximum weight loss decomposition rate is 1.77%/°C, and the weight loss is approximately 53.12%. This stage consists mainly of the high-temperature decomposition of AP.

For sample 25#, the first stage is in the temperature range of 167–270 °C, with a maximum weight loss peak temperature at 236.21 °C and a gentle peak shape. The maximum weight loss rate is 0.19%/°C, and the weight loss is approximately 7.50%. The second stage is in the temperature range of 270–325 °C, with a maximum peak temperature at 296.61 °C and a sharp peak shape. The weight loss is approximately 22.77%, and the maximum weight loss decomposition rate is 0.81%/°C. The third stage is in the temperature range of 325–400 °C, with a peak temperature at 379.81 °C and a sharp peak shape. The maximum weight loss decomposition rate is 1.50%/°C, and the weight loss is approximately 52.52%.

For sample 26#, the first stage is in the temperature range of 164–255 °C, with a maximum weight loss peak temperature at 239.67 °C, a maximum weight loss rate of 0.17%/°C, and a weight loss of approximately 5.96%. The second stage is in the temperature range of 255–313 °C, with a peak temperature at 292.47 °C and a sharp peak shape. The weight loss is approximately 19.52%, and the maximum weight loss decomposition rate is 0.55%/°C. The third stage is in the temperature range of 313–390 °C, with a peak temperature at 359.87 °C and a sharp peak shape. The maximum weight loss decomposition rate is 1.25%/°C, and the weight loss is approximately 52.06%.

In contrast with the HTPB/AP mixture, the thermal decomposition processes of all the aforementioned samples can be divided into three stages. The addition of Al power will not affect its decomposition process.

SEM was used to analyse the apparent morphology of HTPE/AP mixtures and HTPE/AP/Al propellants at different sampling temperatures, as shown in Figure 10. It can be seen that in the HTPE/AP mixture and HTPE/AP/Al propellant, as the sampling temperature increases, the HTPE binder gradually liquefies and coats the surface of AP particles. From samples 12# and 16#, it can be seen that the HTPE binder at this time is still in a viscoelastic state, but the AP particles have already decomposed.

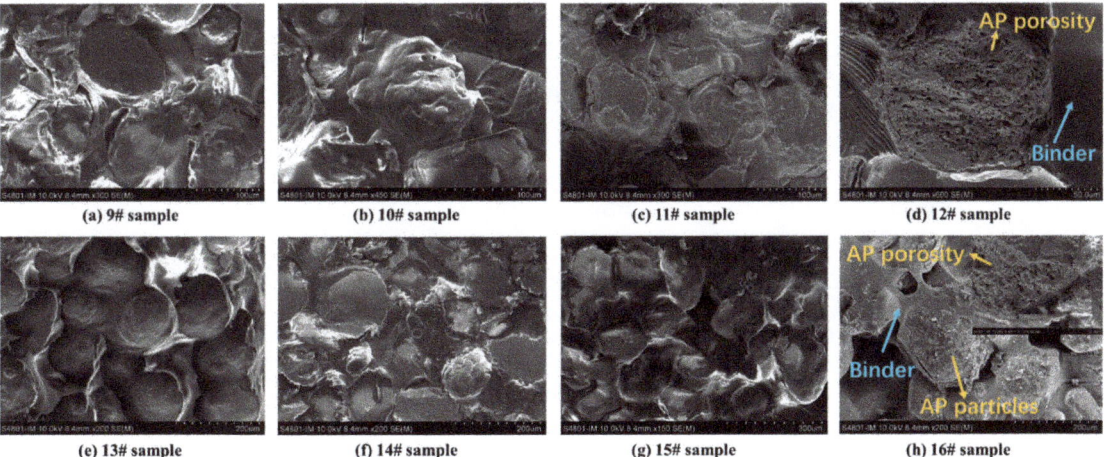

Figure 10. SEM morphology of the HTPE/AP mixtures and HTPE/AP/Al propellants heated to various temperatures.

It can be seen that the interactions between the components of the two binder systems are different. This is because the HTPB binder becomes harder and more brittle as the sample temperature increases, causing it to debond from the AP particles, which in turn weakens the interactions between them. The HTPE binder becomes softer and reaches a certain degree of liquefaction, while continuing to adhere to the AP particles and interact with them.

Figure 9 shows the TG/DTG curves and characteristic temperatures of the HTPE/AP mixtures and HTPE/AP/Al propellants at various temperatures. The heating weight loss processes of the propellants can be divided into two stages. The first stage of the HTPE/AP mixture exhibits a slow weight loss, with a total weight loss of 25–40%, which is higher than the binder content in the propellant, indicating that this stage involves not only the thermal decomposition of the binder but also the low-temperature decomposition of the AP particles. The rapid weight loss in the second stage results mainly from the combustion of the AP oxidant in the propellant, and the weight loss in this stage is approximately 60%. The weight loss of the HTPE/AP/HTPB propellant in the first stage is between 23% and 38%, which is also higher than the binder content in the propellant. This stage includes the thermal decomposition of the binder and the low-temperature decomposition of the AP particles. The weight loss in the second stage is between 45% and 53%. The material that remains after the second stage consists of Al powder and a reaction residue.

Table 4 shows that the S index of sample 19# is 3.78×10^{-8}, and S increases to 4.05×10^{-8} as the sampling temperature increases to 160 °C. This occurs because as the sampling temperature increases, the HTPE binder liquefies under heat and adheres more tightly to the AP particles to coat their surfaces, resulting in stronger interactions. However, as the sampling temperature increases, S gradually decreases to 2.78×10^{-8} owing to the decomposition of the energetic plasticiser A3 in the binder, which cannot provide the heat released by its decomposition to accelerate the low-temperature decomposition of AP. Therefore, the combustion characteristics are initially violent and then slow. The HTPE/AP/Al propellant exhibits the same trend as the mixture, indicating that the addition of Al powder has no significant effect on the interaction between the binder and AP other than performing a catalytic role.

Table 4. Flammability indices of the HTPE/AP mixture and HTPE/AP/Al propellant at various temperatures.

Propellant Samples	Experiment Number	$\left(\frac{dW}{dt}\right)_{max}$ (%/°C)	$\left(\frac{dW}{dt}\right)_{mean}$ (%/°C)	T_i (°C)	T_h (°C)	$S \times 10^{-8}$
HTPE/AP mixture	19#	2.00	0.97	353.35	411.35	3.78
	20#	1.92	0.99	342.99	399.19	4.05
	21#	1.65	0.87	337.85	408.45	3.10
	22#	1.43	0.87	333.92	405.72	2.78
HTPE/AP/Al propellant	23#	1.83	0.84	339.42	392.22	3.40
	24#	1.77	0.85	335.23	391.03	3.42
	25#	1.50	0.72	333.21	400.81	2.43
	26#	1.25	0.65	308.27	390.87	2.19

3.3. Study of the Interaction of Propellant Component

3.3.1. Interaction between the HTPB Binder and AP Particles

To investigate whether there is an interaction between the binder and AP particles, the theoretical TG/DTG curve of the blend was calculated from the average weight of the individual as follows:

$$W = \alpha W_{binder} + \beta W_{AP} \tag{6}$$

where W_{binder} and W_{AP} are the weight loss rates of the binder and AP particles, respectively, and α and β are their respective proportions in the propellant.

The theoretical thermogravimetric curve for the binder mass ratio to AP particles at 18:82 was calculated. The experimental and calculated TG curves are shown in Figure 11. To further clarify the interaction between the HTPB binder and AP particles, ΔW ($\Delta W = TG_{calculated} - TG_{empirical}$) is defined as the difference in weight loss. Figure 12 shows the composition of the HTPB/AP mixture as ΔW changes with temperature.

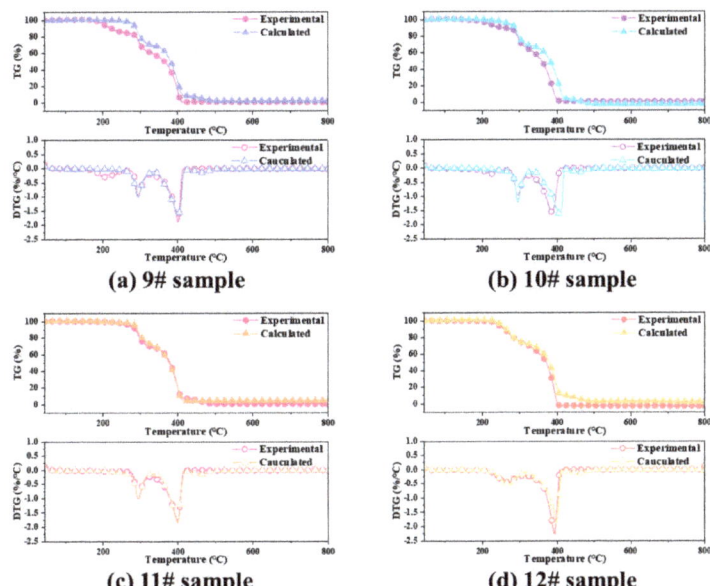

Figure 11. Comparison between the experimental and calculated for HTPB/AP mixture TG and DTG curves.

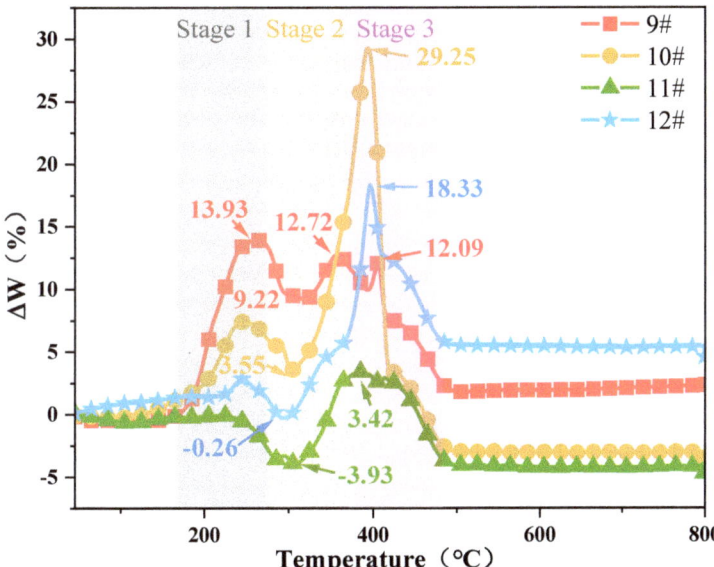

Figure 12. Difference between the HTPB/AP mixture experimental and theoretical weight loss.

The calculated DTG curve of the HTPB/AP mixture almost coincides with the experimental DTG curve within the temperature range below 150 °C. When the temperature of all samples exceeds 150 °C, the calculated TG curve lags behind the experimental TG curve. All the interactions between HTPB and the AP particles are positive and occur at all stages. At 200 °C, the calculated maximum weight loss is 0.30%/°C higher than the experimental value, indicating that the HTPB binder and AP particles interact at low temperatures. Compared with the calculated DTG curve, the experimental DTG curve shifts in the 260–420 °C region. This further confirms the significant interaction between the HTPB binder and AP particles.

Three maximum peaks exist in sample 9#, notably at 262 °C, with a deviation value as high as 13.93. The deviations are 12.72 and 12.09 at 357 and 405 °C, respectively. The maximum deviation for samples 10#, 11#, and 12# are 29.25, 3.42, and 18.33, respectively. These deviations indicate that a promoted interaction occurs between the HTPB binder and AP particles during both the thermal decomposition and combustion stages. This can be attributed to the exothermic heating effect of the HTPB decomposition process, which causes the AP to dissociate at low temperatures and release highly oxidising products in advance. The effect is more significant during the thermal decomposition stage of the unheated HTPB/AP mixture. Above 500 °C, ΔW is stable due to the combustion process of the blend being almost complete.

The interaction between the HTPB binder and AP differs from the mixed HTPB/AP system. Figure 13 shows the combustion characteristic index S of the HTPB and HTPE binder systems. As can be seen from Figures 12 and 13, the impact of ΔW is divided into three stages, which have varying degrees of impact. The first weight loss stages of samples 9#, 10#, and 11# show a gradually decreasing ΔW, and as the sampling temperature increases, the HTPB binder gradually decomposes. Thus, the AP particles promote the decomposition of the HTPB binder at this stage. The second weight loss stage also shows a gradually decreasing ΔW, but the interaction is weaker than that in the first weight loss stage, and ΔW has a negative promoting effect. In the third weight loss stage, ΔW first increases and then decreases as the sampling temperature increases. This is due to the HTPB binder gradually decomposing while weight loss continues, leaving behind substances that are difficult to decompose, thus gradually weakening the interaction. However, the combustion characteristic index, S, of sample 12# is the largest, and the interaction between the HTPB binder and AP particles is the largest in the third stage. The main interaction between the HTPB binder and AP particles occurs in the combustion stage. The interaction of various components in the mixed HTPB/AP system is not only related to its thermal decomposition stage but also affected by its decomposition products and many other factors, such as pore structure.

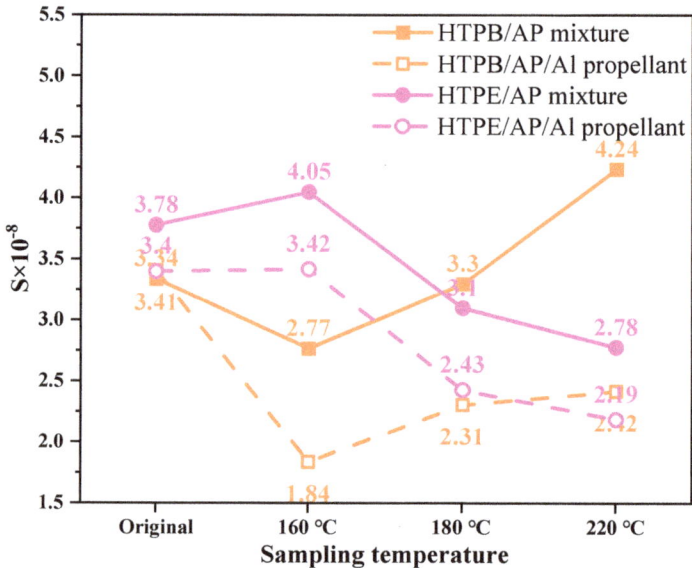

Figure 13. Combustion characteristic index S of HTPB and HTPE binder system.

3.3.2. Interaction between the HTPE Binder and AP Particles

The experimental and calculated TG curves of the HTPE/AP mixture at various sampling temperatures are shown in Figure 14 (the theoretical calculated values in samples 21# and 22# were calculated by using sample 18#). Figure 15 shows the composition of the HTPE/AP mixture at various sampling temperatures, and ΔW changes along with temperature. It can be seen that when the temperature is below 160 °C, the calculated DTG curve of the HTPE/AP mixture is almost consistent with the experimental DTG curve. When the temperature exceeds 160 °C, the calculated TG curve of sample 19# lags behind the experimental TG curve. When the temperature is between 150 and 320 °C, the experimental TG curves of samples 20#, 21#, and 22# lag behind the calculated TG curve. When the temperature exceeds 320 °C, the calculated TG curves of 20#, 21#, and 22# lag behind the experimental TG curve. There are three peaks in samples 19#, 20#, and 21#, of which sample 19# has a value of 4.05 at 206 °C and 3.48 at 220 °C, and a deviation value of up to 25.67 at 385 °C. Sample 20# is −3.10 at 226 °C, −2.26 at 304 °C, and 27.95 at 381 °C. Sample 21# is −1.39 at 229 °C, −3.34 at 299 °C, and 14.62 at 386 °C. Sample 22# has two peaks, ranging from −10.37 at 281 °C to 27.41 at 372 °C. These deviations indicate that when unheated, a positive promoting effect exists between HTPE and the AP particles, whereas when the sampling temperature exceeds 160 °C, a blocking effect exists between HTPE and the AP particles in the first and second weight loss stages of the HTPE/AP mixture and a positive promoting effect in the third weight loss stage. The main interaction between the HTPE binder and AP particles is in the thermal decomposition stage.

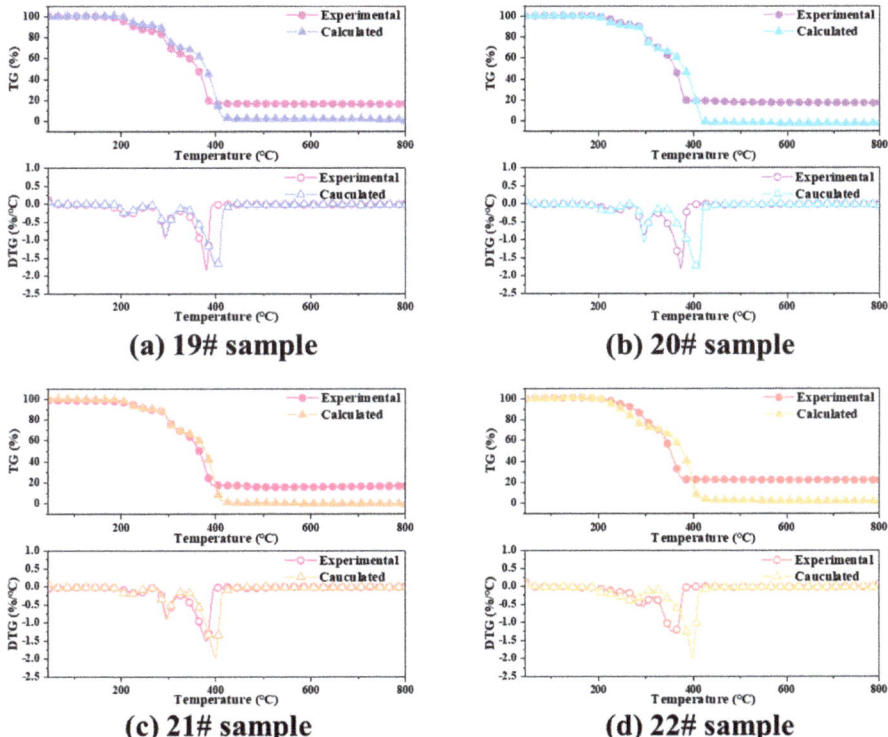

Figure 14. Comparison between the experimental and calculated for HTPE/AP mixture TG and DTG curves.

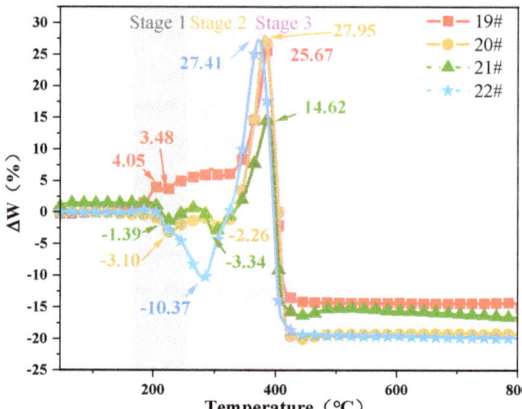

Figure 15. Difference between the HTPE/AP mixture experimental and theoretical weight loss.

As can be seen from Figures 13 and 15, the impact of ΔW is divided into three stages; yet, these three stages have varying degrees of impact, while the third stage has the least. The interaction between the binder and AP is different in the mixed system.

The unheated HTPE binder was beneficial to the low- and high-temperature decompositions of AP. This is related to the structural and thermal decomposition characteristics of HTPE. When AP is promoted, the HTPE binder can decompose in advance to produce short-chain polyethers. Moreover, as the temperature increases, the $HClO_4$ produced by decomposition consumes a large amount of the HTPE binder owing to its decomposition and participation in oxidation reactions. However, when the HTPE propellant is heated, the short-chain polyether produced by the partial decomposition of the binder in it will fill the holes formed by the decomposition of the AP surface, thus acting as a coating and insulation, slowing down further decomposition of AP and inhibiting the concentrated and rapid release of decomposition heat. Therefore, the heated HTPE/AP mixture has a blocking effect, which can be achieved from the first stage when a decrease in ΔW is observed. It can be seen that the interaction of various elements in the mixed HTPE/AP system is not only related to its thermal decomposition stage but also affected by the binder decomposition products and many other factors, such as pore structure. Furthermore, it is also very beneficial for solid propellants to slow down the reaction of the AP oxidation products with Al powder under heating conditions, thereby reducing the responsiveness of solid propellants.

4. Conclusions

By studying the interactions between the components of two binder systems at various temperatures, the conclusions are as follows:

(1) The first and second weight loss decomposition peak temperatures of the HTPB binder are 85.34 and 55.74 °C higher, respectively, than those of the HTPE binder. Therefore, compared to the HTPB binders, the HTPE binders are more easily decomposed.

(2) As the sampling temperature increases, the S index of the HTPB/AP mixture initially decreases from 3.34×10^{-8} to 2.77×10^{-8}, then increases to 4.24×10^{-8}, indicating that its combustion characteristics are initially mild and then intensify. In contrast, the S index of the HTPE/AP mixture from 3.78×10^{-8} first increases to 4.05×10^{-8}, then decreases to 2.78×10^{-8}, indicating that its combustion characteristics are initially rapid and then slow down.

(3) The ΔW deviation between the heated HTPB binder and AP particles is positive, and the maximum deviations are 13.93, 29.25, 3.42, and 18.33, respectively. This indicates a promoting interaction between the HTPB binder and AP particles during the thermal

decomposition and combustion stages. The ΔW deviation between the heated HTPE binder and AP particles is negative in the first and second weight loss stages, but positive in the third weight loss stage, with maximum deviations of 25.67, 27.95, 14.62, and 27.41, respectively. During the first and second weight loss stages of the HTPE/AP mixture, there is a blocking effect between the HTPE and AP particles on the surface, and a positive promoting effect appears in the third weight loss stage. The main interaction between the HTPE binder and AP particles occurs in the thermal decomposition stage.

The study of the interaction of the propellant component after heating is an important influencing factor for mastering and understanding the slow burning mechanism and response severity of propellants. In addition, the influence of the microstructure of propellants after heating cannot be ignored, and it is significant for comprehensively understanding the thermal safety of propellants.

Author Contributions: Conceptualization, J.N. and H.Z.; methodology and formal analysis, J.L.; investigation, J.L. and H.Z.; resources, X.G.; data curation, M.H.; writing—original draft preparation, J.L.; writing—review and editing, J.N.; supervision, S.Y.; funding acquisition, X.G. All authors have read and agreed to the published version of the manuscript.

Funding: This study was supported by the National Natural Science Foundation of China [grant number 22175026].

Institutional Review Board Statement: Not applicable.

Data Availability Statement: The raw/processed data required to reproduce these findings cannot be shared at this time as the data also forms part of an ongoing study.

Acknowledgments: Peng-fei Wang (Xi'an Changfeng Research Institute of Mechanical-Electrical, Xi'an 710065, China).

Conflicts of Interest: The authors declare no conflict of interest.

References

1. Zhang, L.-K.; Zheng, X.-Y. Experimental study on thermal decomposition kinetics of natural ageing AP/HTPB base bleed composite propellant. *Def. Technol.* **2018**, *14*, 422–425. [CrossRef]
2. Wang, Z.-J.; Qiang, H.-F. Mechanical properties of thermal aged HTPB composite solid propellant under confining pressure. *Def. Technol.* **2022**, *18*, 618–625. [CrossRef]
3. Ye, Q.; Yu, Y.-G. Numerical simulation of cook-off characteristics for AP/HTPB. *Def. Technol.* **2018**, *14*, 451–456. [CrossRef]
4. Wang, Z.-J.; Qiang, H.-F.; Wang, G.; Geng, B. Strength criterion of composite solid propellants under dynamic loading. *Def. Technol.* **2018**, *14*, 457–462. [CrossRef]
5. Ji, M.; Cao, L.; Li, Z.; Chen, G.; Cao, P.; Liu, T. Numerical Conversion Method for the Dynamic Storage Modulus and Relaxation Modulus of Hydroxy-Terminated Polybutadiene (HTPB) Propellants. *Polymers* **2023**, *15*, 3. [CrossRef] [PubMed]
6. Zhang, Y.; Tian, Y.; Zhang, Y.; Fu, X.; Li, H.; Lu, Z.; Zhang, T.; Hu, Y. Improvement in Migration Resistance of Hydroxyl-Terminated Polybutadiene (HTPB) Liners by Using Graphene Barriers. *Polymers* **2022**, *14*, 5213. [CrossRef]
7. Shi, L.; Fu, X.; Li, Y.; Wu, S.; Meng, S.; Wang, J. Molecular Dynamic Simulations and Experiments Study on the Mechanical Properties of HTPE Binders. *Polymers* **2022**, *14*, 5491. [CrossRef]
8. Weigand, A.; Unterhuber, G.; Kupzik, K.; Eich, T.; Bucher, B. *Solid Propellant Rocket Motor Insensitive Munitions, Testing and Simulation*; Insensitive Munitions and Energetic Materials Symposium (IMEMTS): Munich, Germany, 2010.
9. Ye, Q.; Yu, Y.; Li, W. Study on cook-off behavior of HTPE propellant in solid rocket motor. *Appl. Therm. Eng.* **2020**, *167*, 114798. [CrossRef]
10. SHo, Y.; Ferschl, T.; Foureur, J. *Correlation of Cook-Off Behaviour of Rocket Propellants with Thermal Mechanical and Thermochemical Properties*; MRL Technical Report; Materials Research Labs Ascot Vale: Victoria, Australia, 1993; pp. 131–132.
11. Wu, X.; Li, J.; Ren, H.; Jiao, Q. Comparative Study on Thermal Response Mechanism of Two Binders during Slow Cook-Off. *Polymers* **2022**, *14*, 3699. [CrossRef]
12. Essel, J.T.; Nelson, A.P.; Smilowitz, L.B.; Henson, B.F.; Merriman, L.R.; Turnbaugh, D.; Gray, C.; Shermer, K.B. Investigating the effect of chemical ingredient modifications on the slow cook-off violence of ammonium perchlorate solid propellants on the laboratory scale. *J. Energetic Mater.* **2020**, *38*, 127–141. [CrossRef]
13. Yan, Q.-L.; Zhao, F.-Q.; Kuo, K.K.; Zhang, X.-H.; Zeman, S.; DeLuca, L.T. Catalytic effects of nano additives on decomposition and combustion of RDX-, HMX-, and AP-based energetic compositions. *Prog. Energy Combust. Sci.* **2016**, *57*, 75–136. [CrossRef]

14. Trache, D.; Maggi, F.; Palmucci, I.; DeLuca, L.T. Thermal behavior and decomposition kinetics of composite solid propellants in the presence of amide burning rate suppressants. *J. Therm. Anal. Calorim.* **2018**, *132*, 1601–1615. [CrossRef]
15. Wu, W.; Jin, P.; Zhao, S.; Luo, Y. Mechanism of AP effect on slow cook-off response of HTPE propellant. *Thermochim. Acta* **2022**, *715*, 179291. [CrossRef]
16. Wu, W.; Zhang, X.; Jin, P.; Zhao, S.; Luo, Y. Mechanism of PSAN effect on slow cook-off response of HTPE propellant. *J. Energetic Mater.* **2022**, *2*, 1–18. [CrossRef]
17. Luo, Y.-J.; Mao, K.-Z.; Xia, M. Effect of Hydroxyl-Terminated Random Copolyether (PET) and Hydroxyl-Terminated Polybutadiene (HTPB) on Thermal Decomposition Characteristics of Ammonium Perchlorate. *J. Res. Update Polym. Sci.* **2015**, *4*, 42–49. [CrossRef]
18. Mallick, L.; Kumar, S.; Chowdhury, A. Thermal decomposition of ammonium perchlorate—A TGA–FTIR–MS study: Part I. *Thermochim. Acta* **2015**, *610*, 57–68. [CrossRef]
19. Zhu, Y.-L.; Huang, H.; Ren, H.; Jiao, Q.-J. Kinetics of Thermal Decomposition of Ammonium Perchlorate by TG/DSC-MS-FTIR. *J. Energetic Mater.* **2014**, *32*, 16–26. [CrossRef]
20. Kohga, M.; Togashi, R. Mechanical Properties and Thermal Decomposition Behaviors of Hydroxyl-Terminated Polybutadiene/Glycerol Propoxylate Blend and Its Application to Ammonium Nitrate-Based Propellants. *Propellants Explos. Pyrotech.* **2021**, *46*, 1016–1022. [CrossRef]
21. Rocco, J.A.F.F.; Lima, J.E.S.; Frutuoso, A.G.; Iha, K.; Ionashiro, M.; Matos, J.R.; Suárez-Iha, M.E.V. TG studies of a composite solid rocket propellant based on HTPB-binder. *J. Therm. Anal. Calorim.* **2004**, *77*, 803–813. [CrossRef]
22. Tingfa, D.; Junfeng, L. Estimation of major volatile products from the first stage of the thermal decomposition of hydroxy-terminated polybutadiene binder. *Thermochim. Acta* **1991**, *184*, 81–90. [CrossRef]
23. Wibowo, H.B.; Sitompul, H.R.D.; Budi, R.S.; Hartaya, K.; Abdillah, L.H.; Ardianingsih, R.; Wibowo, R.S.M. Hexogen Coating Kinetics with Polyurethane-Based Hydroxyl-Terminated Polybutadiene (HTPB) Using Infrared Spectroscopy. *Polymers* **2022**, *14*, 1184. [CrossRef] [PubMed]
24. Yuan, S.; Zhang, B.; Wen, X.; Chen, K.; Jiang, S.; Luo, Y. Investigation on mechanical and thermal properties of HTPE/PCL propellant for wide temperature range use. *J. Therm. Anal. Calorim.* **2022**, *147*, 4971–4982. [CrossRef]
25. Wang, Y.-H.; Liu, L.-L.; Xiao, L.-Y.; Wang, Z.-X. Thermal decomposition of HTPB/AP and HTPB/HMX mixtures with low content of oxidizer. *J. Therm. Anal. Calorim.* **2015**, *119*, 1673–1678. [CrossRef]
26. Padwal, M.B.; Varma, M. Thermal decomposition and combustion characteristics of HTPB-coarse AP composite solid propellants catalyzed with Fe_2O_3. *Combust. Sci. Technol.* **2018**, *190*, 1614–1629. [CrossRef]
27. Wen, X.; Chen, K.; Sang, C.; Yuan, S.; Luo, Y. Applying modified hyperbranched polyester in hydroxyl-terminated polyether/ammonium perchlorate/aluminium/cyclotrimethylenetrinitramine (HTPE/AP/Al/RDX) composite solid propellant. *Polym. Int.* **2020**, *70*, 123–134. [CrossRef]
28. Kim, K.-H.; Kim, C.-K.; Yoo, J.-C.; Yoh, J.J. Test-Based Thermal Decomposition Simulation of AP/HTPB and AP/HTPE Propellants. *J. Propuls. Power* **2011**, *27*, 822–827. [CrossRef]
29. Zhang, H.; Nie, J.; Wang, L.; Wang, D.; Hu, F.; Guo, X. Effect of preignition on slow cook-off response characteristics of composite propellant. *Explos. Shock. Waves* **2022**, *42*, 102901.
30. Gołofit, T.; Ganczyk-Specjalska, K.; Jamroga, K.; Kufel, L. Rheological and thermal properties of mixtures of hydroxyl-terminated polybutadiene and plasticizer (Rapid communication). *Polimery* **2018**, *63*, 53–56. [CrossRef]
31. Zhang, H.-J.; Nie, J.-X.; Jiao, G.-L.; Xu, X.; Guo, X.-Y.; Yan, S.; Jiao, Q.-J. Effect of the microporous structure of ammonium perchlorate on thermal behaviour and combustion characteristics. *Def. Technol.* **2022**, *18*, 1156–1166. [CrossRef]
32. Qi, N. Thermogravimetric analysis on the combustion characteristics of brown coal blends. *Combust. Sci. Technol.* **2001**, *7*, 72–76.
33. Li, X.-G.; Ma, B.-G.; Xu, L.; Hu, Z.-W.; Wang, X.-G. Thermogravimetric analysis of the co-combustion of the blends with high ash coal and waste tyres. *Thermochim. Acta* **2006**, *441*, 79–83. [CrossRef]
34. Yu, L.J.; Wang, S.; Jiang, X.M.; Wang, N.; Zhang, C.Q. Thermal analysis studies on combustion characteristics of seaweed. *J. Therm. Anal. Calorim.* **2008**, *93*, 611–617. [CrossRef]
35. Xie, Z.; Ma, X. The thermal behaviour of the co-combustion between paper sludge and rice straw. *Bioresour. Technol.* **2013**, *146*, 611–618. [CrossRef] [PubMed]

Disclaimer/Publisher's Note: The statements, opinions and data contained in all publications are solely those of the individual author(s) and contributor(s) and not of MDPI and/or the editor(s). MDPI and/or the editor(s) disclaim responsibility for any injury to people or property resulting from any ideas, methods, instructions or products referred to in the content.

MDPI

St. Alban-Anlage 66

4052 Basel

Switzerland

www.mdpi.com

Polymers Editorial Office

E-mail: polymers@mdpi.com

www.mdpi.com/journal/polymers

Disclaimer/Publisher's Note: The statements, opinions and data contained in all publications are solely those of the individual author(s) and contributor(s) and not of MDPI and/or the editor(s). MDPI and/or the editor(s) disclaim responsibility for any injury to people or property resulting from any ideas, methods, instructions or products referred to in the content.